In Vitro Technology and Micropropagated Plants

In Vitro Technology and Micropropagated Plants

Editors

Alicja Tymoszuk
Dariusz Kulus

MDPI

Basel • Beijing • Wuhan • Barcelona • Belgrade • Novi Sad • Cluj • Manchester

Editors
Alicja Tymoszuk
Bydgoszcz University of Science and Technology
Bydgoszcz
Poland

Dariusz Kulus
Bydgoszcz University of Science and Technology
Bydgoszcz
Poland

Editorial Office
MDPI
St. Alban-Anlage 66
4052 Basel, Switzerland

This is a reprint of articles from the Special Issue published online in the open access journal *Horticulturae* (ISSN 2311-7524) (available at: https://www.mdpi.com/journal/horticulturae/special_issues/In_Vitro_Micropropagated).

For citation purposes, cite each article independently as indicated on the article page online and as indicated below:

Lastname, A.A.; Lastname, B.B. Article Title. *Journal Name* **Year**, *Volume Number*, Page Range.

ISBN 978-3-7258-0279-1 (Hbk)
ISBN 978-3-7258-0280-7 (PDF)
doi.org/10.3390/books978-3-7258-0280-7

Cover image courtesy of Dariusz Kulus

Contents

About the Editors

Alicja Tymoszuk

Dr. Alicja Tymoszuk graduated with a degree in Biotechnology in 2007 from the University of Technology and Life Sciences in Bydgoszcz (Poland) and defended a doctoral dissertation in Biology in 2012 at the Nicolaus Copernicus University in Toruń (Poland). She currently works as a Professor (Assistant) at the Faculty of Agriculture and Biotechnology at the Bydgoszcz University of Science and Technology and conducts research related to plant biotechnology and nanobiotechnology. She has experience in conducting scientific experiments, publishing scientific articles, participation in scientific conferences, and raising funds for research projects related to horticultural plants, plant biotechnology, plant breeding, plant tissue culture, plant molecular biology, plant cryopreservation, nanobiotechnology, nanoparticles in horticulture, plant metabolites, oxidative stress in plants, plant protection, and plant cultivation. Her particular interests include the improvement of micropropagation and breeding techniques of horticultural plants with the use of nanomaterials. Dr. Tymoszuk has published over 30 scientific articles in international and Polish journals, e.g., *Acta Physiologiae Plantarum, Acta Scientiarum Polonorum Hortorum Cultus, Agronomy, Industrial Crops and Products, International Journal of Molecular Sciences, Materials, Plant Cell Tissue and Organ Culture*, and *Scientia Horticulturae.*

Dariusz Kulus

Dr. Dariusz Kulus graduated with a degree in Biotechnology from the UTP University of Technology and Life Sciences in Bydgoszcz, Poland (2011) and received his doctoral degree in Biotechnology (summa cum laude) in 2016 from the University of Life Sciences in Poznań, Poland. In 2022, he obtained his habilitation diploma in agriculture and horticulture. Dr. Kulus is currently working as an Associate Professor at the Bydgoszcz University of Science and Technology, Faculty of Agriculture and Biotechnology. He is the head of the Laboratory of Horticulture at the Department of Biotechnology. Dr. Kulus is the winner of numerous awards for scientific achievements, including the Scholarship of the Polish Minister of Science and Higher Education and the Mayor of the Bydgoszcz Metropolis Award for Outstanding Scientific Achievements. He is the author of grants from the Polish Ministry of Science and Higher Education and Polish National Science Centre. His current professional interests focus on issues related to cryopreservation and analyses of biochemical and genetic variability in selected ornamental, vegetable, and medicinal plant species, as well as micropropagation of agricultural and horticultural plants. His theoretical knowledge has been supported by the experience gained during training and internships in domestic and foreign laboratories of diverse profiles, including the PAS Laboratory of Cryobiology in Warsaw, the Czech University of Life Sciences in Prague, Agricultural University of Athens, Zagreb, Tbilisi, Córdoba, and others. He is author and co-author of over 70 scientific and popular science articles in the field of plant biotechnology.

Preface

The development of modern horticulture is strongly associated with improvements in plant micropropagation technologies and the production of high-quality ex vitro plants. Tissue cultures are a source of healthy and true-to-type plant material for further cultivation under covers or in the field. Moreover, in vitro cultures of horticultural plants are used for breeding purposes, germplasm conservation, disease elimination, and production of phytochemicals. The reprint of the Special Issue "In Vitro Technology and Micropropagated Plants" aims to present the results of studies focused on the development, improvement, and usage of plant tissue culture systems for horticultural purposes. This collection of articles brings together the latest research and developments in the use of tissue culture and micropropagation techniques, offering a valuable resource for researchers, educators, and practitioners in the area of plant science and biotechnology.

Alicja Tymoszuk and Dariusz Kulus
Editors

horticulturae

Editorial

Advancements in In Vitro Technology: A Comprehensive Exploration of Micropropagated Plants

Dariusz Kulus * and Alicja Tymoszuk

Laboratory of Horticulture, Faculty of Agriculture and Biotechnology, Bydgoszcz University of Science and Technology, Bernardyńska 6, 85-029 Bydgoszcz, Poland; alicja.tymoszuk@pbs.edu.pl
* Correspondence: dariusz.kulus@pbs.edu.pl

Citation: Kulus, D.; Tymoszuk, A. Advancements in In Vitro Technology: A Comprehensive Exploration of Micropropagated Plants. *Horticulturae* **2024**, *10*, 88. https://doi.org/10.3390/horticulturae10010088

Received: 3 January 2024
Accepted: 9 January 2024
Published: 17 January 2024

1. Introduction

In recent decades, the field of plant science has witnessed several breakthrough discoveries, particularly through the application of in vitro technology [1]. The term "micropropagation" refers to the process of producing numerous plants from small fragments of plant tissue. This technique has become helpful in addressing various challenges faced by traditional propagation methods, such as slow growth rates, plant sterility, susceptibility to diseases, and limited availability of planting material [2,3]. Moreover, in vitro propagation methods have improved plant breeding and conservation methods by providing efficient means by which to produce large quantities of genetically uniform plants under controlled conditions [4,5]. Today, in vitro technology covers a wide array of species and cultivars, ranging from economically important crops to endangered and rare plant species crucial for biodiversity conservation [6,7]. Nonetheless, the efficiency of micropropagation is influenced by several key factors, including the selection of an optimal nutrient medium, the precise balance and type of plant growth regulators, and the genetic makeup of the plant species studied. Additionally, environmental factors such as air temperature and light parameters play crucial roles in the success of micropropagation protocols [8]. Meticulous control of these factors is needed to achieve optimal outcomes.

The aim of this Special Issue is to provide a comprehensive overview of the latest advancements in in vitro technology research with a particular focus on micropropagation. This SI has captured the diversity of the studies that focus on the insights of micropropagation protocols, including the optimization of growth media, hormone supplementation, and the control of environmental factors, to achieve the highest success rates in plant regeneration. It also sheds light on the practical implications of in vitro techniques in agriculture, horticulture, and phytotherapy by means of polyploidization and virus elimination. The molecular, structural, and physiological aspects underlying the success of micropropagation that drive the efficient production of plants were discussed. By synthesizing recent research findings and highlighting key developments, this Special Issue serves as a valuable resource for researchers, educators, and practitioners searching for a deeper understanding of the potential offered by in vitro techniques in the realm of plant science.

2. Overview of Published Articles

Since the in vitro culture technique is most often used for the intensification of plant production, most of the articles published in this Special Issue focus on the micropropagation of ornamental, medicinal, and woody species. Basile et al. (contribution 1) performed research on *Cannabis sativa* subsp. *ruderalis* cv. "Finola", a dioecious Finnish cultivar with a remarkably short cultivation cycle of less than three months. The study aimed to establish an effective micropropagation protocol for in vitro multiplication, rooting, and ex vitro acclimatization, utilizing different explant sources: seed-derived in vitro explants and nodal segments with axillary buds. The optimal results were achieved by combining the Murashige and Skoog medium with sucrose, thidiazuron, and 1-naphthalenacetic acid for shoot proliferation, while rooting induction proved successful under red/blue lights on a medium enriched with sucrose and indol-3-butyric acid, ensuring a high survival rate of over 90% upon transplantation into a controlled greenhouse environment.

The study of Pourhassan et al. (contribution 2) focuses on the development of a comprehensive in vitro propagation protocol for the black-leaved Raven® ZZ plant (*Zamioculcas zamiifolia*), a drought-resistant ornamental foliage plant originating from Africa. The researchers employed meticulous disinfection techniques using sodium hypochlorite and mercury chloride to initiate an axenic culture. The optimized micropropagation method involved culturing leaf explants on Murashige and Skoog medium supplemented with 6-benzyladenine and/or α-naphthalene acetic acid, resulting in increased shoot and root production. The successful acclimatization of rooted plantlets in greenhouse conditions underscores the significance of this study as the first report on a complete micropropagation protocol for the black-leaved zamioculcas, offering valuable insights for the floricultural industry.

The article by Kaviani et al. (contribution 3) outlined an efficient protocol for the in vitro multiplication and root induction of narrow-leaf firethorn (*Pyracantha angustifolia*) using various basal culture media supplemented with 6-benzylaminopurine (BAP) and indole-3-butyric acid (IBA). Successful axillary bud disinfection, with a 70.23% success rate, was achieved on a Murashige and Skoog (MS) basal medium augmented with 0.5 $mg \cdot L^{-1}$ gibberellic acid (GA_3). The resulting micro-shoots, cultivated under optimized conditions, exhibited a high multiplication coefficient (2.4) on MS medium with 2.5 $mg \cdot L^{-1}$ BAP, followed by successful rooting in the presence of 1 $mg \cdot L^{-1}$ IBA. The acclimatization of rooted plantlets to greenhouse conditions demonstrated a notable 92.84% survival rate, highlighting the successful application of this protocol for the mass propagation of *P. angustifolia*.

Sage is a plant genus of medicinal and ornamental significance. The study by Papafotiou et al. (contribution 4) aimed to develop efficient micropropagation protocols for five Mediterranean sage species: *Salvia fruticosa*, *S. officinalis*, *S. ringens*, *S. tomentosa*, and *S. pomifera* ssp. *pomifera*. Through systematic experimentation with various growth regulators, the research identified optimal conditions for shoot multiplication, with *S. officinalis* demonstrating the highest efficiency among the studied species. Despite encountering hyperhydricity challenges, successful rooting and subsequent acclimatization were achieved, providing essential protocols for the propagation of these sage species, valuable in both pharmaceutical and floriculture industries.

Ficus carica, or common fig, is a highly nutritional fruit, well-known for its medicinal and economic values. The study of Ling et al. (contribution 5) aimed to establish an efficient protocol for the mass propagation of *F. carica* cv. "Violette de Solliès". The researchers optimized shoot induction (15.2 shoots per explant) using 6-benzylaminopurine in Murashige and Skoog medium and achieved robust rooting (93.33%) with indole-3-butyric acid in Woody Plant Medium. The resulting plantlets exhibited genetic and morphological stability over six subculture cycles, and successful acclimatization in biochar soil with a 100% survival rate further supports the efficiency of this protocol for commercial propagation of uniform and true-to-type *F. carica* "Violette de Solliès" plant stocks.

Haida et al. (contribution 6) focused on the endangered *Curcuma caesia*, commonly known as Kali Haldi or black turmeric, addressing its decline in natural habitats due to overharvesting for various purposes, including pharmaceutical demands. By employing plant tissue culture techniques, the researchers established an effective protocol for the propagation of *C. caesia*, emphasizing the optimization of shoot induction, multiplication, and rooting stages. The results demonstrated the suitability of the MSB5 medium for shoot induction, a combination of 6-benzylaminopurine and indole-3-butyric acid for enhanced shoot multiplication, and the full-strength MSB5 medium supplemented with indole-3-acetic acid for successful rooting, ultimately providing a valuable protocol for large-scale raw material production, conservation, or bioactive compound extraction.

The next research published in this SI by Clapa et al. (contribution 7) utilized micropropagation in enhancing the large-scale production of blackberry (*Rubus fruticosus*) plant material. The research aimed to assess the proliferative capacity of blackberry in vitro by comparing wheat starch-gelled culture medium to the conventional agar-gelled medium, while also evaluating the genetic fidelity between the proliferated shoots and their parent plants. Results revealed that wheat starch-gelled culture medium significantly outperformed agar-gelled medium in terms of shoot proliferation and length in all tested blackberry cultivars, and the genetic uniformity of the micropropagated shoots was confirmed through molecular markers.

The study by Khatoon et al. (contribution 8) addressed the longstanding challenge of in vitro propagation of olive plants (*Olea euorpea*), characterized by their woody nature and susceptibility to culture oxidation. Focused on optimizing shoot induction and proliferation protocols, the research investigates the effects of different concentrations of 6-benzylaminopurine (BAP) and pre-cooling regimes on various olive cultivars. The results highlight significant improvements in morphological, physiological, and biochemical attributes, with the 48 h pre-cooling and 2.5 $mg \cdot L^{-1}$ BAP concentration interaction yielding the most favorable results, demonstrating the potential for enhanced olive cultivation through in vitro techniques.

The next published article, that by Faisal et al. (contribution 9), focused on the in vitro regeneration of *Plectranthus amboinicus*, a perennial plant with aromatic leaves and health benefits. Using thidiazuron treatment, the authors achieved optimal shoot regeneration from nodal segments, with the highest number and length of shoots. The in vitro-regenerated plants had higher contents of phenolic, tannin, and flavonoid compounds, along with enhanced antioxidant activity compared to ex-vitro plants. Flow cytometry analysis confirmed the nuclear genome stability of the in vitro-propagated plants. This efficient in vitro multiplication method not only enhances availability but also provides insights into the genetic and phytochemical properties of *P. amboinicus*, contributing to its preservation and sustainable use.

Alongside micropropagation and large-scale production of plants, in vitro cultures can also be used for breeding purposes. Contribution 10 by Jin et al. addressed the importance of transforming the commercial cultivar "YX4" of *Gypsophila paniculata*, a popular cut flower, to produce blue flowers. By successfully introducing the flavonoid 3′5′-hydroxylase gene (PgF3′5′H) from *Platycodon grandiflorus*, the research achieved a transformation efficiency of 13.5%. The optimized protocol, involving shoot apex without meristem, thidiazuron supplementation, and a specific Agrobacterium-mediated transformation system, not only enhances the ornamental value of *G. paniculata* but also provides a valuable tool for studying genes associated with important ornamental traits in this commercially significant species.

The artificial polyploidization of *Ajuga reptans*—a medicinal species, involving genome doubling—was investigated by Navrátilová et al. (contribution 11) to assess its impact on the composition and quantity of biologically active substances from the glycoside and phytoecdysone families. The study revealed increased levels of *trans*-teupolioside, *trans*-verbascoside, and 20-hydroxyecdysone in the aboveground parts of the tetraploid lines compared to diploids. These findings suggest the potential use of Ajuga tetraploids in breeding programs to augment the production of substances with medicinal and industrial applications in pharmaceuticals, cosmetics, and food production, along with potential botanical pesticide effects.

Elimination of viruses is another practical application of plant tissue culture. Viruses have a significant economic impact on agriculture, particularly potato virus M (PVM) and potato virus S (PVS), which collectively cause over 50% of annual potato tuber yield losses. The research by Kereša et al. (contribution 12) aimed to eliminate PVM and PVS from the traditional Croatian potato (*Solanum tuberosum*) cultivar "Brinjak", valued for its economic profitability and genetic potential, to enhance its productivity. For this purpose, the authors used chemotherapy via the application of ribavirin at various concentrations. Despite a 33% success rate in eliminating PVS, ribavirin failed to eradicate PVM, leading to more severe disease symptoms, adversely affecting photochemistry and multispectral parameters, and ultimately reducing the yield of tubers in plants with mixed infections (PVM + PVS) compared to those with a single PVM infection.

Finally, Koufan et al. (contribution 13) prepared a review article on Caper (*Capparis spinosa*), a shrubby plant species known for its difficulty in vegetative propagation and seed germination. Recognizing the limitations in seed germination, root induction from stem cuttings, and plant hardening, the study highlights the potential of tissue culture as a promising alternative for the clonal propagation of caper plants. This comprehensive review outlines various micropropagation methods, including in vitro seed germination, propagation via nodal segmentation, and adventitious organogenesis, highlighting their role in overcoming obstacles and an approach to the large-scale propagation and genetic improvement of caper plants.

3. Conclusions

Micropropagation has emerged as a powerful tool, offering significant advantages in terms of rapid multiplication, disease-free propagation, breeding, and conservation of plant genetic resources. The compilation of diverse studies presented in this Special Issue underlines the pivotal role of in vitro culture techniques, particularly micropropagation, in the advancement of plant production across various species. The presented research on ornamental, medicinal, vegetable, and woody plants, such as *Ajuga reptans, Cannabis sativa, Capparis spinosa, Curcuma caesia, Ficus carica, Gypsophila paniculate, Olea euorpea, Plectranthus amboinicus, Pyracantha angustifolia, Rubus fruticosus, Solanum tuberosum, Zamioculcas zamiifolia*, and *Salvia* species, demonstrate the versatility and applicability of tissue culture methods. These studies provide efficient protocols for shoot proliferation, rooting, and acclimatization, paving the way for large-scale production, genetic improvement, and conservation efforts. Overall, these contributions collectively contribute valuable insights into the optimization of in vitro techniques for plant propagation, offering practical solutions for challenges in agriculture and horticulture.

Funding: This research received no external funding.

Acknowledgments: As Guest Editors of the Special Issue "In Vitro Technology and Micropropagated Plants", we would like to thank and express our deep appreciation to all contributing authors of this SI.

Conflicts of Interest: The authors declare no conflicts of interest.

List of Contributions

1. Basile, G.N.; Tedone, L.; Pulvento, C.; De Mastro, G.; Ruta, C. Establishment of an Efficient In Vitro Propagation Protocol for *Cannabis sativa* L. subsp. *ruderalis* Janish. *Horticulturae* **2023**, *9*, 1241. https://doi.org/10.3390/horticulturae9111241.
2. Pourhassan, A.; Kaviani, B.; Kulus, D.; Miler, N.; Negahdar, N. A Complete Micropropagation Protocol for Black-Leaved *Zamioculcas zamiifolia* (Lodd.) Engl. 'Dowon'. *Horticulturae* **2023**, *9*, 422. https://doi.org/10.3390/horticulturae9040422.
3. Kaviani, B.; Deltalab, B.; Kulus, D.; Tymoszuk, A.; Bagheri, H.; Azarinejad, T. In Vitro Propagation of *Pyracantha angustifolia* (Franch.) C.K. Schneid. *Horticulturae* **2022**, *8*, 964. https://doi.org/10.3390/horticulturae8100964.
4. Papafotiou, M.; Vlachou, G.; Martini, A.N. Investigation of the Effects of the Explant Type and Different Plant Growth Regulators on Micropropagation of Five Mediterranean *Salvia* spp. Native to Greece. *Horticulturae* **2023**, *9*, 96. https://doi.org/10.3390/horticulturae9010096.
5. Ling, W.T.; Tan, L.V.; Khor, S.P.; Sriskanda, D.; Subramaniam, S.; Chew, B.L. Rapid In Vitro Propagation of Fig (*Ficus carica* L.) 'Violette de Solliès' Supported by Molecular and Microscopy Analyses. *Horticulturae* **2022**, *8*, 1025. https://doi.org/10.3390/horticulturae8111025.

6. Haida, Z.; Sinniah, U.R.; Nakasha, J.J.; Hakiman, M. Shoot Induction, Multiplication, Rooting and Acclimatization of Black Turmeric (*Curcuma caesia* Roxb.): An Important and Endangered *Curcuma* Species. *Horticulturae* **2022**, *8*, 740. https://doi.org/10.3390/horticulturae8080740.
7. Clapa, D.; Hârța, M.; Szabo, K.; Teleky, B.-E.; Pamfil, D. The Use of Wheat Starch as Gelling Agent for In Vitro Proliferation of Blackberry (*Rubus fruticosus* L.) Cultivars and the Evaluation of Genetic Fidelity after Repeated Subcultures. *Horticulturae* **2023**, *9*, 902. https://doi.org/10.3390/horticulturae9080902.
8. Khatoon, S.; Liu, W.; Ding, C.-b.; Liu, X.; Zheng, Y.; Zhang, Y.; Chen, X.; Rauf, M.; Alghabari, F.; Shah, Z.H. In Vitro Evaluation of the Effects of BAP Concentration and Pre-Cooling Treatments on Morphological, Physiological, and Biochemical Traits of Different Olive (*Olea euorpea* L.) Cultivars. *Horticulturae* **2022**, *8*, 1108. https://doi.org/10.3390/horticulturae8121108.
9. Faisal, M.; Qahtan, A.A.; Alatar, A.A. Thidiazuron Induced In Vitro Plant Regeneration, Phenolic Contents, Antioxidant Potential, GC-MS Profiles and Nuclear Genome Stability of *Plectranthus amboinicus* (Lour.) Spreng. *Horticulturae* **2023**, *9*, 277. https://doi.org/10.3390/horticulturae9020277.
10. Jin, C.; Sun, D.; Ma, L.; Mo, X.; Yang, C.; Li, F. Ectopic Expression of *PgF3′5′H* in Commercial *Gypsophila paniculata* Cultivar through Optimized Agrobacterium-Mediated Transformation. *Horticulturae* **2023**, *9*, 321. https://doi.org/10.3390/horticulturae9030321.
11. Navrátilová, B.; Ondřej, V.; Vrchotová, N.; Tříska, J.; Horník, Š.; Pavela, R. Impact of Artificial Polyploidization in *Ajuga reptans* on Content of Selected Biologically Active Glycosides and Phytoecdysone. *Horticulturae* **2022**, *8*, 581. https://doi.org/10.3390/horticulturae8070581.
12. Kereša, S.; Vončina, D.; Lazarević, B.; Bošnjak Mihovilović, A.; Pospišil, M.; Brčić, M.; Matković Stanković, A.; Habuš Jerčić, I. Partial Elimination of Viruses from Traditional Potato Cultivar 'Brinjak' by Chemotherapy and Its Impact on Physiology and Yield Components. *Horticulturae* **2022**, *8*, 1013. https://doi.org/10.3390/horticulturae8111013.
13. Koufan, M.; Belkoura, I.; Mazri, M.A. In Vitro Propagation of Caper (*Capparis spinosa* L.): A Review. *Horticulturae* **2022**, *8*, 737. https://doi.org/10.3390/horticulturae8080737.

References

1. Wijerathna-Yapa, A.; Hiti-Bandaralage, J. Tissue Culture—A Sustainable Approach to Explore Plant Stresses. *Life* **2023**, *13*, 780. [CrossRef] [PubMed]
2. Sharma, V.; Ankita; Karnwal, A.; Sharma, S.; Kamal, B.; Jadon, V.S.; Gupta, S.; Sivanasen, I. A Comprehensive Review Uncovering the Challenges and Advancements in the In Vitro Propagation of Eucalyptus Plantations. *Plants* **2023**, *12*, 3018. [CrossRef]
3. Xue, Y.; Hiti-Bandaralage, J.C.A.; Mitter, N. Micropropagation of Duboisia Species: A Review on Current Status. *Agronomy* **2023**, *13*, 797. [CrossRef]
4. Radomir, A.-M.; Stan, R.; Florea, A.; Ciobotea, C.-M.; Bănuță, F.M.; Negru, M.; Neblea, M.A.; Sumedrea, D.I. Overview of the Success of In Vitro Culture for Ex Situ Conservation and Sustainable Utilization of Endemic and Subendemic Native Plants of Romania. *Sustainability* **2023**, *15*, 2581. [CrossRef]
5. Vahdati, K.; Sadeghi-Majd, R.; Sestras, A.F.; Licea-Moreno, R.J.; Peixe, A.; Sestras, R.E. Clonal Propagation of Walnuts (*Juglans* spp.): A Review on Evolution from Traditional Techniques to Application of Biotechnology. *Plants* **2022**, *11*, 3040. [CrossRef] [PubMed]
6. Custódio, L.; Charles, G.; Magné, C.; Barba-Espín, G.; Piqueras, A.; Hernández, J.A.; Ben Hamed, K.; Castañeda-Loaiza, V.; Fernandes, E.; Rodrigues, M.J. Application of In Vitro Plant Tissue Culture Techniques to Halophyte Species: A Review. *Plants* **2023**, *12*, 126. [CrossRef] [PubMed]
7. Duta-Cornescu, G.; Constantin, N.; Pojoga, D.-M.; Nicuta, D.; Simon-Gruita, A. Somaclonal Variation-Advantage or Disadvantage in Micropropagation of the Medicinal Plants. *Int. J. Mol. Sci.* **2023**, *24*, 838. [CrossRef] [PubMed]
8. Kaur, K.; Dolker, D.; Behera, S.; Pati, P.K. Critical Factors Influencing In Vitro Propagation and Modulation of Important Secondary Metabolites in *Withania somnifera* (L.) Dunal. *Plant Cell Tissue Organ Cult.* **2022**, *149*, 41–60. [CrossRef] [PubMed]

horticulturae

Article

Establishment of an Efficient In Vitro Propagation Protocol for *Cannabis sativa* L. subsp. *ruderalis* Janish

Giuseppe N. Basile, Luigi Tedone, Cataldo Pulvento, Giuseppe De Mastro * and Claudia Ruta *

Department of Soil, Plant and Food Sciences, University of Bari Aldo Moro, 70125 Bari, Italy; g.basile18@studenti.uniba.it (G.N.B.); luigi.tedone@uniba.it (L.T.); cataldo.pulvento@uniba.it (C.P.)
* Correspondence: giuseppe.demastro@uniba.it (G.D.M.); claudia.ruta@uniba.it (C.R.)

Abstract: *Cannabis sativa* L., subsp. *ruderalis* Janish., 'Finola' is a dioecious cultivar of Finnish origin. This cultivar is very interesting because its cultivation cycle lasts less than 3 months. The aim of this study was to define an efficient micropropagation protocol to ensure in vitro multiplication and rooting and in vivo acclimatization. Two different explant sources were tested: seed-derived in vitro explants and nodal segments containing axillary buds from selected mother plants. Shoot proliferation was tested on different growth media enriched with cytokinin alone or cytokinin in combination with auxins. Among all combinations, the best results were obtained by combining the Basal Medium (BM—a Murashige and Skoog modified medium) with sucrose (20 g L^{-1}), thidiazuron (TDZ 0.4 mg L^{-1}), and 1-naphthalenacetic acid (NAA 0.2 mg L^{-1}). Regarding rooting induction, the plants developed an extensive root system under red/blue lights on BM enriched with sucrose (30 g L^{-1}) and indol-3 butyric acid (0.1 mg L^{-1}), which allowed the survival of more than 90 percent of the plantlets once transplanted into the climate-controlled greenhouse.

Keywords: Finola; hemp; auto-flowering; clonal propagation; micropropagation

Citation: Basile, G.N.; Tedone, L.; Pulvento, C.; De Mastro, G.; Ruta, C. Establishment of an Efficient In Vitro Propagation Protocol for *Cannabis sativa* L. subsp. *ruderalis* Janish. *Horticulturae* **2023**, *9*, 1241. https://doi.org/10.3390/horticulturae9111241

Academic Editors: Alicja Tymoszuk and Dariusz Kulus

Received: 7 September 2023
Revised: 14 November 2023
Accepted: 15 November 2023
Published: 17 November 2023

1. Introduction

In recent years, there has been renewed interest in hemp (*Cannabis sativa* L., genus *Cannabis* family Cannabaceae), a multipurpose crop cultivated in many parts of the world for its nutritional, medicinal, and industrial uses [1–4].

The taxonomic determinations of *Cannabis* genus have been deeply studied and discussed. Based on genetic evaluations, *Cannabis sativa* can be subdivided into three subspecies: *sativa*, *indica*, and *ruderalis* [5]. *C. sativa* subsp. *ruderalis* is a lesser-known hemp that originated in regions with harsh climates, such as Russia and Eastern Europe. This subspecies is typically smaller in size and contains very low Δ9-tetrahydrocannabinol (THC), the psychoactive compound in the plant, but is often rich in CBD (cannabidiol, a non-intoxicating compound), known for its potential therapeutic benefits [6]. Furthermore, *Ruderalis* strains are auto-flowering, which means that they are photoperiod-insensitive. For this reason, the transition from the vegetative stage to the flowering stage is based on the plant's age, rather than relying on light cycles [7]. Conversely, the cultivation cycle of the other subspecies of *Cannabis sativa* is typically photo-dependent, strongly linked to seasonal trends and, consequently, influenced by the number of hours of light and darkness per day [5].

The above-mentioned characteristics of *ruderalis* make vegetative reproduction difficult, while seed propagation can lead to uneven growth and maturity because the flowering process can vary among individual plants. Indeed, because auto-flowering strains containing *ruderalis* genetics are renowned for their ability to flower without strict light schedules [8], some plants may initiate flowering earlier or later than desired. This factor can be challenging for growers seeking a consistent and synchronized harvest [9]. Belonging to *ruderalis*, 'Finola' is a hemp cultivar of Finnish origin, characterized by high cannabidiol (CBD) and low Δ9-tetrahydrocannabinol (THC) contents. This cultivar is dioecious, independent of the photoperiod, and interesting due to its very short cycle of fewer than 3 months, with early sowing and flowering needed. This cultivar is characterized by its low branching, small size, small seeds, and low biomass recovery [10]. Indeed, 'Finola' has high potential at a commercial level and could be used in super-intensive production models of protected environments, allowing significant water savings [11].

To obtain uniform plant production, the micropropagation technique could be a useful tool. Indeed, in vitro propagation has great potential in rapidly producing many true to type clonal and high-quality plants using synthetic growth media in a controlled environment [12–15]. Micropropagation requires few selected mother plants as explant sources to initiate an in vitro culture [15]. Furthermore, the in vitro plants are grown in reduced spaces under sterile conditions to avoid the transmission of pests, diseases, and pathogens and the influence of weather or soil [16,17]. Due to all these advantages, in recent years, several researchers have developed different protocols to define the in vitro cultures of *C. sativa* subspecies *sativa* and *indica* derived from the leaves, axillary nodes, cotyledons, shoot tips, and epicotyls of explants with the aim to obtain a high multiplication rate, produce disease-free plants, select elite clones, and overcome the limits of heterozygosity from cross-pollination [15,18–21]. To the best of our knowledge, only two studies focused on the in vitro culture of the subspecies *ruderalis*. These studies underline the low branching tendency, strong apical dominance [22,23], and influence of the genotype on the explant's behavior and multiplication rate. The difficulties of in vitro growth and the maintenance of a vegetative state may be due to auto-flowering varieties and photoperiod independence [11,24]. These factors could also be other limits for the micropropagation of this subspecies. Therefore, the main challenge in developing an effective micropropagation protocol is the specificity of species, varieties, and genotypes, resulting in different responses in vitro depending on the explant, nutrient media, and growth regulators, such as the culture conditions [21].

Considering the above-mentioned information, the focus of this experimental research was to develop the most suitable protocol for the in vitro direct propagation of *C. sativa* L., subsp. *ruderalis* Janish., cultivar 'Finola'.

2. Materials and Methods

2.1. Plant Materials and Explant Source of *C. sativa* L., subsp. *ruderalis* Janish., Cultivar 'Finola'

This research began with certified 'Finola' seeds. The seeds were sown directly in vitro or under controlled conditions to obtain mother plants. The nodes of the seedlings were then excised to initiate micropropagation via axillary shoots.

2.1.1. In Vitro Germinated Seeds

After initial rinsing under running water to remove impurities, the effects of the sterilant and immersion times on the surface sterilization of 'Finola' seeds were evaluated based on the following two treatments:

(a) Sodium hypochlorite (NaClO with 14% active chlorine) at 1.4% for 20 min;
(b) Mercuric chloride (HgCl) at 0.1% for 15 min.

One-hundred seeds in each treatment were then rinsed five times in sterile distilled water. For the in vitro sowing, 50 mL glass tubes containing 10 mL of the nutrient medium (BM [25]: Murashige and Skoog (MS) macro nutrients [26], Nitsch and Nitsch (NN) micronutrients [27], FeNaEDTA (40 mg L^{-1}), and thiamine HCl (0.4 mg L^{-1})), adding agar (7 g L^{-1}) as a gelling agent, were used. The pH of the medium was adjusted to 5.6–5.8 before autoclaving. The tubes were maintained in a growth chamber equipped with broad-spectrum LED lights between 400 and 700 nm and a light intensity of 50 $\mu m\ s^{-1}\ m^{-2}$. The photoperiod was set to 16 h/8 h (day/night–d/n) and the temperature to 22 ± 1 °C.

2.1.2. Mother Plant Production

Thirty seeds were first soaked in distilled water for 48 h at a temperature of 22 ± 1 °C in total darkness. Then, seeds showing open seed coats and the emergence of coleorhiza (root sheath) were sown in polystyrene trays filled with a substrate of pH 5.5–6.5 consisting of a mix of blond peat (fraction 20–40 mm), German black peat (fraction 0–15 mm), and perlite (3:1 v/v). Once the seedlings had emerged, and the first pair of true leaves were grown, the plants were transplanted into pots 9 cm in diameter and 11 cm deep, which were filled with peat and expanded clay and perlite (3:1:1).

Ten selected mother plants were grown in a climate-controlled greenhouse, and the same number of plants was grown in a growth cabinet to compare their behaviors and growth responses under two different conditions. Starting from these plants, five mother plants were selected to collect the nodal buds. The controlled greenhouse, located at the Department of Soil, Plant, and Food Sciences (DiSSPA) (University of Bari Aldo Moro, Bari, Italy), had a temperature control system maintained between 18 °C and 25 °C, with a 15 h/9 h (d/n) photoperiod of natural light per day (in Summer) and relative humidity of 85%. On the other hand, the thermostat-controlled cabinet (F-Cell 707 comfort Blue-Line—MMM Group) featured illumination using white LED lights (VIS LED light Cool White), with a spectrum between 400 nm and 700 nm at an intensity of 80 µm s^{-1} m^{-2}; the photoperiod set to 24 h/0 h (d/n), and a temperature of 22 ± 1 °C was used.

The growth of the seedlings was monitored to record the response to the growth conditions up to the time of the appearance of flower buds in order to define the time of the end of growth and determine the best method of cultivation to obtain a high number of axillary buds per female plant.

2.2. *In Vitro Multiplication*

One month after the seeds were planted in test tubes, flower buds appeared at the apex of the seedlings. The nodal buds below the apex of female plants were taken and immediately transferred to jars containing 30 mL of BM culture medium, solidified with agar (7 g L^{-1}) ((Agar No. 1) Oxoid Agar Bacteriological), with the addition of sucrose (20 g L^{-1}) and growth phytoregulators of the cytokinin group to stimulate multiplication. The cytokinins used were Metatopolin (MT) [6-(3-hydroxybenzylaminopurine) at a concentration of 0.5 mg L^{-1} and 6-Benzylaminopurine (BAP) at a concentration of 0.05 mg L^{-1}. The test was carried out on 80 jars for each cytokinin, with a total of 160 jars containing a single bud. After 21 days, the first subculture was performed, maintaining the same starting cytokinins with equal concentrations.

For the mother plants, when the first female flower buds appeared, the nodal buds below (1 cm in length), that were not yet induced to flower, were excised and sterilized with sodium hypochlorite (NaClO, active chlorine 14%) at 1.4% for 20 min following extensive washing with sterile-distilled water to remove residues of the sterilizing agent. The nodal buds were then placed in 220 mL glass jars containing 30 mL of nutrient media and the growth regulators reported below to enable the in vitro introduction and subsequent induction of multiplication.

Two different growth media were tested:

(1) BM [25], regularly used at the Laboratory of Micropropagation and Microscopy, DiSSPA, University of Bari Aldo Moro (Bari, Italy), for routine culture produced by stocks.
(2) Driver and Kuniyuki (DKW) (1984) [28].

Both media, agarized with 7 g L^{-1} Agar (No. 1 Oxoid Agar Bacteriological), were enriched with sucrose (20 g L^{-1}).

The tested growth regulators were selected based on several studies exploring in vitro cultures of *Cannabis* [29–32] or based on the best results of previous research on other subspecies carried out at the Laboratory of Micropropagation and Microscopy, DiSSPA, University of Bari Aldo Moro (unpublished data):

- Benzylaminopurine (BAP) (0.05 mg L^{-1});
- Metatopoline (MT) (0.5 mg L^{-1});
- Thidiazuron (TDZ) (0.4 mg L^{-1}) + 1 naphthalenacetic acid (NAA) (0.2 mg L^{-1});

- Thidiazuron (TDZ) (0.4 mg L^{-1}) + 2,3,5- Triiodobenzoic acid (TIBA 1 g L^{-1}).

For each growth medium, 80 buds were considered, with 20 buds for each of the four growth regulators tested. When miniaturization of the shoots was recorded (for BAP, MT, and TDZ in combination with TIBA), with lengths less than 2 cm and shortened internodes, the shoots were subjected to an elongation phase by adding gibberellic acid (GA_3 0.25 mg L^{-1}) to achieve greater extension between nodes and increased leaf growth. The pH level was adjusted to 5.7 prior to autoclaving at 121 °C for 20 min.

Three subsequent subcultures (lasting three weeks) were performed on the same growth media, and several parameters (number of axillary buds that formed shoots, shoot length, and number of nodes per each shoot) were recorded for each subculture. In each subculture, the apical bud was always removed to reduce apical dominance, promote lateral shoots as much as possible, and delay flowering. During the trial, the explants were maintained in a growth chamber at 22 ± 1 °C, with a photoperiod of 16 h/8 h (d/n).

2.3. *In Vitro Rooting*

Around 200 shoots that had reached a minimum length of 3 cm were transplanted to the agarized BM supplemented with sucrose (30 g L^{-1}) and indol-3-butyric acid (IBA) at two different concentrations (0.5 mg L^{-1} and 0.1 mg L^{-1}) to induce rooting. The pH level was adjusted to 5.7 prior to autoclaving at 121 °C for 20 min.

To evaluate the influence of different kinds and combinations of light on the photomorphogenic and physiological responses in the last in vitro step, 100 plantlets for each concentration of IBA were divided under two different light conditions. The first light condition used white (Osram Lumilux De Luxe T8 36 W/954)/pink (Osram Fluora T8 36 W/77) light (LEDs with a broad spectrum between 400 and 700 nm, a white/pink ratio of 2:1, and a light intensity of 50 µm s^{-1} m^{-2}), as white light is traditionally used as the light source in micropropagation, while pink light enhances plant growth, increasing fresh and dry weights compared to monochromatic light [33]. The second light condition used in the growth cabinet (F-Cell 707 comfort Blue-MMM-Group) was red/blue light (VQ-GLSF0014 W 1:3) LEDs with a wavelength of 700 nm and 400 nm. LEDs with a red/blue ratio of 3:1 at an intensity of 50 µm s^{-1} m^{-2} were used for the possible positive induction of root formation [34]. For both groups, the photoperiod was a 16 h/8 h d/n cycle.

After 20 days, the effects of the different light types and phytoregulators on the rooting rate, in terms of root percentage, and the average length and number of roots per shoot were evaluated.

2.4. *Ex Vitro Acclimatization*

Approximately 50 rooted microplants were removed from the in vitro containers. Then, the roots were gently rinsed in distilled water and the microplants were transplanted for acclimatization in organic Jiffypots® square planting pots (Ø 8 cm^2) filled with sterile peat (46% organic carbon, 1–2% organic nitrogen, and 80% organic matter, pH 6.5) and perlite (2:1, v/v ratio) and covered with transparent plastic. Acclimatization conditions in the greenhouse were 18–25 °C and humidity that was reduced from 85–90% to 50–60% over 20 days. At this time, the survival of the plants was also evaluated.

2.5. *Statistical Analysis*

Data were subjected to an analysis of variance (ANOVA) test using the CoStat software, Version 6.40. The means of the different treatments were compared to each other in terms of significance using the Student Newman Keuls (SNK) Test ($p \leq 0.001$). Before the ANOVA analysis, the percentage data were subjected to angular transformation. All the experiments were repeated three times.

3. Results

3.1. Germination of Seeds In Vitro

The sterilization test, carried out on a sample of 100 seeds for each sterilizing agent, demonstrated that exposure to sodium hypochlorite (with 14% active chlorine) at a concentration of 1.4% for 20 min was able to obtain highly contamination-free vital seeds. The percentage of in vitro germinated seeds, not contaminated seeds, was 80%. Among those seeds, only 42% were female, as identified visually by the inflorescence sex as soon as flower buds appeared. Thirty days after in vitro sowing, the morphological parameters measured on 'Finola' female seedlings indicated good development of the shoot length (5.3 cm on average), with the shoots containing about 3 nodes (3.1 on average). The shoots were then excised and transferred to a culture medium (BM) enriched with BAP and MT to induce multiplication.

The growth and multiplication responses of the shoots of 'Finola' in the presence of the two different phytoregulators tested were recorded after 21 days (Table 1) (Figure 1). The results showed no significant differences in the parameters evaluated, although the data were higher after adding MT to the growth medium for both shoot length (5.8 cm vs. 2.8 cm) and the number of nodes (5 vs. 4) compared to the average multiplication index (5 vs. 3) (Table 1). When the development parameters were recorded, the sufficiently grown shoots were subcultured for the first time to increase the number of shoots obtained from each explant. After a further 21 days from the start of the first subculture, although the growth conditions remained identical, all explants responded to the second cut with completely arrested development (Figure 1b), showing necrosis of the apices, diffuse hyperhidricity, or early flowering.

Table 1. Influence of different phytoregulators on morphological parameters and mean multiplication index (MMI) measured on 'Finola' shoots grown from nodal buds of seedlings 21 days after the multiplication induction.

Phytoregulator	Concentration (mg L^{-1})	Length (cm)	Nodes (n)	M.M.I. * (n)
BAP	0.05	2.8 a	4.0 a	3.0 a
MT	0.5	5.8 a	5.0 a	5.0 a

M.M.I. * = Mean multiplication index was calculated as the mean number of shoots obtained 21 days after the multiplication induction. Different letters for each column indicate significant differences (SNK test at $p \leq 0.001$).

Figure 1. 'Finola' shoots multiplied on BM enriched with BAP, scale bar 1 cm (**a**); decline in the subcultures of 'Finola' shoots developed from nodal buds of seedlings sowed directly in vitro (**b**).

3.2. Mother Plant Production and Micropropagation by Axillary Buds

Table 2 presents the values of the different growth parameters measured under the two different conditions of mother plant development at the time of the first flower's appearance.

Table 2. Influence of different growing conditions on the growth of the stem and appearance of female and male flowers in mother plants of 'Finola' cultivar.

	Stem			Flower		
Growing Conditions	Length (cm)	Nodes (n)	Diameter (mm)	Female Plants (%)	Female Flower Appearance (dd *)	Male Flower Appearance (dd *)
Controlled greenhouse	15.6 b	3.2 b	2.4 b	57.0 a	32.2 a	25.7 b
Growth cabinet	47.1 a	4.8 a	4.5 a	66.7 a	35.6 a	32.6 a

* dd number of days from sowing to flowering. Different letters for each column indicate significant differences (SNK test at $p \leq 0.001$).

Plants obtained in the growth cabinet showed significant differences (Table 2, Figure 2) in vigor compared to the vigor of plants obtained in the climate-controlled greenhouse. The length was more than two times greater, and the stem diameter was almost two times greater, for the mother plants obtained in the growth cabinet. To complete the cycle, the time required for the appearance of female flowers differed by only three days between the two growth conditions (32.2 days in the climate-controlled greenhouse vs. 35.6 days in the growth cabinet) and was not statistically significant. Moreover, the nodal buds from the mother plants (Figure 3) grown in the climate-controlled greenhouse did not readily respond to the in vitro conditions due to both their smaller sizes and the difficulty in producing pathogen-free plants. Indeed, the plants required a long period of sterilization, which, due to their slender size, caused burning, oxidation, and arrested the development of excise materials once inoculated in the culture media chosen for multiplication. On the other hand, by reducing the sterilization time, the explants presented widespread contaminants. For this reason, the aseptic culture was initiated considering only nodal buds cut from the mother plants grown inside the growth cabinet.

Figure 2. Mother plant in climate-controlled greenhouse, scale bar 10 cm (**a**); mother plant in growth cabinet, scale bar 10 cm (**b**).

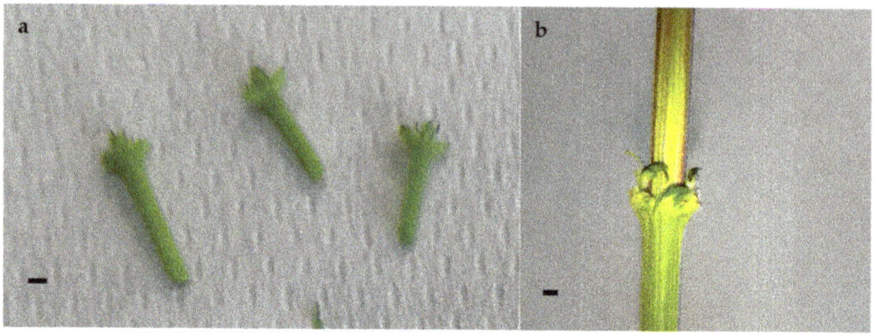

Figure 3. Nodal segments containing axillary buds excised from mother plants grown in climate-controlled greenhouse, scale bar 2 mm (**a**); nodal segments containing axillary buds from mother plants grown in growth cabinet, scale bar 2 mm (**b**).

Table 3 shows the effects of nutrient media and different phytoregulators on in vitro shoots of 'Finola'. A comparison between the two nutrient media tested, the Basal Medium and Driver and Kuniyuki (main effect NM), indicated no significant differences in the influence of media on the measured parameters, even if the very high presence of hyperhydrated explants (90%) grown on DKW fully limited the use of this nutrient medium. Considering the phytoregulator's main effects, the best results for the growth and multiplication index were obtained from the shoots grown using a combination of TDZ and NAA, in terms of length, the number of nodes, and the mean multiplication index. All these data were confirmed and found to increase based on an analysis of the following interaction: BM x TDZ + NAA (3.4 cm, 3.1, and 3.0, respectively, for the length, number of nodes, and number of shoots per explant) (Figure 4). No significant increase in shoot length was obtained by adding gibberellic acid (GA_3) to the nutrient media enriched with TDZ and TIBA, BAP, or MT.

Table 3. Effect of nutrient media and different phytoregulators on growth, mean multiplication index, and hyperhidricity of 'Finola' shoots, measured 21 days after the beginning of in vitro culture.

Nutrient Medium (NM)	Phytoregulator * (P)	Length (cm)	Nodes (n)	M.M.I. ** (n)	Hyperhydricity (%)
Interaction NM × P					
MB	TDZ + NAA	3.4 a	3.1 a	3.0 a	0.0 c
	TDZ + GA_3 + TIBA	2.8 b	2.5 b	1.8 b	30.0 b
	BAP + GA_3	1.9 c	2.0 c	1.5 b	100.0 a
	MT + GA_3	2.3 c	1.5 c	1.3 b	100.0 a
DKW	TDZ + NAA	2.9 a	2.7 b	2.0 a	80.0 a
	TDZ + GA_3 + TIBA	2.6 a	2.0 a	1.7 a	70.0 b
	BAP + GA_3	1.5 b	1.0 b	1.0 b	100.0 a
	MT + GA_3	1.9 b	1.0 b	1.0 b	100.0 a
Main effect NM					
MB		2.6 a	2.3 a	1.5 a	57.5 b
DKW		2.2 a	1.7 a	1.4 a	90.0 a
Main effect P					
	TDZ + NAA	3.2 a	2.5 a	2.5 a	40.0 b
	TDZ + GA_3 + TIBA	2.7 b	2.2 ab	1.8 b	50.0 b
	BAP + GA_3	1.7 c	1.5 b	1.3 bc	100.0 a
	MT + GA_3	2.1 c	1.2 b	1.2 bc	100.0 a

* Phytoregulator: TDZ 0.4 mg L^{-1}; NAA 0.2 mg L^{-1}; GA_3 0.25 mg L^{-1}; TIBA 1 g L^{-1}; BAP 0.05 mg L^{-1}; MT 0.5 mg L^{-1}. ** M.M.I. = Mean multiplication index was calculated as the mean number of shoots obtained during one subculture. Data were a mean of three subcultures. Different letters for each column in each treatment or interaction of treatments indicate significant differences (SNK test at $p \leq 0.001$).

Figure 4. 'Finola' shoot proliferation induced by TDZ and NAA, scale bar 1 cm.

3.3. Results for In Vitro Rooting and Ex Vitro Acclimatization

After 20 days, the effects of different types of light and two different concentrations of IBA on the rooting induction were evaluated (Table 4). The best results were obtained under red/blue lights using IBA at a lower concentration (Figure 5), whereas no rooting was obtained under white/pink lights. Once transplanted into the climate-controlled greenhouse, the plantlets showed a survival rate of over 90 percent (Figure 6).

Table 4. Effect of lights and different concentrations of IBA on the induction of rooting on 'Finola' shoots detected after 20 days.

Lights	IBA (mg L^{-1})	Rooting (%)	Length (cm)	Roots (n.)
White/Pink	0.1	0.0	-	-
	0.5	0.0	-	-
Red/Blue	0.1	60.0 a	3.5 a	2.5 a
	0.5	45.0 b	2.3 a	1.5 a

Different letters for each column indicate significant differences (SNK test at $p \leq 0.001$).

Figure 5. Rooted shoot of 'Finola' obtained under red/blue light, using IBA at the lower concentration (0.1 mg L^{-1}), scale bar 1 cm.

Figure 6. Detail of rooted propagule of 'Finola' shoots obtained on MB × TDZ × NAA, scale bar 1 cm (**a**); plantlet during acclimatization phase, scale bar 1 cm (**b**); plant from in vitro culture, fully adapted to live conditions, scale bar 8 cm (**c**).

4. Discussion

Several authors underscored strong cultivar-specific responses to treatments in *Cannabis* tissue culture [35–37]. In this study, to define the most suitable protocol for the micropropagation of *C. sativa* L. subsp. *ruderalis* Janish., cv. 'Finola', different experimental procedures were tested. The first step of this experimental research was to select the most suitable explant to establish an aseptic culture. The next step involved drafting a functional in vitro propagation protocol while exploring the choice of nutrient substrates with different known combinations of nutrients and growth regulators, Finally, rooting trials were carried out, also evaluating the influence of different kinds and combinations of light to obtain microplants ready for transplantation in the greenhouse.

To determine the most suitable explant for micropropagation initiation, both in vitro germinated seeds and nodal explants derived from mother plants were evaluated. To ensure a successful in vitro culture, it is crucial to optimize explant sterilization protocols [38]. The comparative seed sterilization test conducted using two different sterilizing agents indicated that sodium hypochlorite was the most effective in reducing contamination values, with results of less than 10 percent. On the other hand, mercuric chloride was too harsh on the seeds, resulting in arrested germination or reduced development and slightly oxidized shoots during in vitro growth. This strong toxicity, causing irreversible damage to plants, was reported in several studies on seeds of different species [39–41]. Furthermore, NaClO is recommended for seed sterilization due to it is lower costs, higher availability, and decreased toxicity compared to $HgCl_2$. Evaluating the effects of different growth regulators on inducing the multiplication of seedlings germinated under in vitro conditions yielded the best results when using metatopolin (MT) 0.5 mg L^{-1}, although the differences were not statistically significant (Table 1). Unfortunately, 21 days after the start of the first subculture, under identical growth conditions, the explants responded by arresting growth, showing signs of necrosis of the apices, diffuse hyperhydricity, or early flowering. Starting directly from in vitro germinated seeds is not the best option for this hemp cultivar, due to the inability to differentiate the starting material based on sex and the high difficulty in cutting for subculturing, which effectively limits the possibility of mass propagation of the selected clone. This culture decline, also observed by Page et al. [35] in *Cannabis sativa*, limits long-term culture proliferation. Similarly, in 'Finola', this decline was already observed in the first subculture through hyperhydricity and/or the death of the shoots.

The results relating to the two climate-controlled environments used to grow the mother plants showed different morphological developments achieved by the plants (Table 2) that strongly affected adaptation to the aseptic culture of the nodal explants excised for in vitro micropropagation. Only the plants grown in the thermostatically controlled cabinet became suitable (Table 3).

'Finola' showed the pronounced presence of hyperhydrated shoots when grown on Driver and Kuniyuki (DKW). Indeed, the latter resulted in the presence of vitrescence in almost all explants (90%), causing the alteration of tissues that were found to be thickened, turgid, and glassy in texture, with leaves featuring lightened coloration, very slow or no growth, a poor multiplication index, and brittle leaves. This behavior is different from the results reported by Page et al. [35], which suggested the use of DKW nutrient medium for multiple commercial varieties of *C. sativa*. These divergent results could be due to the different subspecies evaluated in this study.

Based on the results shown in Table 3, most of the phytoregulators tested, alone or in combination, produced strong shortening and clustering of the shoots. Although several authors [29–32] suggest using phytoregulators like BAP and MT to improve in vitro culture of *Cannabis*, in the case of 'Finola', the results achieved using these phytoregulators have been very disappointing. Nevertheless, the addition of GA_3 to the nutrient substrate to promote stem elongation, which is common in many micropropagation protocols [42–44], resulted only in the excessive growth of leaves, while the internodes on the stem remained very short. This factor limited the number of microcuttings that could be produced. Moreover, to overcome the apical dominance of in vitro shoots that was strongly linked to the induction of in vitro flowering in 'Finola', 2,3,5-triiodobenzoic acid (TIBA) was tested in combination with TDZ and GA_3. TIBA is a synthetic inhibitor of auxin flux and has been widely applied in studies on in vitro shoot regeneration, as reported in *Rosa hybrida* [45], *Cucumis sativus* [46], and 'Finola' in combination with TDZ [22]. The results of the present study showed an increase of about 1 cm in the length of the shoots compared with the combination of BAP + GA_3 or MT + GA_3, albeit with a lower multiplication rate than the best mean value achieved with the combination TDZ + NAA (Figure 6a). Instead, the best results were found by adding TDZ combined with NAA. TDZ is a phenylurea-substituted compound that is highly effective in the tissue culture morphogenesis of many plant species, including *C. sativa* variety MX-1 [47–49]. Indeed, several authors added this compound in different concentrations, alone or in combination with NAA, to induce the multiplication or regeneration of shoots from different explants of *Cannabis sativa* [9,23,32,50], achieving high multiplication indices for each explant. In contrast to these results, 'Finola' yielded a much lower multiplication rate, caused by the high degree of apical dominance and low levels of branching.

To prevent or delay flowering, which represents the greatest limit in maintaining a vegetative state for day-neutral genotypes such as 'Finola' [21], the apical tips were regularly eliminated in each subculture to break apical dominance and promote axillary bud growth. This technique was already found to be effective in promoting the shoot regeneration of *C. sativa* var. Epsilon 68 [51] and *Piper sarmentosum* [50].

Lastly, the different LED lighting systems used in the phase of rooting induction were discussed. In recent years, several studies have been conducted to determine the best light source to improve the quality of micropropagated plants, and, at the same time, reduce costs [33,52]. In particular, Budiarto (2010) [34] reported that the in vitro root activity increased under red and blue LED lights. In this study, 'Finola' shoots exhibited good rooting susceptibility under red/blue lights when treated with IBA at lower concentrations.

However, IBA is an auxin frequently used in the rooting induction of *C. sativa* [53–55]. This application resulted in the development of an extensive root system, in terms of both length and number of roots, which facilitated the survival of all plantlets after transplantation into the climate-controlled greenhouse (Figure 6).

5. Conclusions

'Finola' (*C. sativa* subsp. *ruderalis*) is an in vitro hemp recalcitrant cultivar because it is day-neutral and flowers independent of the photoperiod. For this reason, 'Finola' is difficult to maintain in a vegetative state. The ambitious aim of this research was to overcome this limitation and define an efficient protocol for micropropagation by evaluating many different factors and combining them. The positive results obtained in this work could contribute to developing a tissue culture system suitable for other recalcitrant hemp varieties.

In the future, the selection of in vitro elite plant materials could help to realize the super intensive production of 'Finola' under protected environments in very short cycles. This approach can offer a viable option for maximizing production yields while maintaining optimal conditions for in vitro plant growth.

Author Contributions: Conceptualization, C.R.; methodology, C.R. and G.N.B., validation, L.T., C.R. and G.D.M.; investigation, G.N.B.; data curation C.R. and G.N.B.; writing—original draft preparation, C.R. and G.N.B.; writing—review and editing, L.T., C.R., G.D.M. and C.P.; project administration, G.D.M.; funding acquisition, G.D.M. All authors have read and agreed to the published version of the manuscript.

Funding: This research was funded by Puglia Region, 6 June 2017, n. 21 "Promotion of the cultivation of hemp for productive and environmental purposes". Project "Micropropagation Apulian Autoflowering".

Data Availability Statement: Not applicable.

Acknowledgments: The support of the project leader Fattoria della Canapa and the partner Rete Etruscum for the success of the results of this research is gratefully acknowledged.

Conflicts of Interest: The authors declare no conflict of interest.

References

1. Salami, S.A.; Martinelli, F.; Giovino, A.; Bachari, A.; Arad, N.; Mantri, N. It is our turn to get cannabis high: Put cannabinoids in food and health baskets. *Molecules* **2020**, *25*, 4036. [CrossRef]
2. Zhao, J.; Xu, Y.; Wang, W.; Griffin, J.; Wang, D. High Ethanol Concentration (77 g/L) of Industrial Hemp Biomass Achieved Through Optimizing the Relationship between Ethanol Yield/Concentration and Solid Loading. *ACS Omega* **2020**, *5*, 21913–21921. [CrossRef]
3. Shen, P.; Gao, Z.; Fang, B.; Rao, J.; Chen, B. Ferreting out the secrets of industrial hemp protein as emerging functional food ingredients. *Trends Food Sci. Technol.* **2021**, *112*, 1–15. [CrossRef]
4. Amaducci, S.; Scordia, D.; Liu, F.H.; Zhang, Q.; Guo, H.; Testa, G.; Cosentino, S.L. Key evaluation techniques for hemp in Europe and in China. *Ind. Crops Prod.* **2015**, *69*, 2–16. [CrossRef]
5. Zhang, Q.; Chen, X.; Guo, H.; Trindade, L.M.; Salentijn, E.M.J.; Guo, R.; Guo, M.; Xu, Y.; Yang, M. Latitudinal Adaptation and Genetic Insights into the Origins of *Cannabis sativa* L. *Front. Plant Sci.* **2018**, *9*, 1876. [CrossRef]
6. Gloss, D. An overview of products and bias in research. *Neurotherapeutics* **2015**, *12*, 731–734. [CrossRef]
7. McPartland, J.M.; Clarke, R.C.; Watson, D.P. *Hemp Diseases and Pests: Management and Biological Control: An Advanced Treatise*; CABI: Wallingford, UK, 2000.
8. Dowling, C.A.; Shi, J.; Toth, J.A.; Quade, M.A.; Smart, L.B.; McCabe, P.F.; Melzer, R.; Schilling, S. A FLOWERING LOCUS T ortholog is associated with photoperiod-insensitive flowering in hemp (*Cannabis sativa* L.). *bioRxiv* **2023**. *preprint*.
9. Hillig, K.W. Genetic evidence for speciation in Cannabis (Cannabaceae). *Genet. Resour. Crop Evol.* **2005**, *52*, 161–180. [CrossRef]
10. Fadel, D.; Assaad, N.; Alghazal, G.; Hamouche, Z.; Lazari, D. "Finola" Cannabis Cultivation for Cannabinoids Production in Thessaloniki-Greece. *J. Agric. Sci.* **2020**, *12*, 172–181. [CrossRef]
11. Pagnani, G.; Pellegrini, M.; Galieni, A.; D'Egidio, S.; Matteucci, F.; Ricci, A.; Stagnari, F.; Sergi, M.; Lo Sterzo, C.; Pisante, M.; et al. Plant growth-promoting rhizobacteria (PGPR) in *Cannabis sativa* 'Finola' cultivation: An alternative fertilization strategy to improve plant growth and quality characteristics. *Ind. Crops Prod.* **2018**, *123*, 75–83. [CrossRef]

12. George, E.F.; Hall, M.A.; De Klerk, G.J. (Eds.) *Plant Propagation by Tissue Culture: Volume 1. The Background*; Springer Science & Business Media: Berlin/Heidelberg, Germany, 2007.
13. Cardoso, J.C.; Lee Tseng Sheng, G.; Teixeira da Silva, J.A. Micropropagation in the twenty-first century. In *Plant Cell Culture Protocols*; Springer: Berlin/Heidelberg, Germany, 2018; pp. 17–46.
14. Boonsnongcheep, P.; Benyakan, P. Factors affecting micropropagation of *Cannabis sativa* L.: A review. *Pharm. Sci. Asia* **2020**, *47*, 21–29. [CrossRef]
15. Adhikary, D.; Kulkarni, M.; El-Mezawy, A.; Mobini, S.; Elhiti, M.; Gjuric, R.; Ray, A.; Polowick, P.; Slaski, J.J.; Jones, M.; et al. Medical cannabis and industrial hemp tissue culture: Present status and future potential. *Front. Plant Sci.* **2021**, *12*, 627240. [CrossRef]
16. Chadipiralla, K.; Gayathri, P.; Rajani, V.; Reddy, P.V.B. Plant Tissue Culture and Crop Improvement. In *Sustainable Agriculture in the Era of Climate Change*; Roychowdhury, R., Choudhury, S., Hasanuzzaman, M., Srivastava, S., Eds.; Springer: Cham, Switzerland, 2020; pp. 391–412.
17. El-Sherif, N.A. Impact of plant tissue culture on agricultural sustainability. In *Sustainability of Agricultural Environment in Egypt: Part II: Soil-Water-Plant Nexus*; The Handbook of Environmental Chemistry Series; Springer: Berlin/Heidelberg, Germany, 2019; pp. 93–107.
18. Lata, H.; Chandra, S.; Khan, I.A.; ElSohly, M.A. Micropropagation of *Cannabis sativa* L.—An update. In *Cannabis sativa L.—Botany and Biotechnology*; Springer: Berlin/Heidelberg, Germany, 2017; pp. 285–297.
19. Monthony, A.S.; Bagheri, S.; Zheng, Y.; Jones, A.M.P. Flower power: Floral reversion as a viable alternative to nodal micropropagation in *Cannabis sativa*. *In Vitr. Cell. Dev. Biol. Plant* **2021**, *57*, 1018–1030. [CrossRef]
20. Schilling, S.; Melzer, R.; Dowling, C.A.; Shi, J.; Muldoon, S.; McCabe, P.F. A protocol for rapid generation cycling (speed breeding) of hemp (*Cannabis sativa*) for research and agriculture. *Plant J.* **2023**, *113*, 437–445. [CrossRef]
21. Monthony, A.S.; Page, S.R.; Hesami, M.; Jones, A.M.P. The Past, Present and Future of *Cannabis sativa* Tissue Culture. *Plants* **2021**, *10*, 185. [CrossRef] [PubMed]
22. Dreger, M.; Szalata, M. The Effect of TIBA and NPA on Shoot Regeneration of *Cannabis sativa* L. Epicotyl Explants. *Agronomy* **2022**, *12*, 104. [CrossRef]
23. Galán-Ávila, A.; García-Fortea, E.; Prohens, J.; Herraiz, F.J. Development of a Direct in vitro Plant Regeneration Protocol from *Cannabis sativa* L. Seedling Explants: Developmental Morphology of Shoot Regeneration and Ploidy Level of Regenerated Plants. *Front. Plant Sci.* **2020**, *11*, 645. [CrossRef]
24. ElSohly, M.A.; Radwan, M.M.; Gul, W.; Chandra, S.; Galal, A. Phytochemistry of *Cannabis sativa* L. In *Phytocannabinoids. Progress in the Chemistry of Organic Natural Products*; Kinghorn, A., Falk, H., Gibbons, S., Kobayashi, J., Eds.; Springer: Cham, Switzerland, 2017; Volume 103, pp. 1–36.
25. Ancona, S.; De Mastro, G.; Jenderek, M.M.; Ruta, C. Micropropagation Supports Reintroduction of an Apulian Artichoke Landrace in Sustainable Cropping Systems. *Agronomy* **2021**, *11*, 1169. [CrossRef]
26. Murashige, T.; Skoog, F. A revised medium for rapid growth and bio assays with tobacco tissue cultures. *Physiol. Plant.* **1962**, *15*, 473–497. [CrossRef]
27. Nitsch, J.P.; Nitsch, C. Haploid plants for pollen grains. *Science* **1969**, *63*, 85–87. [CrossRef]
28. Page, S.R.G.; Monthony, A.S.; Jones, A.M.P. DKW basal salts improve micropropagation and callogenesis compared with MS basal salts in multiple commercial cultivars of *Cannabis sativa*. *Botany* **2021**, *99*, 269–279. [CrossRef]
29. Lata, H.; Chandra, S.; Techen, N.; Khan, I.A.; ElSohly, M.A. In vitro mass propagation of *Cannabis sativa* L.: A protocol refinement using novel aromatic cytokinin meta-topolin and the assessment of eco-physiological, biochemical and genetic fidelity of micropropagated plants. *J. Appl. Res. Med. Aromat. Plants* **2016**, *3*, 18–26. [CrossRef]
30. Mehdi, M.; Vali-Ollah, G.-O.; Sepide, T. The effect of different concentrations of TDZ and BA on in vitro regeneration of Iranian cannabis (*Cannabis sativa*) using cotyledon and epicotyl explants. *J. Plant Mol. Breed.* **2015**, *3*, 20–27.
31. Mestinšek Mubi, Š.; Svetik, S.; Flajšman, M.; Murovec, J. In Vitro tissue culture and genetic analysis of two high-CBD medical Cannabis (*Cannabis sativa* L.) breeding lines. *Genetika* **2020**, *52*, 925–941. [CrossRef]
32. Chaohua, C.; Gonggu, Z.; Lining, Z.; Chunsheng, G.; Qing, T.; Jianhua, C.; Xinbo, G.; Dingxiang, P.; Jianguang, S. A rapid shoot regeneration protocol from the cotyledons of hemp (*Cannabis sativa* L.). *Ind. Crops Prod.* **2016**, *83*, 61–65. [CrossRef]
33. Dutta Gupta, S.; Jatothu, B. Fundamentals and applications of light-emitting diodes (LEDs) in in vitro plant growth and morphogenesis. *Plant Biotechnol. Rep.* **2013**, *7*, 211–220. [CrossRef]
34. Budiarto, K. Spectral quality affects morphogenesis on Anthurium plantlet during in vitro culture. *Agrivita* **2010**, *32*, 234–240.
35. Page, S.R.G.; Monthony, A.S.; Jones, A.M.P. Basal media optimiza-tion for the micropropagation and callogenesis of *Cannabis sativa* L. *bioRxiv* **2020**. *preprint*.
36. Campbell, L.G.; Naraine, S.G.U.; Dusfresne, J. Phenotypic plasticity influences the success of clonal propagation in industrial pharmaceutical *Cannabis sativa*. *PLoS ONE* **2019**, *14*, e0213434. [CrossRef]
37. Codesido, V.; Meyer, S.; Casano, S. Influence of media composition and gen-otype for successful *Cannabis sativa* L. In Vitro introduction. *Acta Hortic.* **2020**, *1285*, 75–80. [CrossRef]
38. Dodds, J.H.; Roberts, L.W. *Experiments in Plant Tissue Culture*, 2nd ed.; Cambridge University Press: Cambridge, UK, 1985; pp. 21–35.

39. Barampuram, S.; Allen, G.; Krasnyanski, S. Effect of Various Sterilization Procedures on the in Vitro Germination of Cotton Seeds. *Plant Cell Tissue Organ Cult.* **2014**, *118*, 179–185. [CrossRef]
40. Lu, L.M.; An, Y. Effects of Different Disinfectant on Sterilization Effect and Germination of Tobacco Seeds. *Seed* **2012**, *31*, 93–95.
41. Yuan, Y. Selection and Disinfection of Tissue Culture Explants of Toona sinensis Roem. *Anhui Agric.* **2020**, *26*, 19–20.
42. Camara, M.; Vandenberghe, L.; Rodríguez, C.; Oliveira, J.; Faulds, C.; Bertrand, E.; Soccol, C. Current advances in gibberellic acid (GA3) production, patented technologies and potential applications. *Planta* **2018**, *248*, 1049–1062. [CrossRef]
43. Ayano, M.; Kani, T.; Kojima, M.; Sakakibara, H.; Kitaoka, T.; Kuroha, T.; Angeles-Shim, R.; Kitano, H.; Nagai, K.; Ashikari, M. Gibberellin biosynthesis and signal transduction is essential for internode elongation Subburaman in deepwater rice. *Plant Cell Environ.* **2014**, *37*, 2313–2324. [CrossRef] [PubMed]
44. Geng, F.; Moran, R.; Day, M.; Halteman, W.; Zhang, D. Increasing In Vitro Shoot Elongation and Proliferation of 'G.30' and 'G.41' Apple by Chilling Explants and Plant Growth Regulators. *HortScience* **2016**, *51*, 899–904. [CrossRef]
45. Sing, S.K.; Syamal, M.M. Anti-auxin enhance *Rosa hybrida* L. micropropagation. *Biol. Plant.* **2000**, *43*, 279–281. [CrossRef]
46. Shukla, P.S.; Das, A.K.; Jha, B.; Agarwal, P.K. High-frequency in vitro shoot regeneration in *Cucumis sativus* by inhibition of endogenous auxin. *In Vitr. Cell. Dev. Biol. Plant* **2014**, *50*, 729–737. [CrossRef]
47. Donna, I.L.; John, E.P. Thidiazuron stimulates adventitious shoot production from *Hydrangea quercifolia* Bartr., leaf explants. *Sci. Hortic.* **2004**, *101*, 121–126.
48. Murthy, B.N.S.; Murch, S.J.; Saxena, P.K. Thidiazuron: A potent regulator of in vitro plant morphogenesis. *In Vitr. Cell. Dev. Biol. Plant* **1998**, *34*, 267–275. [CrossRef]
49. Lata, H.S.; Chandra, I.K.; ElSohly, M.A. Thidiazuron-induced high-frequency direct shoot organogenesis of *Cannabis sativa* L. *In Vitr. Cell. Dev. Biol. Plant* **2009**, *45*, 12–19. [CrossRef]
50. Stephin, S.; Gangaprasad, A.; Mathew, S.P.; Muthukrishnan, S. Enhanced In Vitro Shoot Multiplication of *Piper sarmentosum* by Suppression of Apical Dominance. *Proc. Natl. Acad. Sci. India Sect. B Biol. Sci.* **2020**, *90*, 87–94. [CrossRef]
51. Wróbel, T.; Dreger, M.; Wielgus, K.; Słomski, R. Modified nodal cuttings and shoot tips protocol for rapid regeneration of *Cannabis sativa* L. *J. Nat. Fibers* **2022**, *19*, 536–545. [CrossRef]
52. Batista, D.S.; Felipe, S.H.S.; Silva, T.D.; Motta de Castro, K.; Mamedes-Rodrigues, T.C.; Miranda, N.A.; Ríos-Ríos, A.-M.; Vidal Faria, D.; Fortini, E.A.; Chagas, K.; et al. Light quality in plant tissue culture: Does it matter? *In Vitr. Cell. Dev. Biol. Plant* **2018**, *54*, 195–215. [CrossRef]
53. Lata, H.; Chandra, S.; Khan, I.; ElSohly, M.A. High frequency plant regeneration from leaf derived callus of high Δ9-tetrahydrocannabinol yielding *Cannabis sativa* L. *Planta Med.* **2010**, *76*, 1629–1633. [CrossRef] [PubMed]
54. Stephen, C.; Zayas, V.A.; Galic, A.; Bridgen, M.P. Micropropagation of hemp (*Cannabis sativa* L.). *HortScience* **2023**, *58*, 307–316. [CrossRef]
55. Wang, R.; He, L.S.; Xia, B.; Tong, J.F.; Li, N.; Peng, F. A micropropagation system for cloning of hemp (*Cannabis sativa* L.) by shoot tip culture. *Pak. J. Bot.* **2009**, *41*, 603–608.

horticulturae

Article

The Use of Wheat Starch as Gelling Agent for In Vitro Proliferation of Blackberry (*Rubus fruticosus* L.) Cultivars and the Evaluation of Genetic Fidelity after Repeated Subcultures

Doina Clapa [1], Monica Hârța [1,*], Katalin Szabo [2,*], Bernadette-Emőke Teleky [3] and Doru Pamfil [1,4]

1 Faculty of Horticulture and Business in Rural Development, BIOCERA Research Centre, University of Agricultural Sciences and Veterinary Medicine Cluj-Napoca, 3-5 Manastur Street, 400372 Cluj-Napoca, Romania; doina.clapa@usamvcluj.ro (D.C.); doru.pamfil@usamvcluj.ro (D.P.)
2 Technological Transfer Center COMPAC, University of Agricultural Sciences and Veterinary Medicine Cluj-Napoca, 3-5 Manastur Street, 400372 Cluj-Napoca, Romania
3 Life Science Institute, University of Agricultural Sciences and Veterinary Medicine Cluj-Napoca, 3-5 Manastur Street, 400372 Cluj-Napoca, Romania; bernadette.teleky@usamvcluj.ro
4 Romanian Academy, Cluj-Napoca Branch, Republicii St. 9, 400015 Cluj-Napoca, Romania
* Correspondence: monica.harta@usamvcluj.ro (M.H.); katalin.szabo@usamvcluj.ro (K.S.)

Abstract: Micropropagation has an important role in the large-scale production of blackberry plant material, given the high proliferation rates of this species. The aim of the present study was to evaluate the proliferative capacity of blackberry grown in vitro on wheat starch-gelled culture medium compared to classical agar-gelled medium and to assess the genetic fidelity between the proliferated shoots in starch-gelled culture medium and their mother plants. Six blackberry varieties ('Čačanska Bestrna', 'Chester Thornless', 'Driscoll's Victoria', 'Loch Ness', 'Polar', and 'Karaka Black') were tested. For the in vitro shoots proliferation, Murashige and Skoog (MS) medium supplemented with 0.5 mg dm^{-3} 6-benzyladenine (BA) was used. The conventional medium was gelled with 0.5% plant agar, and wheat starch was used as an alternative gelling agent in a concentration of 5%. The results showed that for all blackberry cultivars, the highest number of shoots/inoculum was obtained in wheat starch-gelled culture medium, with a maximum value of 54.42 ± 4.18 presented by 'Karaka Black'. Considering the length of the proliferated shoots, all tested cultivars presented outstanding results on the culture medium gelled with 5% wheat starch. The highest values regarding shoots length were observed on the 'Chester Thornless' followed by 'Čačanska Bestrna', and 'Loch Ness' with values of 5.55 ± 0.04 cm, 5.46 ± 0.06 cm, and 5.37 ± 0.09 cm, respectively. The genetic uniformity of the micropropagated shoots in relation to their mother plants was confirmed by sequence-related amplified polymorphism (SRAP) and start codon targeted (SCoT) molecular markers.

Keywords: gelling agent; proliferation rate; rheological analyses; SRAP; SCoT

Citation: Clapa, D.; Hârța, M.; Szabo, K.; Teleky, B.-E.; Pamfil, D. The Use of Wheat Starch as Gelling Agent for In Vitro Proliferation of Blackberry (*Rubus fruticosus* L.) Cultivars and the Evaluation of Genetic Fidelity after Repeated Subcultures. *Horticulturae* **2023**, 9, 902. https://doi.org/10.3390/horticulturae9080902

Academic Editors: Alicja Tymoszuk, Dariusz Kulus and Michailidis Michail

Received: 4 July 2023
Revised: 28 July 2023
Accepted: 4 August 2023
Published: 8 August 2023

1. Introduction

Blackberry (*Rubus fruticosus* L.) is a well-known shrub from the Rubus genus with edible fruits having a delicious taste and a pleasant aroma due to its specific biochemical composition. Fruits are widely consumed as they have a high content in vitamins, minerals, antioxidants, and dietary fibers that are beneficial for human health and well-being [1–4].

Similar to other *Rubus* species with edible fruits, blackberry varieties have been commercially propagated by classical methods of vegetative propagation, e.g., by hardwood and softwood cuttings, by layering, and/or by bush division [5–8]. Another successful strategy for large-scale commercial micropropagation of blackberry plant material includes in vitro tissue culture-based techniques [9–14].

The micropropagation process involves several distinct stages: initiation or the establishment of culture, proliferation of the shoots, elongation of the shoots, and rooting, followed by the acclimatization of in vitro-grown plants [15–18]. Among the aforementioned stages, shoot proliferation is considered crucial for successful mass propagation linked to the quantity and quality of the obtained in vitro plant material [19,20].

Prior research on the multiplication of *Rubus* species has demonstrated that blackberry varieties can be successfully propagated using a wide range of culture media. Among these, the Murashige and Skoog (MS) medium have emerged as the most commonly employed options [21–25], followed by McCown's Woody Plants medium [4,13,25–27], Gamborg's B5 medium [4,28], Juglans Medium [29–32], Quoirin–Lepoivre [25,33,34], Anderson medium [25,35,36] and Murashige and Tucker medium [17,37].

The multiplication of blackberry shoots is usually made on media supplemented with cytokinins, in concentrations between 0.3 and 2.0 mg dm^{-3} as the major plant growth regulator, and with smaller concentrations of auxins, and occasionally with gibberellins [13,31,32]. Thus, the highest values of shoot proliferation rates were reported in the presence of 6-benzylaminopurine (BAP) alone or in combination with indole-3-butyric acid (IBA) and/or gibberellic acid (GA3) [12,14,38–42].

The large-scale gelling agent used for in vitro propagation of blackberry is agar [13,14,43,44]. Given the relatively high cost of agar, several alternatives were tested in the blackberry multiplication phase previously. For example, Gelcarin GP-812, Isubgol, and Guar gum were used for the in vitro multiplication phase of 'Thornless evergreen' [31]. The culture media gelled with agents such as wheat starch and corn starch were also successfully tested on the blackberry varieties 'Navaho', and 'Čačanska Bestrna' [12].

During tissue culture for the large-scale propagation of commercially important plants, somaclonal variations might occur at any stage of the plantlet's development, especially in the multiplication stage. These variations can be caused by environmental conditions, explant type, the number of successive subcultures, and culture media type [45,46]. First of all, the use of plant growth regulators in high doses, combined with the number of subcultures, causes stress that leads to cellular instability, triggering genetic or epigenetic variations in plants in vitro [47]. Another aspect to consider is the morphogenetic pathway utilized for clone production. Obtaining plants through axillary branching typically does not lead to the generation of variants, whereas cultures that undergo a callus phase are the ones that theoretically promote a higher mutation rate [46,48]. Therefore, it is important to check the genetic uniformity of the multiplied plants in relation to the mother plants to confirm their quality for commercial use [49].

The assessment of genetic fidelity between the micropropagated plantlets and their mother plants using molecular markers such as Sequence-related amplified polymorphism (SRAP) and Start codon targeted polymorphism (SCoT) serve as valuable tools to test the uniformity of plant material [50–53].

In this context, the first objective of the present study was to investigate the in vitro proliferative capacity of six *Rubus fruticosus* L. cultivars grown on wheat starch-gelled culture medium in comparison with the same cultivars grown on the classical agar-gelled medium. The second objective of the study was to evaluate the genetic fidelity between shoots propagated in starch-gelled culture medium after twelve successive subcultures and their mother plants using SRAP and SCoT molecular markers.

2. Materials and Methods

2.1. Plant Material and In Vitro Shoot Cultures

The studied blackberry cultivars were 'Čačanska Bestrna', 'Chester Thornless', 'Driscoll's Victoria', 'Loch Ness', 'Polar' (thornless blackberry), and 'Karaka Black' (thorny blackberry), having a tetraploid genetic structure (2n = 4x = 28).

In vitro culture initiation was carried out on annual shoots at the end of October on MS medium supplemented with 3% sucrose and 0.5 mg dm^{-3} 6-benzyladenine (BA), gelled with 0.5% plant agar, and pH = 5.8. The small cuttings were washed with tap water and rinsed with sterile deionized water. Disinfection was made with 20% bleach (ACE Procter and Gamble, București, Romania; <5% active ingredient) for 20 min and rinsed three times with sterile water. The axillary and apical buds were excised and inoculated into the culture medium, one bud/test tube. After six weeks, the regenerated shoots were transferred to a culture medium with the same composition for in vitro culture stabilization. Further, eleven successive subcultures were carried out (each subculture lasting for 10 weeks) on the same agar-gelled culture medium, and in parallel, on culture media with 5% wheat starch.

In this experiment, specifically during the twelfth subculture, the in vitro culture of blackberry was conducted under identical conditions. Two different variants of the culture media were prepared: one was solidified with plant agar at a concentration of 0.5%, while the other utilized wheat starch at a concentration of 5%. The plants were cultured in 720 mL jars with a diameter of 9 cm and a height of 13.5 cm, with screw caps equipped with leukopore tape filters, using 100 mL culture medium/jar.

The culture medium gelled with agar was sterilized for 20 min in an autoclave at 121 °C and 0.11 MPa, while the other one gelled with wheat starch was sterilized for 30 min under the same conditions. The medium gelled with starch needs a longer sterilization period to avoid infections. All the components of the culture media were added prior to sterilization, as well the pH was adjusted to 5.8.

In each jar, four shoot fragments of approx. 2 cm were inoculated.

The in vitro cultures were incubated in the growth room at 16 h photoperiod, 32.4 μmol $m^{-2}s^{-1}$ light intensity (Philips CorePro LEDtube 1200 mm 16W865 CG, 1600 lm Cool Daylight) at 22 $\pm$ 3 °C and 50 $\pm$ 5% humidity.

All the necessary components of the culture media were purchased from Duchefa (Biochemie B.V, Haarlem, Netherlands), and the wheat starch was purchased from SanoVita (https://sanovita.ro/, accessed on 15 February 2023).

2.2. Estimation of the Costs per Liter of Culture Medium

The cost per liter of medium was calculated based on the current prices of various types of agar and wheat starch (Table 1).

Table 1. Comparing the costs of various gelling agents per liter of culture medium.

Gelling Agent	Price/kg (Euro)	g/L Culture Medium	Price/L Culture Media (Euro)	Price Difference (Euro)
Wheat starch SANOVITA (used in the present study)	2	50	0.10	-
Plant agar * (used in the present studies)	87.4	5	0.44	0.34
Daishin agar *	206.5	7	1.45	1.35
Micro agar *	111.7	6	0.67	0.55
Phyto agar *	101	6	0.61	0.51
Gelrite *	174.4	3	0.52	0.42
Agar Sigma A1296 **	438	5.6	2.45	2.35

Catalog prices * Duchefa and ** Sigma.

2.3. Rheological Analyses

Dynamic rheological analyses were carried out on both wheat starch-gelled culture medium and agar-gelled medium using an Anton Paar MCR 72 modular compact rheometer (Anton Paar, Graz, Austria). The rheometer was equipped with a Peltier plate-plate system (P-PTD 200/Air) that allowed temperature control. However, for this study, the samples' viscosity was measured at room temperature of 22 $\pm$ 3 °C. Each sample, approximately 3 mL in volume, was placed between two plates. The upper plate had a smooth parallel plate geometry with a diameter of 50 mm (PP-50-67300), while the lower plate was at a gap of 1 mm [54]. Prior to measurement, any excess sample was removed, and the samples were left undisturbed for 5 min to ensure thermal equilibrium before testing. Duplicate measurements were conducted using a linearly increasing shear rate, ranging from 5 to 300 1/s [55].

2.4. Genetic Fidelity Analysis Using SRAP and SCoT Markers

Each analyzed cultivar was represented by 12 samples obtained from three jars, each containing four initial inoculums. From each inoculum, a shoot was randomly selected, and the leaves were used for DNA isolation. The harvested leaves were dried, ground into a fine powder (TissueLyser II, Qiagen, Hilden, Germany), and stored at 4 °C until DNA isolation.

2.4.1. DNA Isolation

Total genomic DNA was isolated from 0.1 g of dried powder using a protocol based on the CTAB (cetyltrime-thylammonium bromide) method according to Lohdi et al., 1994 [56] and slightly modified by Pop et al., 2003 [57] and Bodea et al., 2016 [58]. The DNA purity and concentration were evaluated using a NanoDrop 1000 spectrophotometer (Thermo Fisher Scientific, Waltham, MA, USA). Prior to performing the polymerase chain reaction (PCR) amplifications, all the DNA samples were diluted to 50 ng/µL using sterile double distilled water.

2.4.2. SCoT and SRAP Analysis

For the SRAP analysis, the PCR amplification reactions were carried out according to Li and Quiros' (2001) [59] protocol, and the reaction volumes were adjusted to 15 µL. The reaction mixture consisted of 50 ng/µL of gDNA, 5X Green Go Taq flexi buffer, 1.5 mM $MgCl_2$, 0.2 mM of dNTPs, 0.3 µM of both forward and reverse primer, nuclease-free water and 1 U of Taq DNA polymerase (Promega, Madison, WI, USA).

The DNA amplification was carried out in a Gradient thermal cycler, SuperCycler Trinity (Kyratec, Mansfield, Australia), programmed for 1 cycle of 5 min at 94 °C for initial denaturation, followed by five cycles of 1 min of denaturation at 94 °C, 1 min of annealing at 35 °C and 1 min of elongation at 72 °C and then 35 cycles (94 °C for 1 min; 50 °C for 1 min and 72 °C for 1 min) with a final elongation step of 10 min at 72 °C.

For the SCoT analysis, the PCR amplification reactions were performed using the protocol described by Collard and Mackill (2009) [60]. The reaction mixture (a total volume of 15 µL) consisted of 50 ng/µL of gDNA, distilled H_2O for the PCR reactions, 5X GoTaq Flexi Green buffer (Promega, Madison, WA, USA), 1.5 mM $MgCl_2$ (Promega, Madison, WA, USA), 0.2 mM of dNTP mix (Promega, Madison, WA, USA), 1 µM SCoT primer (GeneriBiotech, Hradec Králové, Czechia), and 1U of GoTaq polymerase (Promega, Madison, WA, USA). The PCR temperature cycling conditions were: (a) 1 cycle of 5 min at 94 °C for initial denaturation, (b) 35 cycles of denaturation at 94 °C for 1 min, annealing at 50 °C for 1 min and elongation at 72 °C for 2 min, and (c) the final elongation step of 7 min at 72 °C.

The list of SRAP primer combinations and SCoT primers used in this study is shown in Table S1. The PCR amplifications were repeated twice for each SCoT primer and each SRAP primer combination to ensure the reproducibility of the results. Separation of the PCR amplicons for both techniques was performed by electrophoresis on 1.4% agarose gels (Promega, Madison, WA, USA) stained with RedSafeTM Nucleic Acid staining solution (iNtRON Biotech, Seoul, Republic of Korea) in 1X TBE (Tris Borate-EDTA buffer), at 110 V and 136 mA for 2.5–3 h. The electrophoretic profiles were visualized under UV (ultraviolet in UVP Biospectrum AC Imaging System (UVP BioImaging Systems, Upland, CA, USA).

2.5. Data Collection and Statistical Analysis

Data regarding shoot number (SN) and shoot length (SL) were collected after twelve subcultures of the blackberry cultivars, and each subculture's duration was 10 weeks.

The in vitro experiments were carried out in a completely randomized design (CRD) in factorial with two factors (two gelling agents and six cultivars), and two-way ANOVA was performed to check the differences between the experimental variants. When the null hypothesis was rejected, ANOVA was completed with Duncan's test ($\alpha < 0.05$) to separate and highlight the differences between means [61]. The presented values are means ± S.E.

For the SCoT and SRAP analysis, the gel images were analyzed using TotalLab120 software (Nonlinear Dynamics, Newcastle upon Tyne, UK) to determine the number and molecular weight range of amplified bands. The number and size range in base pairs (bp) of the PCR-amplified bands were recorded with the statement that the low intensity of some amplified bands in gels was not considered an eliminative factor while scoring. The genetic distances between analyzed blackberry genotypes were calculated using the Euclidean coefficient of similarity. Cluster analysis was performed with the UPGMA algorithm using PAST software (PAle-ontological STatistics Version 4.11, Natural History Museum, Oslo, Norway) [62]. Its consistency was assessed using the bootstrap method with 10,000 repetitions.

3. Results

3.1. In Vitro Shoot Cultures in Starch-Gelled Culture Media and Agar-Gelled Culture Media

The wheat starch-gelled culture media compared to the classical agar-gelled medium (Figure 1) had a distinct influence on the number of shoots/inoculum and the shoot length for all the studied cultivars, as presented in Figures 2 and 3.

Figure 1. In vitro cultures of blackberry, cultivated on MS medium supplemented with 0.5 mg dm^{-3} BA, gelled with 5% wheat starch (left-sided jar) and 0.5% plant agar (right-sided jar): (**a**) 'Chester Thornless'; (**b**) 'Čačanska Bestrna'; (**c**) 'Loch Ness'; (**d**) 'Polar'; (**e**) 'Driscoll's Victoria'; and (**f**) 'Karaka Black'.

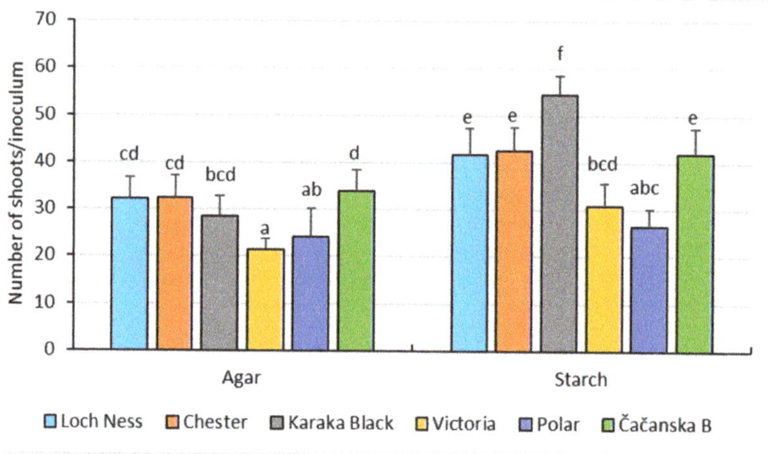

Figure 2. The number of shoots/inoculum of the in vitro blackberry cultivars 'Čačanska Bestrna', 'Chester Thornless', 'Driscoll's Victoria', 'Loch Ness', 'Polar', and 'Karaka Black' grown on MS medium supplemented with 0.5 mg dm^{-3} BA, gelled with 0.5% plant agar (agar), and 5% wheat starch (starch). Different lowercase letters above the bars indicate significant differences between the means of the proliferation rate among cultivars, according to Duncan's test ($\alpha < 0.05$).

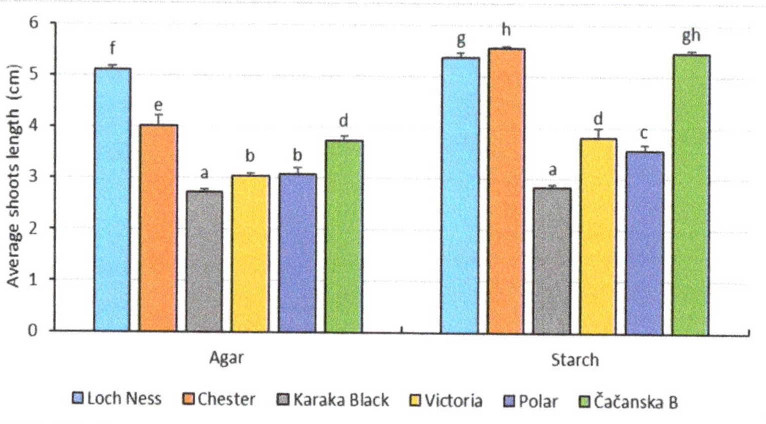

Figure 3. Shoot length of the in vitro blackberry cultivars 'Čačanska Bestrna', 'Chester Thornless', 'Driscoll's Victoria', 'Loch Ness', 'Polar', and 'Karaka Black' grown on MS medium supplemented with 0.5 mg dm^{-3} BA, gelled with 0.5% plant agar (agar), and 5% wheat starch (starch). Different lowercase letters above the bars indicate significant differences between the means of the proliferation rate among cultivars, according to Duncan's test ($\alpha < 0.05$).

On the agar-gelled culture medium, the highest SN was shown by 'Čačanska Bestrna' with 33.92 ± 4.44, followed by 'Chester Thornless' (32.42 ± 4.62) and 'Loch Ness' (32.08 ± 4.70). A decreased SN was observed in cultivars 'Driscoll's Victoria' (21.42 ± 2.24) and 'Polar' (24.25 ± 6.08), as shown in Figure 2

On wheat starch-gelled culture medium, the highest SN was attained by 'Karaka Black' (54.42 ± 4.18), followed by 'Chester Thornless' (42.58 ± 4.92), as shown in Figure 2. The 'Polar' variety had the lowest SN on the culture medium gelled with starch, respectively 26.50 ± 3.71.

Considering the length of the obtained shoots, it can be observed that all of the tested cultivars presented outstanding results on culture medium gelled with 5% wheat starch. The highest values regarding shoots length were observed on the 'Chester' cultivar, followed by 'Čačanska Bestrna', and 'Loch Ness' with values of 5.55 ± 0.08 cm, 5.46 ± 0.07 cm and 5.37 ± 0.09, respectively, as illustrated in Figure 3. On the culture media gelled with agar, the longest shoots were recorded by the 'Loch Ness' cultivar (5.11 ± 0.08 cm), statistically significant compared to the other studied varieties (Figure 3).

3.2. Rheological Analyses

The rheological properties of the two samples (wheat starch-gelled culture medium and agar-gelled medium) can be seen in Figure 4, assessed across shear rates ranging from 5 to 300 s^{-1}. This evaluation aimed to determine the relationship between viscosity and shear rate for the samples.

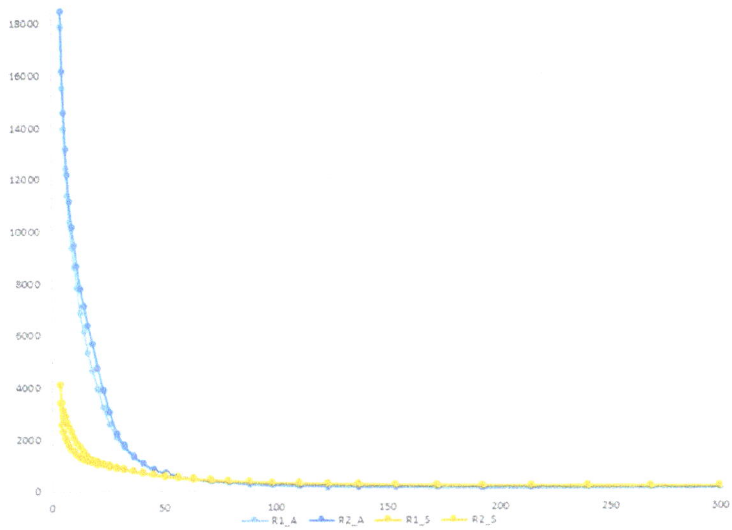

Figure 4. The viscosity of wheat starch-gelled medium (yellow) and agar-gelled medium (blue) samples (in duplicates).

The results showed that the agar-gelled medium samples had a higher viscosity than the wheat starch-gelled medium samples. In the case of the agar-gelled medium samples, the viscosity started from 18,160 ± 29 and decreased to 59 ± 21, while in the case of the wheat starch-gelled medium samples, it started from 3739 ± 28 and decreased to 120 ± 30. Both samples presented a shear-thinning (pseudo-plastic) behavior.

3.3. Genetic Fidelity Evaluation Using SRAP and SCoT Molecular Markers

In the present study, SRAP and SCoT markers were used to assess the genetic uniformity of blackberry shoots proliferated on wheat starch-gelled medium and their mother plants.

SRAP analysis. Ten SRAP primer combinations were used for the genetic fidelity analysis. However, only eight primer combinations produced clear and scorable PCR bands (Table 2).

Table 2. Number and size range of SRAP amplified bands in the analyzed *R. fruticosus* L. cultivars.

Primer Combinations	Size Range of Bands (bp)	No. of Scorable Monomorphic Bands					
		Čačanska Bestrna	Chester Thornless	Loch Ness	Polar	Driscoll's Victoria	Karaka Black
Em4-Me6	180–1450	12	10	12	15	16	13
Em7-Me1	150–1350	14	15	11	10	10	11
Em3-Me8	200–2000	11	12	11	13	12	10
Em5-Me4	150–1800	10	11	8	10	12	8
Em2-Me8	300–2000	12	9	10	12	10	11
Em8-Me5	150–2000	15	15	13	15	16	14
Em1-Me7	200–1900	10	10	12	11	10	13
Em6-Me3	200–1700	11	11	9	12	12	10
Total no. of bands/cultivar		95	93	86	98	98	90

Each SRAP primer combination generated monomorphic bands that ranged in size from 150 bp (Em7-Me1; Em5-Me4; Em8-Me5) to 2000 bp (Em3-Me8; Em2-Me8; Em8-Me5). The lowest number of monomorphic bands was recorded by 'Loch Ness' and 'Karaka Black' generated by the Em5-Me4 primer combination, and the highest number of bands was detected in 'Driscoll's Victoria' cultivar with the primer combinations Em4-Me6 and Em8-Me5, as presented in Table 2.

Regarding the total number of monomorphic bands/cultivar, the highest score was detected in 'Polar' and 'Driscoll's Victoria' cultivars with 98 PCR bands, and the lowest number of bands (86) was detected in 'Loch Ness' cultivar (Table 2).

Despite the fact that no differences were revealed between the mother plants and their in vitro proliferated shoots of each cultivar, the six analyzed cultivars had different SRAP profiles, as shown in Figure 5.

SCoT analysis. From nine SCoT primers used for the initial screening of the genetic fidelity between the mother plants and the proliferated shoots, only seven SCoT primers generated clear and reproducible bands. The number and size range of SCoT amplified bands in the analyzed *R. fruticosus* L. cultivars are presented in Table 3.

Table 3. Number and size range of SCoT amplified bands in the analyzed *R. fruticosus* L. cultivars.

Primer Name	Size Range of Bands (bp)	No. of Scorable Monomorphic Bands					
		Čačanska Bestrna	Chester Thornless	Loch Ness	Polar	Driscoll's Victoria	Karaka Black
SCoT-1	250–1900	8	10	8	11	11	9
SCoT-2	320–2100	7	9	9	8	8	9
SCoT-3	210–2200	11	11	9	7	9	11
SCoT-4	300–2300	8	9	11	9	8	12
SCoT-5	250–1800	13	12	13	11	12	11
SCoT-6	280–2350	12	13	11	13	12	11
SCoT-9	220–1900	15	14	10	16	9	16
Total no. of bands/cultivar		74	78	71	75	69	79

The generated PCR bands ranged in size from 210 bp (SCoT-3) to 2350 bp (SCoT-6). It is noteworthy that no polymorphism was observed between the in vitro proliferated shoots and their mother plants (Figure 6). The number of scorable monomorphic amplified bands varied between 7 and 16. A total of 7 bands were observed for SCoT-2 in samples 'Čačanska Bestrna' and 16 bands for SCoT-9 in samples 'Karaka Black' and 'Polar' (Table 3). Regarding the total number of monomorphic bands generated after the PCR amplifications, the highest number of amplified bands were detected at 'Karaka Black' (79), and the lowest number of bands was recorded at 'Driscoll's Victoria' (69) (Table 3).

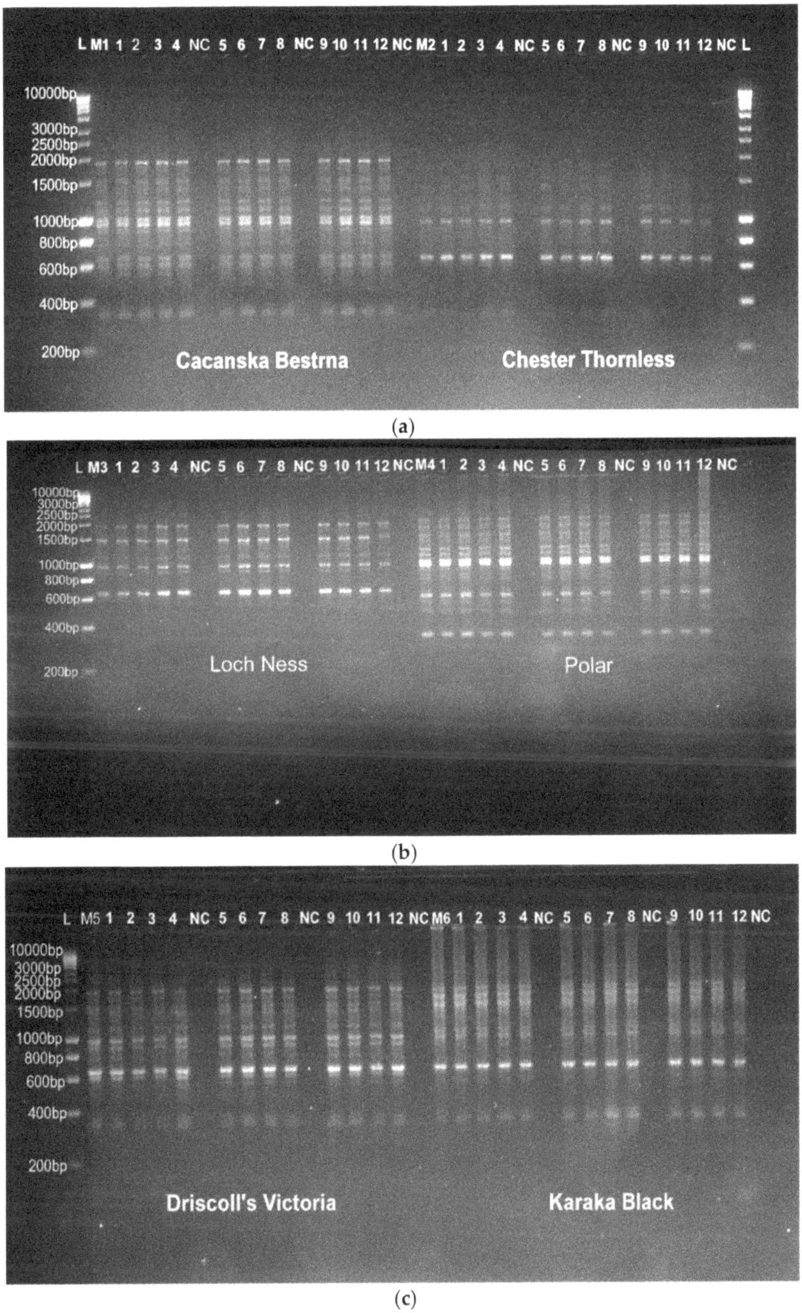

Figure 5. The monomorphic SRAP profiles of mother plants (M1–M6) from each of the six *Rubus fruticosus* L. cultivars and their in vitro proliferated shoots (1–12) generated by the primer combination Me4-Em5; (**a**) 'Čačanska Bestrna', 'Chester Thornless'; (**b**) 'Loch Ness', 'Polar'; (**c**) 'Driscoll's Victoria', 'Karaka Black'. Lane L—indicates the molecular marker (100 bp Ladder, Promega, USA); NC—sample controls without DNA.

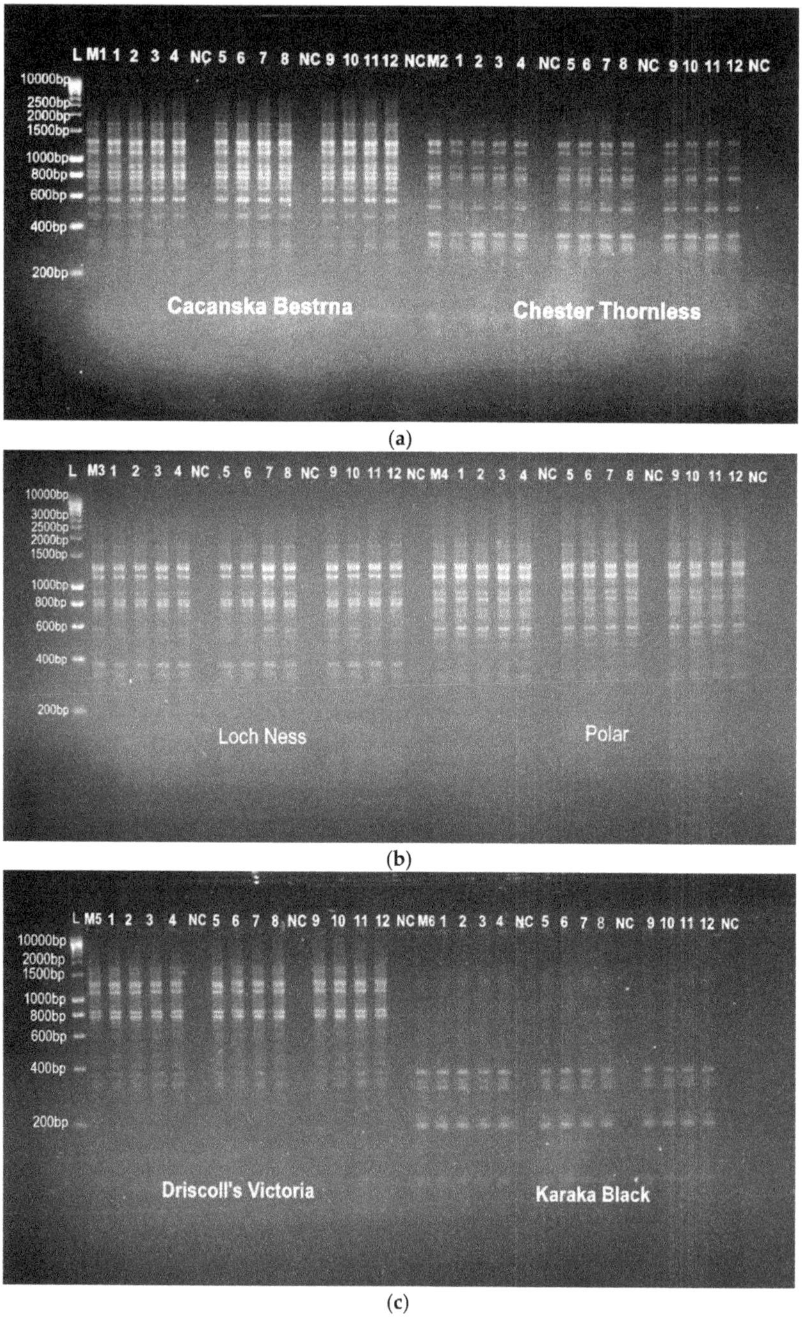

(c)

Figure 6. The monomorphic SCoT profiles of mother plants (M1–M6) from each of the six *Rubus fruticosus* L. cultivars and their in vitro proliferated shoots (1–12) generated by the primer SCoT 9, (**a**) 'Čačanska Bestrna', 'Chester Thornless', (**b**) 'Loch Ness', 'Polar' (**c**) 'Driscoll's Victoria', 'Karaka Black'. Lane L—indicates the molecular marker (100 bp Ladder, Promega, USA); NC—sample controls without DNA.

The results of our study confirm that the SRAP and SCoT markers can be successfully used to rapidly assess the genetic stability of blackberry in vitro grown plants at the proliferation stage. These findings are confirmed by UPGMA cluster analysis. Thus, the constructed dendrograms based on SRAP and SCoT analysis revealed the genetic uniformity between the proliferated shoots and their mother plants. Moreover, the UPGMA cluster analysis showed that there were differences between the genetic profiles of the analyzed blackberry cultivars (Figures 7 and 8).

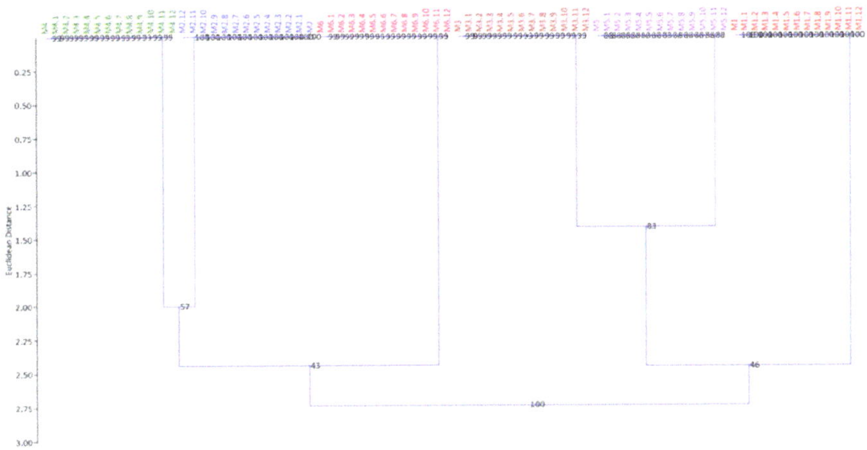

Figure 7. UPGMA dendrogram based on SRAP analysis, showing the genetic fidelity between in vitro proliferated shoots from the analyzed cultivars (M1.1–M1.12; M2.1–M2.12; M3.1–M3.12; M4.1–M4.12; M5.1–M5.12; M6.1–M6.12) and their mother plants (M1–M6). Numbers on the branches show bootstrap values computed from 10,000 replications. M1—'Čačanska Bestrna'; M2—Chester Thornless'; M3—'Loch Ness'; M4—'Polar'; M5—'Driscoll's Victoria'; M6—'Karaka Black'.

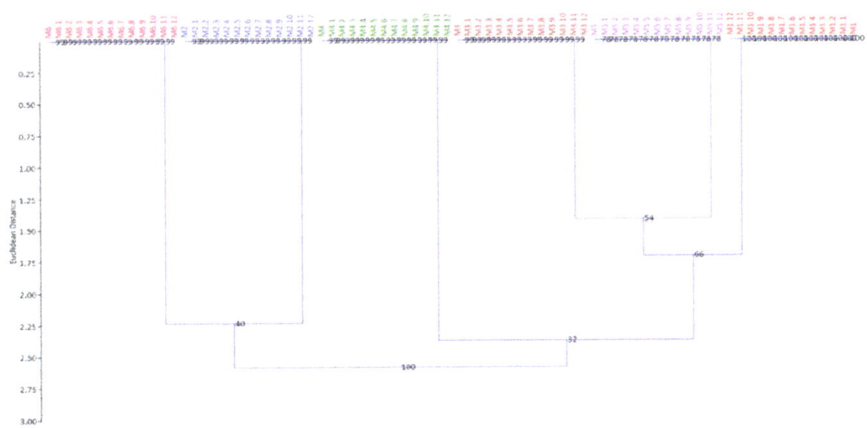

Figure 8. UPGMA dendrogram based on SCoT analysis, showing the genetic fidelity between in vitro proliferated shoots from the analyzed cultivars (M1.1–M1.12; M2.1–M2.12; M3.1–M3.12; M4.1–M4.12; M5.1–M5.12; M6.1–M6.12) and their mother plants (M1–M6). Numbers on the branches show bootstrap values computed from 10,000 replications. M1—'Čačanska Bestrna'; M2—'Chester Thornless'; M3—'Loch Ness'; M4—'Polar'; M5—'Driscoll's Victoria'; M6—'Karaka Black'.

4. Discussion

Generally, the mass propagation of berry fruit plants can be achieved by using plant tissue culture techniques to produce high-quality plant material characterized by genetic uniformity and free of contamination [63]. Furthermore, micropropagation strategies aim to achieve two goals simultaneously: to increase the proliferation rate of the targeted plant material and to reduce production costs at the same time. Replacing agar, the most expensive component of the culture medium, with alternative low-priced gelling agents can lead to reduced production costs of the in vitro cultures [64].

In this context, the outcomes of our study demonstrated two positive aspects when replacing agar with wheat starch in the in vitro multiplication stage of six blackberry cultivars: (1) reduction in cost per liter of culture medium and (2) increased proliferation rate and shoot length. According to the data presented in Table 3, substituting agar with wheat starch can lead to a cost reduction of up to 2.45 euros per liter of culture medium. In our case, replacing plant agar (0.5%) with 5% wheat starch reduced the cost of one liter of the medium by 0.34 euro.

Aside from cost reduction, achieving the appropriate hardness of the medium is crucial for successful plant tissue cultures. If the medium is excessively soft, it can lead to hyperhydricity in tissue cultures, while if it is excessively hard, it can impede shoot development. [64].

In our study, the viscosity of the culture medium gelled with 5% wheat starch was almost five times lower than that of the medium gelled with 0.5% plant agar.

The distinctive structural properties of agar molecules and their interactions within the culture medium account for the variation in viscosity and shear-thinning behavior. Conversely, in the case of starch, these characteristics can be attributed to the molecular structure and interactions within the substrate [65].

Agar is a polysaccharide derived from seaweed, consisting of long chains of repeating sugar units. In its natural state, agar forms a gel-like structure due to the formation of physical cross-links between the agar molecules. These cross-links contribute to the viscosity and elasticity of the agar gel [66]. When shear is applied to the agar-based substrate, such as during stirring or flow, the shearing forces disrupt the physical cross-links between the agar molecules. This results in the temporary reduction of viscosity, allowing the substrate to flow more easily. As the shearing forces increase, the structural network of agar molecules continues to break down, leading to a further reduction in viscosity [67].

Starch is a complex carbohydrate composed of glucose units organized into two main types of polymers: amylose and amylopectin. The amylose component consists of linear chains, while amylopectin has a branched structure [68]. These molecular arrangements contribute to the unique rheological properties of starch-based substrates. When subjected to shear stress, the starch molecules undergo various changes that result in shear thinning behavior, like molecular alignment, disruption of granule structure, and solvent interactions [69].

Our results show that the lower viscosity of the wheat starch-gelled medium caused the inocula to sink into the culture medium as they developed and became heavier. Consequently, the proliferated shoots had a larger surface in contact with the culture medium, which could explain their better development in starch-gelled media compared to agar-gelled media. This phenomenon was observed in a study conducted by Jain and Babbar (2002) [70] when katira gum was used as a gelling agent in culture media for in vitro shoot formation and rooting at *Syzygium cuminii*. In this case, the viscosity of the culture medium gelled with katira gum was less than one-sixth of the viscosity of the medium gelled with agar, and frequent shaking of the culture vessel often caused the explants to sink, with submerged explants showing a positive response.

Similar results were obtained in a previous study conducted by Amlesom et al. in 2021 [71], who evaluated the efficacy of three types of starch (corn, potato, and barley) in both laboratory and commercial grades as alternatives to agar for potato micropropagation.

In terms of physical parameters, including plant height, root length, fresh weight, and dry weight, the media containing laboratory-grade potato starch, commercial corn starch, and laboratory-grade corn starch yielded superior results compared to the control medium gelled with agar. Both laboratory and commercial-grade starch-based media led to cost reductions of 15–22% and 61–66%, respectively. These findings indicate that both corn and potato starches can be considered reliable and cost-effective substituents for agar in potato micropropagation. In a recent study, plantain and banana explants were propagated on 16 starch-based substrates (mung bean, sago, xanthan, Isabgol, guar gum, pear sago, cassava starch, tapioca starch and their combinations with agar) to evaluate their suitability as tissue culture gelling agents [72]. Two of the substrates, mung bean and Isabgol, had suitable gelling properties and cost less than one euro, and were more economical than agar.

In the present study, the type of gelling agent had a significant effect on the proliferation rate and shoot length in all six blackberry cultivars, as shown in Figures 2 and 3. The highest proliferation rate was obtained in the culture medium gelled with wheat starch in all blackberry cultivars. In this regard, the 'Karaka Black' cultivar exhibited the highest proliferation rate, nearly twice as high as that on the agar-gelled medium. The smallest differences in proliferation rate were observed in the 'Polar' variety, with only a 1.09-fold increase in the starch-gelled medium compared to the agar-gelled medium. Also, even if the consistency of the culture media gelled with wheat starch was softer, no hyperhydrated cultures or other physiological disorders were observed.

Since the culture medium gelled with 5% wheat starch is opaque, possible endogenous bacterial contamination cannot be observed. For this reason, this medium is indicated to be used only in the in vitro multiplication stage of blackberry. Our previous investigations regarding in vitro multiplication of *Rubus* sp. showed that the culture medium, solidified with different alternative gelling agents, including wheat starch and potato starch, generated high proliferation rates and vigorous shoots suitable for concomitant acclimation and ex vitro rooting to different blackberry varieties: 'Thornless Evergreen' [33], 'Loch Ness', 'Čačanska bestrna', 'Chester Thornless', and 'Navaho' [12].

Previous research results show that the assessment of genetic fidelity is one of the most important requirements in tissue culture-based propagation of any fruit plant species [50] due to the possible presence of somaclonal variations. Chromosomal rearrangements are an important source of somaclonal variations that can occur between the in vitro plants regenerated from one parental line as a result of the successive subculture of plant organs, tissues, and cells [46]. In this particular context, molecular marker techniques have emerged as valuable tools for two purposes: verifying the genetic fidelity of in vitro propagated plants in comparison to their mother plants and confirming their uniformity for commercial applications [73–76].

During the last decades, several research articles have been published regarding the evaluation of the genetic uniformity of in vitro propagated plants by SRAP markers on different fruit species like *Vitis vinifera* L. [77], *Musa* sp. [78], *Rubus fruticosus* L. [50], *Ribes grossularia* L. [79], *Aronia melanocarpa* (Michx.) [80]. The results of these studies are consistent with the findings of the present research, showing that SRAP markers were adequate to assess the genetic fidelity of the proliferated shoots after 12 consecutive subcultures. The experimental design included the genetic fidelity analysis for the in vitro proliferated shoots in the starch-gelled medium and confirmed the uniformity of the proliferated shoots with their mother plant, as illustrated by the DNA fingerprinting profiles (Figure 6).

It is worth mentioning that the SRAP technique is an easily accessible tool for different laboratories to assess the genetic fidelity of in vitro grown plant material, and the reproducibility of the results is satisfactory [79].

In the present study, SCoT markers were used to confirm the SRAP results, given that this type of DNA-based molecular marker is highly reproducible. SCoT technique amplifies the short, conserved regions around the ATG translation start codon located in plant genes, and previous studies confirm the complementary validation of the two techniques regarding genetic fidelity on other micropropagated plant species [51]. Unlike

RAPD or ISSR markers that amplify DNA fragments from non-coding regions of the genome, the advantage of using SCoT markers is associated with functional genes and their corresponding traits [81–83]. The results presented in Figure 7 confirm the genetic fidelity of the in vitro grown shoots after 12 concomitant subcultures and their mother plant and represents a complementary validation of the SRAP results at the same time. To the best of our knowledge, this study is the first report on blackberry that applies SCoT markers to assess the genetic uniformity of the in vitro propagated shoots.

Furthermore, the results of this study show that both the SRAP and SCoT marker systems are useful for cultivar differentiation at the DNA molecular level.

5. Conclusions

In conclusion, our study on in vitro shoot proliferation of six blackberry cultivars ('Čačanska Bestrna', 'Chester Thornless', 'Driscoll's Victoria', 'Loch Ness', 'Polar', and 'Karaka Black') indicates that wheat starch is an efficient gelling agent. The use of wheat starch for agar replacement in blackberry culture media resulted in a cost reduction of 0.34 euro per liter (compared to plant agar). Additionally, it promoted a greater number of shoots, longer and physiologically suitable for acclimation. The genetic uniformity of the micropropagated shoots with their mother plants was confirmed by SRAP and SCoT molecular markers, thus, revealing the sustainability of the wheat starch as an effective gelling agent for the production of clonal blackberry planting material.

Supplementary Materials: The following supporting information can be downloaded at: https://www.mdpi.com/article/10.3390/horticulturae9080902/s1, Table S1: Sequences of SCoT primers and SRAP primer combinations used in this study.

Author Contributions: Conceptualization, D.C., K.S. and M.H.; methodology, D.C. and M.H.; software, D.C. and M.H.; validation, D.P.; investigation, D.C., M.H., B.-E.T. and K.S.; resources, D.C.; data curation, D.C.; writing—original draft preparation, D.C., M.H. and B.-E.T.; writing—review and editing, K.S. and D.P.; supervision, D.P.; project administration, D.C.; funding acquisition, D.C. All authors have read and agreed to the published version of the manuscript.

Funding: This research was funded by National Research Development Projects to finance Excellence (PFE)—14/2022-2024, granted by the Romanian Ministry of Research and Innovation.

Data Availability Statement: Not applicable.

Conflicts of Interest: The authors declare no conflict of interest.

References

1. Cho, M.J.; Howard, L.R.; Prior, R.L.; Clark, J.R. Flavonol glycosides and antioxidant capacity of various blackberry and blueberry genotypes determined by high-performance liquid chromatography/mass spectrometry. *J. Sci. Food Agric.* **2005**, *85*, 2149–2158. [CrossRef]
2. Zia-Ul-Haq, M.; Riaz, M.; De Feo, V.; Jaafar, H.Z.E.; Moga, M. *Rubus fruticosus* L.: Constituents, Biological Activities and Health Related Uses. *Molecules* **2014**, *19*, 10998–11029. [CrossRef] [PubMed]
3. Oszmiański, J.; Nowicka, P.; Teleszko, M.; Wojdyło, A.; Cebulak, T.; Oklejewicz, K. Analysis of phenolic compounds and antioxidant activity in wild blackberry fruits. *Int. J. Mol. Sci.* **2015**, *16*, 14540–14553. [CrossRef] [PubMed]
4. Fathy, H.M.; Abou El-Leel, O.F.; Amin, M.A. Micropropagation and Biomass Production of *Rubus fruticosus* L. (*Blackberry*) plant. *Middle East J. Appl. Sci.* **2018**, *8*, 1215–1228.
5. Bray, M.; Rom, C.C.; Clark, J.R. Propagation of thornless Arkansas blackberries by hardwood cuttings. *Discov. Stud. J. Dale Bump. Coll. Agric. Food Life Sci.* **2003**, *4*, 9–13.
6. Takeda, F.; Soria, J. Method for producing long-cane blackberry plants. *HortTechnology* **2011**, *21*, 563–568. [CrossRef]
7. Hussain, I.; Roberto, S.R.; Colombo, R.C.; Assis, A.; Koyama, R. Cutting types collected at different seasons on blackberry multiplication. *Rev. Bras. Frutic.* **2017**, *39*. [CrossRef]
8. Gomes, H.T.; Bartos, P.M.C.; Andrade, M.T.D.; Almeida, R.F.; Lacerda, L.F.D.; Scherwinski-Pereira, J.E. In vitro conservation of blackberry genotypes under minimal growth conditions and subsequent large-scale micropropagation. *Pesqui. Agropecu. Bras.* **2017**, *52*, 1286–1290. [CrossRef]
9. Broome, O.C.; Zimmerman, R.H. In vitro propagation of blackberry. *Hortic. Sci.* **1978**, *13*, 151–153.
10. Najaf-Abadi, A.J.; Hamidoghli, Y. Micropropagation of thornless trailing blackberry ('*Rubus* sp.') by axillary bud explants. *Aust. J. Crop Sci.* **2009**, *3*, 191–194.

11. Vujović, T.; Ružić, D.J.; Cerović, R. In vitro shoot multiplication as influenced by repeated subculturing of shoots of contemporary fruit rootstocks. *Hortic. Sci.* **2012**, *39*, 101–107. [CrossRef]
12. Fira, A.; Clapa, D.; Simu, M. Studies Regarding the Micropropagation of Some Blackberry Cultivars. *Bull. UASVM (H)* **2014**, *71*, 29–37.
13. Kefayeti, N.; Kafkas, E.; Ercişli, S. Micropropagation of 'Chester thornless' Blackberry Cultivar using Axillary Bud Explants. *Not. Bot. Horti Agrobot. Cluj-Napoca* **2019**, *47*, 162–168. [CrossRef]
14. Kolarević, T.; Milinčić, D.D.; Vujović, T.; Gašić, U.M.; Prokić, L.; Kostić, A.Ž.; Cerović, R.; Stanojevic, S.P.; Tešić, Ž.L.; Pešić, M.B. Phenolic Compounds and Antioxidant Properties of Field-Grown and In vitro Leaves, and Calluses in Blackberry and Blueberry. *Horticulturae* **2021**, *7*, 420. [CrossRef]
15. Murashige, T. Plant Propagation through Tissue Cultures. *Annu. Rev. Plant Physiol.* **1974**, *25*, 135–166. [CrossRef]
16. Soumare, A.; Diédhiou, A.G.; Arora, N.K.; Tawfeeq Al-Ani, L.K.; Ngom, M.; Fall, S.; Hafidi, M.; Ouhdouch, Y.; Kouisni, L.; Sy, M.O. Potential role and utilization of plant growth promoting microbes in plant tissue culture. *Front. Microbiol.* **2021**, *12*, 649878. [CrossRef]
17. Abdalla, N.; El-Ramady, H.; Seliem, M.K.; El-Mahrouk, M.E.; Taha, N.; Bayoumi, Y.; Shalaby, T.A.; Dobránszki, J. An Academic and Technical Overview on Plant Micropropagation Challenges. *Horticulturae* **2022**, *8*, 677. [CrossRef]
18. Viswanath, M.; Ravindra Kumar, K.; Chetanchidambar, N.M.; Mahesh, S.S.N.M. Regeneration mechanisms in plant tissue culture: A. *J. Pharm. Innov.* **2023**, *12*, 2948–2952.
19. Hussain, A.; Qarshi, I.A.; Nazir, H.; Ullah, I. Plant Tissue Culture: Current Status and Opportunities. In *Recent Advances in Plant In Vitro Culture*; IntechOpen: London, UK, 2012; pp. 2–29. [CrossRef]
20. Monja-Mio, K.M.; Olvera-Casanova, D.; Herrera-Alamillo, M.Á.; Sánchez-Teyer, F.L.; Robert, M.L. Comparison of conventional and temporary immersion systems on micropropagation (multiplication phase) of *Agave angustifolia* Haw 'Bacanora'. *3 Biotech* **2021**, *11*, 77. [CrossRef]
21. Murashige, T.; Skoog, F. A revised medium for rapid growth and bio assays with tobacco tissue cultures. *Physiol. Plant.* **1962**, *15*, 473–497. [CrossRef]
22. Bobrowski, V.L.; Mello-Farias, P.; Petters, J. Micropropagation of blackberries (*Rubus* sp.) cultivars. *J. Agric. Sci. Technol.* **1996**, *2*, 17–20.
23. Gonzalez, M.V.; Lopez, M.; Valdes, A.E.; Ordas, R.J. Micropropagation of three berry fruit species using nodal segments from field-grown plants. *Ann. Appl. Biol.* **2000**, *137*, 73–78. [CrossRef]
24. Pasa, M.D.S.; Carvalho, G.L.; Schuch, M.W.; Schmitz, J.D.; Torchelsen, M.D.M.; Nickel, G.K.; Sommer, L.R.; Lima, T.S.; Camargo, S.S. Qualidade de luz e fitorreguladores na multiplicação e enraizamento in vitro da amoreira-preta 'Xavante'. *Cienc. Rural* **2012**, *42*, 1392–1396. [CrossRef]
25. Schuchovski, C.S.; Biasi, L.A. Development of an efficient protocol for 'Brazos' blackberry in vitro multiplication. *Acta Hortic.* **2017**, *1224*, 157–164. [CrossRef]
26. Llyod, G.; McCown, B. Commercially feasible micropropagation of mountain loure, *Kalmia latifolia*, by using of shoot tip culture. *Comb. Proceed. Int. Plant Prop. Soc.* **1980**, *30*, 421–427.
27. Villa, F.; Fráguas, C.B.; Dutra, L.F.; Pio, L.A.S.; Pasqual, M. Multiplicacao in vitro de amoreira-preta cultivar Brazos. *Ciênc. Agrotec.* **2006**, *30*, 266–270. [CrossRef]
28. Gamborg, O.L. Cells, Protoplasts and Plant Regeneration in Culture. In *Manual of Industrial Microbiology and Biotechnology*; Demain, A.L., Salomon, N.A., Eds.; American Society for Microbiology: Washington, DC, USA, 1986; pp. 263–273.
29. Driver, J.A.; Kuniyuki, A.H. In vitro propagation of Paradox walnut rootstock. *HortScience* **1984**, *19*, 507–509. [CrossRef]
30. Fira, A.; Clapa, D.; Plopa, C. Micropropagation of blackberry thornless cultivars. *Fruit Grow. Res.* **2009**, *25*, 213–221.
31. Fira, A.; Clapa, D.; Plopa, C. New aspects regarding the micropropagation of blackberry cultivar 'Thornless evergreen'. *Bull. UASVM (H)* **2010**, *67*, 106–114.
32. Vescan, L.A.; Clapa, D.; Fira, A.; Pamfil, D. Micropropagation of cold resistant blackberry cultivar 'Gazda'. *Bull. USAMV Anim. Sci. Biotechnol.* **2012**, *69*, 282–289.
33. Quoirin, M.; Lepoivre, P. Improved media for in vitro culture of *Prunus* sp. *Acta Hortic.* **1977**, *78*, 437–442. [CrossRef]
34. Borodulina, I.D.; Plaksina, T.V.; Panasenko, V.N.; Sokolova, G.G. Optimization of blackberry clonal micropropagation. *Ukr. J. Ecol.* **2019**, *9*, 339–345. [CrossRef]
35. Anderson, W.C. Tissue culture propagation of red and black raspberries, *Rubus idaeus* and *R. occidentalis*. *Acta Hortic.* **1980**, *112*, 13–20. [CrossRef]
36. Reed, B.M. Multiplication of Rubus germplasm in vitro: A screen of 256 accessions. *Fruit Var. J.* **1990**, *44*, 141–148.
37. Murashige, T. Growth factor requirements of citrus tissue culture. *Proc. First Int. Citrus Symp.* **1969**, *3*, 1151–1161.
38. Wu, J.H.; Miller, S.A.; Hall, H.K.; Mooney, P.A. Factors affecting the efficiency of micropropagation from lateral buds and shoot tips of Rubus. *PCTOC* **2009**, *99*, 17–25. [CrossRef]
39. Erig, A.C.; De Rossi, A.; Fortes, G.R. 6-benzilaminopurina e ácido indolbutírico na multiplicação in vitro da amoreira-preta (*Rubus idaeus* L.), cv. Tupy. *Ciência Rural* **2002**, *32*, 765–770. [CrossRef]
40. Ružić, D.; Lazić, T. Micropropagation as means of rapid multiplication of newly developed blackberry and black currant cultivars. *Agric. Conspec. Sci.* **2006**, *71*, 149–153.

41. Lepse, L.; Laugale, V. Micropropagation, Rooting and Acclimatization of Blackberry 'Agavam'. *Acta Hortic.* **2008**, *839*, 43–49. [CrossRef]
42. Hunkova, J.; Libiakova, G.; Gajdošová, A. Shoot proliferation ability of selected cultivars of Rubus spp. as influenced by genotype and cytokinin concentration. *J. Cent. Eur. Agric.* **2016**, *17*, 252–259. [CrossRef]
43. Hunková, J.; Gajdošová, A.; Szabóová, M. Effect of Mesos Components ($MgSO_4$, $CaCl_2$, KH_2PO_4) on in vitro Shoot Growth of Blackberry, Blueberry, and Saskatoon. *Plants* **2020**, *9*, 935. [CrossRef] [PubMed]
44. Muñoz-Concha, D.; Quintero, J.; Ercişli, S. Media and hormones influence in micropropagation success of blackberry cv. 'Chester'. *Res. J. Biotechnol.* **2021**, *16*, 103–108.
45. Bairu, M.W.; Aremu, A.O.; Van Staden, J. Somaclonal variation in plants: Causes and detection methods. *Plant Growth Regul.* **2011**, *63*, 147–173. [CrossRef]
46. Krishna, H.; Alizadeh, M.; Singh, D.; Singh, U.; Chauhan, N.; Eftekhari, M.; Sadh, R.K. Somaclonal variations and their applications in horticultural crops improvement. *3 Biotech* **2016**, *6*, 54. [CrossRef] [PubMed]
47. Ferreira, M.D.S.; Rocha, A.d.J.; Nascimento, F.d.S.; Oliveira, W.D.d.S.; Soares, J.M.d.S.; Rebouças, T.A.; Morais Lino, L.S.; Haddad, F.; Ferreira, C.F.; Santos-Serejo, J.A.d.; et al. The Role of Somaclonal Variation in Plant Genetic Improvement: A Systematic Review. *Agronomy* **2023**, *13*, 730. [CrossRef]
48. Smulders, M.J.M.; de Klerk, G.J. Epigenetics in plant tissue culture. *Plant Growth Regul.* **2011**, *63*, 137–146. [CrossRef]
49. Moharana, A.; Das, A.; Subudhi, E.; Naik, S.K.; Barik, D.P. High frequency shoot proliferation from cotyledonary node of *Lawsonia inermis* L. and validation of their molecular finger printing. *J. Crop Sci. Biotechnol.* **2017**, *20*, 405–416. [CrossRef]
50. Borsai, O.; Hârța, M.; Szabo, K.; Kelemen, C.D.; Andrecan, F.A.; Codrea, M.M.; Clapa, D. Evaluation of genetic fidelity of in vitro-propagated blackberry plants using RAPD and SRAP molecular markers. *Hortic. Sci.* **2020**, *47*, 21–27. [CrossRef]
51. Clapa, D.; Hârța, M. Establishment of an Efficient Micropropagation System for *Humulus lupulus* L. cv. Cascade and Confirmation of Genetic Uniformity of the Regenerated Plants through DNA Markers. *Agronomy* **2021**, *11*, 2268. [CrossRef]
52. Longchar, T.B.; Deb, C.R. Optimization of in vitro propagation protocol of *Dendrobium heterocarpum* Wall. ex. Lindl. and clonal genetic fidelity assessment of the regenerates: An orchid of horticultural and medicinal importance. *S. Afr. J. Bot.* **2022**, *149*, 67–78. [CrossRef]
53. Rai, M.K. Start codon targeted (SCoT) polymorphism marker in plant genome analysis: Current status and prospects. *Planta* **2023**, *257*, 34. [CrossRef]
54. Teleky, B.-E.; Mitrea, L.; Plamada, D.; Nemes, S.A.; Călinoiu, L.-F.; Pascuta, M.S.; Varvara, R.-A.; Szabo, K.; Vajda, P.; Szekely, C.; et al. Development of Pectin and Poly(vinyl alcohol)-Based Active Packaging Enriched with Itaconic Acid and Apple Pomace-Derived Antioxidants. *Antioxidants* **2022**, *11*, 1729. [CrossRef]
55. Mitrea, L.; Teleky, B.E.; Leopold, L.F.; Nemes, S.A.; Plamada, D.; Dulf, F.V.; Pop, I.D.; Vodnar, D.C. The physicochemical properties of five vegetable oils exposed at high temperature for a short-time-interval. *J. Food Compos. Anal.* **2022**, *106*, 104305. [CrossRef]
56. Lodhi, M.A.; Guang-Ning, Z.; Weeden, F.N.F.; Reisch, B.I. A simple and efficient method for DNA extraction from grapevine cultivars and Vitis species. *Plant Mol. Biol. Rep.* **1994**, *12*, 6–13. [CrossRef]
57. Pop, R.; Ardelean, M.; Pamfil, D.; Gaboreanu, I.M. The efficiency of different DNA isolation and purification in ten cultivars of *Vitis vinifera*. *Bul. USAMV CN(ZB)* **2003**, *59*, 259–261.
58. Bodea, M.; Pamfil, D.; Pop, R.; Sisea, R.C. DNA isolation from desiccated leaf material from plum tree (*Prunus domestica* L.) molecular analysis. *Bul. UASVM CN (H)* **2016**, *1*, 1–2. [CrossRef]
59. Li, G.; Quiros, C. Sequence-related amplified polymorphism (SRAP), a new marker system based on a simple PCR reaction: Its application to mapping and gene tagging in Brassica. *Theor. Appl. Genet.* **2001**, *103*, 455–461. [CrossRef]
60. Collard, B.C.Y.; Mackill, D.J. Start Codon Targeted (SCoT) Polymorphism: A Simple, Novel DNA Marker Technique for Generating Gene-Targeted Markers in Plants. *Plant Mol. Biol. Rep.* **2009**, *27*, 86–93. [CrossRef]
61. *ANOVA and Duncan's Test PC Program for Variant Analyses Made for Completely Randomized Polyfactorial Experiences*; Poli-Fact 2020 Version 4; University of Agricultural Sciences and Veterinary Medicine: Cluj-Napoca, Romania, 2020.
62. Hammer, Ø.; Harper, D.A.T.; Ryan, P.D. Past: Paleontological statistics software package for education and data analysis. *Palaeontol. Electron.* **2001**, *4*, 9.
63. AbdAlla, M.M.; Mostafa, R.A.A. In vitro Propagation of Blackberry (*Rubus fruticosus* L.). *Assiut J. Agric. Sci.* **2015**, *46*, 88–99.
64. Fira, A.; Magyar-Tábori, K.; Hudák, I.; Clapa, D.; Dobránszky, J. Effect of gelling agents on in vitro development of *Amelanchier canadensis* 'Rainbow Pillar'. *Int. J. Hort. Sci.* **2013**, *19*, 75–79. [CrossRef]
65. Xu, Y.; Stokes, J.R. Soft lubrication of model shear-thinning fluids. *Tribol. Int.* **2020**, *152*, 106541. [CrossRef]
66. Zhang, L.; Che, L.; Zhou, W.; Chen, X.D. Rheological behavior of agar solution in relation to the making of instant edible bird's nest products. *Int. J. Food Eng.* **2012**, *8*. [CrossRef]
67. Alam, K.; Iqbal, M.; Hasan, A.; Al-Maskari, N. Rheological characterization of biological hydrogels in aqueous state. *J. Appl. Biotechnol. Rep.* **2020**, *7*, 172–176. [CrossRef]
68. You, K.M.; Murray, B.S.; Sarkar, A. Rheology and tribology of starch + κ-carrageenan mixtures. *J. Texture Stud.* **2021**, *52*, 16–24. [CrossRef]
69. Iskakova, J.; Smanalieva, J.; Methner, F.J. Investigation of changes in rheological properties during processing of fermented cereal beverages. *J. Food Sci. Technol.* **2019**, *56*, 3980–3987. [CrossRef]

70. Jain, N.; Babbar, S.B. Gum katira–a cheap gelling agent for plant tissue culture media. *Plant Cell Tissue Organ Cult.* **2002**, *71*, 223–229. [CrossRef]
71. Amlesom, W.S.; Mehari, T.; Saleh, B.K. Evaluation of Different Starches as Gelling Agents for Micropropagation of Potato. *J. Agric. Sci.* **2021**, *13*, 144. [CrossRef]
72. Ebile, P.A.; Opata, J.; Hegele, S. Evaluating suitable low-cost agar substitutes, clarity, stability, and toxicity for resource-poor countries' tissue culture media. *In Vitro Cell. Dev. Biol.-Plant* **2022**, *58*, 989–1001. [CrossRef]
73. Manish, C.; Dhawni, S.; Chandra, V.; Garima, V. Molecular markers: An important tool to assess genetic fidelity in tissue culture grown long-term cultures of economically important fruit plants. *Asian J. Bio Sci.* **2015**, *10*, 101–105.
74. Abdolinejad, R.; Shekafandeh, A.; Jowkar, A.; Gharaghani, A.; Alemzadeh, A. Indirect regeneration of *Ficus carica* by the TCL technique and genetic fidelity evaluation of the regenerated plants using flow cytometry and ISSR. *Plant Cell Tssue Organ Cult.* **2020**, *143*, 131–144. [CrossRef]
75. Tyagi, S.; Rajpurohit, D.; Sharma, A. Genetic Fidelity Studies for Testing True-to-Type Plants in Some Horticultural and Medicinal Crops Using Molecular Markers. In *Agricultural Biotechnology: Latest Research and Trends*; Kumar Srivastava, D., Kumar Thakur, A., Kumar, P., Eds.; Springer: Singapore, 2021; pp. 147–170.
76. Kessel-Domini, A.; Pérez-Brito, D.; Guzmán-Antonio, A.; Barredo-Pool, F.A.; Mijangos-Cortés, J.O.; Iglesias-Andreu, L.G.; Cortés-Velázquez, A.; Canto-Flick, A.; Avilés-Viñas, S.A.; Rodríguez-Llanes, Y.; et al. Indirect Somatic Embryogenesis: An Efficient and Genetically Reliable Clonal Propagation System for *Ananas comosus* L. Merr. Hybrid "MD2". *Agriculture* **2022**, *12*, 713. [CrossRef]
77. Guo, D.; Zhang, J.; Liu, C.; Zhang, G.; Li, M.; Zhang, Q. Genetic variability and relationships between and within grape cultivated varieties and wild species based on SRAP markers. *Tree Genet. Genomes* **2012**, *8*, 789–800. [CrossRef]
78. Khatab, I.; Youssef, M. Micropropagation and assessment of genetic stability of *Musa* sp. cv. Williams using RAPD and SRAP markers. *Egypt. J. Bot.* **2018**, *58*, 371–380. [CrossRef]
79. Wójcik, D.; Trzewik, A.; Kucharska, D. Field Performance and Genetic Stability of Micropropagated Gooseberry Plants (*Ribes grossularia* L.). *Agronomy* **2021**, *11*, 45. [CrossRef]
80. Borsai, O.; Clapa, D.; Fira, A.; Hârța, M.; Szabo, K.; Dumitraș, A.F.; Pamfil, D. In vitro propagation of *Aronia melanocarpa* (Michx.) Elliott. *Acta Hortic.* **2021**, *1308*, 213–222. [CrossRef]
81. Bekheet, S.A.; Gabr, A.M.M.; Reda, A.A.; El Bahr, M.K. Micropropagation and assessment of genetic stability of in vitro raised jojoba (*Simmondsia chinensis* link.) plants using SCoT and ISSR markers. *Plant Tissue Cult. Biotechnol.* **2015**, *25*, 165–179. [CrossRef]
82. Kudikala, H.; Jogam, P.; Sirikonda, A.; Mood, K.; Allini, V.R. In vitro micropropagation and genetic fidelity studies using SCoT and ISSR primers in *Annona reticulata* L.: An important medicinal plant. *Vegetos* **2020**, *33*, 446–457. [CrossRef]
83. Thakur, M.; Sharma, V.; Chauhan, A. Genetic fidelity assessment of long term in vitro shoot cultures and regenerated plants in Japanese plum cvs Santa Rosa and Frontier through RAPD, ISSR and SCoT markers. *S. Afr. J. Bot.* **2021**, *140*, 428–433. [CrossRef]

horticulturae

MDPI

Communication

A Complete Micropropagation Protocol for Black-Leaved *Zamioculcas zamiifolia* (Lodd.) Engl. 'Dowon'

Amirhossein Pourhassan [1], Behzad Kaviani [1,*], Dariusz Kulus [2], Natalia Miler [2,*] and Naser Negahdar [1,3]

[1] Department of Horticultural Science, Rasht Branch, Islamic Azad University, Rasht 4147654919, Iran
[2] Laboratory of Ornamental Plants and Vegetable Crops, Faculty of Agriculture and Biotechnology, Bydgoszcz University of Science and Technology, Bernardyńska 6, 85-029 Bydgoszcz, Poland; dkulus@gmail.com
[3] Hyrcan Agricultural Sciences and Biotechnology Research Institute, Amol 1697846139, Iran
* Correspondence: kaviani@iaurasht.ac.ir (B.K.); nmiler@pbs.edu.pl (N.M.); Tel.: +98-9111777482 (B.K.)

Abstract: *Zamioculcas zamiifolia*, a drought-resistant plant in the family Araceae, is a popular ornamental potted foliage plant originating from tropical east and subtropical southeast Africa. The growth and propagation rate of this species is low in conventional propagation methods. Therefore, the current study aimed at developing a complete in vitro propagation protocol of black-leaved Raven® ZZ plant (Z. *zamiifolia* 'Dowon')—a novelty on the floricultural market. In order to initiate an axenic culture, the disinfection of leaf explants was performed with sodium hypochlorite and mercury chloride. Next, leaf segments were cultured on the Murashige and Skoog (MS) medium with the addition of 6-benzyladenine (BA) and/or α-naphthalene acetic acid (NAA) at various concentrations. The highest number of shoots (11) and leaves (22) per explant was obtained in a medium enriched with 2 mg·L^{-1} BA together with 0.5 mg·L^{-1} NAA. The maximum number of roots (3.33) was produced in microshoots cultured on the medium supplemented with 2 mg·L^{-1} NAA. On the other hand, the longest roots (2.66 cm) were produced on a medium containing 2 mg·L^{-1} NAA and 0.5 mg·L^{-1} BA. The combination of 0.5 mg·L^{-1} BA and 0.5 mg·L^{-1} NAA was most effective in stimulating callus formation (78.33%). Rooted plantlets were transferred to plastic pots filled with coco peat and acclimatized to ambient greenhouse conditions with an average 68.19% survival rate. This is the first report on a complete micropropagation protocol of black-leaved zamioculcas.

Keywords: Araceae; in vitro propagation; ornamental plants; plant growth regulators

Citation: Pourhassan, A.; Kaviani, B.; Kulus, D.; Miler, N.; Negahdar, N. A Complete Micropropagation Protocol for Black-Leaved *Zamioculcas zamiifolia* (Lodd.) Engl. 'Dowon'. *Horticulturae* **2023**, *9*, 422. https://doi.org/10.3390/horticulturae9040422

Academic Editor: Ute Albrecht

Received: 23 February 2023
Revised: 17 March 2023
Accepted: 22 March 2023
Published: 24 March 2023

1. Introduction

Zamioculcas is a monotypic genus in the family Araceae (aroids), comprising the single species *Zamioculcas zamiifolia* (Lodd.) Engl. It is a stem-less tropical herbaceous monocotyledonous originating from eastern Africa, from southern Kenya to northeastern South Africa, growing on the stony ground or tropical moist forest floor [1]. Zamioculcas (commonly named as 'ZZ plant') differs from most other aroids in possessing pinnate leaves [2]. It is grown as a medicinal-ornamental plant, mostly for its attractive glossy foliage and some pharmaceutical metabolites. Moreover, ZZ plant has the potential to reduce the concentration of pollutant gases such as benzene, ethylbenzene, xylene, and toluene from contaminated indoor air [1]. It is drought-resistant, can grow under low light conditions, and generates short sprouts from a thick underground tuber-like rhizome that stores water [3,4]. Consequently, it is described as a "unique" indoor foliage plant. The species is conventionally propagated through leaf cuttings (it rarely flowers), but this method is inefficient and time-consuming as ZZ plant is slow-growing, even under optimal production conditions [3]. For example, in a study by Prathibha et al. [5] an average of 2.2 shoots had formed per propagule at 180 days under in vivo conditions. Consequently, stock plants are expensive [2].

The development of an efficient micropropagation protocol for ZZ plant could increase the plant yield and decrease the production costs. Tissue culture techniques provide an opportunity for the large-scale production of an elite material and plants of commercial interest. The micropropagation of some Araceae genera such as *Aglaonema, Alocasia, Anthurium, Dieffenbachia, Homalomena, Philodendron, Spathiphyllum*, and *Syngonium* has been carried out [6,7]. In these studies, seedling fragments, leaves, shoots, inflorescences, and corms were used as explant sources and inoculated usually on the MS [8] medium. As for the plant growth regulators (PGRs), thidiazuron (TDZ) and 6-benzyladenine (BA) were extensively used as cytokinins, while α-naphthalene acetic acid (NAA), indole-3-acetic acid (IAA), and 2,4-D (2,4-dichlorophenoxyacetic acid) were the most common auxins [7]. Among these PGRs, BA is the most effective in stimulating organogenesis in aroids. For example, Chen et al. [9] found that a 5 mg·L^{-1} BA treatment could induce the formation of adventitious shoot in *Aglaonema* 'White Tip'. In *Philodendron*, more shoots were regenerated on the BA- than kinetin-supplemented media [10]. The best conditions for in vitro multiplication of *Homalomena pineodora* were found in the MS medium with 0.5 mg·L^{-1} BA [11]. Nonetheless, there may be considerable cultivar differences in response to the type and concentration of cytokinins used.

Despite the popularity of the species, information on in vitro culture systems in *Z. zamiifolia* is limited to green-leaved cultivars and published in non-English articles [5,12]. Papafotiou and Martini [13] studied the effect of the inoculation position of leaf explants concerning PGRs during the micropropagation of ZZ plant. They found that medium supplementation with 2,4-D resulted in the formation of somatic embryos, NAA combined with BA induced mainly roots, and NAA alone produced tubers and roots, while BA alone resulted mainly in leaf development [13]. According to Sayadi Nejad and Sadeghi [14], the combination of BA and NAA was optimal for callus induction, while NAA alone stimulated shoot elongation if rhizome or petiole explants were inoculated in a half-strength (1/2) MS medium. Vanzie-Canton and Leonhardt [15] developed a protocol for the oryzalin treatment of *Z. zamiifolia* callus in a study aimed at producing a tetraploid plant in vitro.

Recently, the black-leaved attractive cultivar 'Dowon' has been created and patented by the South Korean breeder Hyuk Jin Lee [16] and offered on sale under the tradename Raven®. This cultivar has lime green new growth that turns a purple-black color as it ages, which brings an outstanding ornamental effect and gives the plant high popularity. As with other ZZ plants, it stores water in its thick petioles and a tuber-like rhizome, making it an exceptionally easy-to-grow houseplant that withstands even dark conditions. In some areas, this cultivar can be grown outdoors if the temperature does not fall below 15 °C [17].

Currently, there is no published information on the in vitro propagation of the black-leaved *Zamioculcas zamiifolia*. Considering the increasing demand for ZZ plants in the market, the present study aimed to evaluate the effect of BA and NAA on the mass proliferation of 'Dowon' cultivar through tissue culture technique.

2. Materials and Methods

2.1. Plant Material and Initiation of In Vitro Culture

Plant material (*Zamioculcas zamiifolia* (Lodd.) Engl. 'Dowon') was obtained from a greenhouse in Amol city, Mazandaran province, the northern part of Iran. The geographical coordinates of Amol are as follows: latitude: 36°28′10″ N, longitude: 52°21′02″ E, and elevation above sea level: 96 m. Leaf segments (15 mm × 15 mm) were used as primary explants. *Z. zamiifolia* 'Dowon' is a stem-less species and has underground rhizome. Therefore, the most appropriate organ for explant preparation is the leaf. To eliminate surface contamination, the leaves were washed with dishwashing liquid for 20 min. Next, the explants were placed under running tap water for half an hour and treated with 10% (*v/v*) sodium hypochlorite ($NaOCl/H_2O$ ± 3% active chloride, Chem-Lab, Zedelgem, Belgium) for 10 min. Then, the samples were disinfected with 0.1 mg·L^{-1} (*w/v*) mercury chloride ($HgCl_2$, Merck, Darmstadt, Germany) for 10 min followed by ethanol 70% for one min. Finally, the explants were washed thrice with sterile distilled water (5 min each). The margins of the explants were cut off to remove the parts damaged in the disinfection treatment. Each leaf was cut into four segments of 15 mm × 15 mm.

2.2. Organogenesis Induction

Disinfected explants were placed inside sterile glass-jars. Three explants were cultured in each jar filled with 50 mL of basal MS culture medium with 3% (*w/v*) sucrose (Merck, Darmstadt, Germany) and 0.7% (*w/v*) agar (SIGMA Aldrich, St. Louis, MO, USA). The media contained different concentrations of BA (0, 0.5, 1, and 2 mg·L^{-1}; i.e., 0, 2.22, 4.44, 8.88 μM) and NAA (0, 0.5, 1, and 2 mg·L^{-1}; i.e., 0, 2.69, 5.37, 10.75 μM) for the induction of organogenesis. Our aim was to evaluate the effect of 16 treatments on the performance of explants in vitro (direct or indirect organogenesis for both shoot multiplication and rooting). The PGRs combinations (treatments) stimulating simultaneous caulogensis and rhizogenesis were used in the further acclimatization step. The pH of the medium was adjusted to 5.6–5.8 after adding all components and before autoclaving at 121 °C and 105 kPa for 20 min. Cultures were kept in a growth room at 24 ± 2 °C with a 16 h photoperiod and light intensity of 50–60 μmol·m^{-2}·s^{-1} provided by cool-white fluorescent tubes and adjusted by a spectroradiometer. Plantlet number and length, leaf number, callus induction percentage, mean root length, and longest root length, all per explant, were measured after 12 weeks. To measure the traits and take pictures, the samples were removed from the glass containers and placed on Petri dishes.

2.3. Experimental Design and Data Analysis

The experiment was conducted in a completely randomized block design with three replications. Each experimental unit consisted of five glass-jars and three explants were cultured in each jar. Means were subjected to the analysis of variance and compared by Duncan's test at $p < 0.05$ using the SAS ver. 9.1 software (SAS Institute, 2003).

Considering that the interaction effects became significant, instead of comparing the averages of the main treatments, the average comparison between the levels of treatment compounds has been done. Therefore, the interaction effect of each treatment at different levels of the other treatment is well represented.

2.4. Acclimatization Process

All surviving and rooted microshoots (120-day-old) were removed from the culture jars and placed in the coco peat in 18 cm plastic pots (with a capacity of 4 kg) and watered with sterile water. In each pot, 3 plantlets obtained from the rooting step were planted (more than 100 pots were used in total). The pots were placed in a greenhouse with high humidity (70–80%), controlled photoperiod (16/8 light/dark with a light intensity not exceeding 100 μmol·m^{-2}·s^{-1}), and temperature (24 ± 2 °C) for acclimatization for 1.5–2 months. Irrigation was conducted by the rainy method and from the roof of the greenhouse. Plantlets were watered in this way every two weeks to prevent them from drying out. Light intensity was reduced by using two layers of canopy or nets. After 45 days of placing the plantlets in the adaptation pots, fully developed plants were transferred to a substrate (plot) containing cocopeat, peat moss, perlite, sand, and leaf soil in equal proportion for acclimatization and supply to the sales market. The survival (%) of the plants was evaluated after 8 weeks.

3. Results

3.1. Microshoot Proliferation

After disinfection (the efficiency of which reached over 85%), the explants were transferred to the proliferation culture medium. About 90% of explants regenerated shoots (Figure 1). A mean comparison of the data (Table 1) showed that the longest plantlets (2.06 cm) were obtained in MS medium augmented with 1 mg·L^{-1} NAA and 2 mg·L^{-1} BA. Their length was two- to four-fold higher than that of the other treatments. The shortest plantlets (0.4 cm), on the other hand, were produced in a medium containing 0.5 mg·L^{-1} BA. Plantlet length was also low in the control treatment, i.e., medium without PGRs.

Figure 1. In vitro propagation of *Z. zamiifolia* 'Dowon'. Callus production in a medium augmented with 2 mg·L^{-1} NAA + 1 mg·L^{-1} BA (**A**) and the PGRs-free control (**B**); development and multiplication of shoots in a medium containing 0.5 mg·L^{-1} NAA ((**C**), left), 1 mg·L^{-1} NAA + 2 mg·L^{-1} BA

((**C**), medial), and 0.5 mg·L^{-1} NAA + 2 mg·L^{-1} BA ((**C**), right); comparison of plantlets produced in a medium containing 1 mg·L^{-1} NAA + 2 mg·L^{-1} BA ((**D**), down) and the control medium ((**D**), up); leaf development in a medium containing 2 mg·L^{-1} NAA + 2 mg·L^{-1} BA (**E**); root development in a medium containing 2 mg·L^{-1} NAA + 0.5 mg·L^{-1} BA (**F**) and the control treatment (**G**); leaves developed in a medium supplemented with 0.5 mg·L^{-1} NAA + 2 mg·L^{-1} BA (**H**); plantlets produced in a medium enriched with 1 mg·L^{-1} NAA + 2 mg·L^{-1} BA (**I**); plantlets developed in a medium supplemented with 0.5 mg·L^{-1} NAA + 2 mg·L^{-1} BA, ready for acclimatization (**J**); plantlets cultivated ex vitro in a plastic pot filled with cocopeat (**K**); fully developed plants after one year in a plot containing cocopeat, peat moss, perlite, sand, and leaf soil in equal proportions (**L**). Age of cultures: (**A–J**): 60 days, (**K**): 105 days, (**L**): one year. Scale bars: (**A**,**B**,**D**,**I**) = 10 mm, (**C**,**F**) = 12 mm, (**E**,**H**) = 8 mm, (**G**) = 2 mm, and (**J**) = 5 mm.

Table 1. Effect of different concentrations of NAA and BA on the measured characteristics of *Zamioculcas zamiifolia* (Lodd.) Engl. 'Dowon' grown in vitro in the multiplication step.

NAA (mg·L^{-1})	BA (mg·L^{-1})	Plantlet Length (cm)	Plantlet Number per Explant	Leaf Number per Explant	Callus Induction (%)
0	0	0.54 ± 0.44 ^{c}	2.88 ± 0.19 ^{f}	5.66 ± 0.72 ^{f}	32.33 ± 5.80 ^{f}
0	0.5	0.40 ± 0.31 ^{d}	4.66 ± 0.28 ^{d,e}	9.33 ± 0.30 ^{d}	70.00 ± 20.00 ^{a,b}
0	1	0.56 ± 046 ^{c}	3.00 ± 0.36 ^{f}	7.33 ± 0.38 ^{e,f}	46.66 ± 15.30 $^{c–e}$
0	2	1.20 ± 0.30 ^{b}	4.33 ± 0.25 ^{d,e}	6.00 ± 0.27 ^{f}	33.33 ± 5.80 ^{f}
0.5	0	0.60 ± 0.11 ^{b,c}	3.66 ± 0.72 ^{e,f}	8.66 ± 0.482 ^{d}	45.00 ± 13.20 $^{c–e}$
0.5	0.5	0.53 ± 0.22 ^{c,d}	4.00 ± 0.15 ^{e}	7.33 ± 0.74 ^{e,f}	78.33 ± 7.60 ^{a}
0.5	1	0.60 ± 0.16 ^{b,c}	5.66 ± 0.17 ^{b,c}	8.00 ± 0.22 ^{e,f}	51.66 ± 7.60 ^{c,d}
0.5	2	1.16 ± 0.42 ^{b}	11.00 ± 0.42 ^{a}	22.00 ± 0.63 ^{a}	41.66 ± 7.60 ^{d,e}
1	0	0.66 ± 0.18 ^{c}	5.00 ± 0.53 $^{b–d}$	10.00 ± 0.62 ^{d}	40.00 ± 5.00 ^{d,e}
1	0.5	0.50 ± 0.37 ^{c,d}	3.66 ± 0.55 ^{e,f}	7.33 ± 0.29 ^{e,f}	56.66 ± 2.90 ^{b,c}
1	1	0.73 ± 0.44 ^{b,c}	4.66 ± 0.18 ^{d,e}	9.33 ± 0.28 ^{d}	46.66 ± 5.80 $^{c–e}$
1	2	2.06 ± 0.38 ^{a}	9.00 ± 0.34 ^{a,b}	18.00 ± 0.15 ^{b}	45.00 ± 5.00 $^{c–e}$
2	0	0.66 ± 0.11 ^{c}	5.00 ± 0.12 $^{b–d}$	10.00 ± 0.42 ^{d}	36.66 ± 2.90 ^{e,f}
2	0.5	0.63 ± 0.22 ^{c}	5.66 ± 0.15 ^{b,c}	11.33 ± 0.22 ^{c}	56.66 ± 5.80 ^{b,c}
2	1	0.66 ± 0.32 ^{c}	5.00 ± 0.44 $^{b–d}$	10.00 ± 0.15 ^{d}	45.00 ± 13.20 $^{c–e}$
2	2	0.93 ± 0.23 ^{b,c}	9.00 ± 0.23 ^{a,b}	18.00 ± 0.42 ^{b}	38.30 ± 2.90 ^{d,e}

Means with different letters in the same column are significantly different ($p < 0.05$) based on Duncan's test.

The highest number of plantlets per explant (11) was produced in the MS medium supplemented with 0.5 mg·L^{-1} NAA and 2 mg·L^{-1} BA and it was approximately four-fold higher than in the control object. Media with 1 mg·L^{-1} NAA together with 2 mg·L^{-1} BA and 2 mg·L^{-1} NAA with 2 mg·L^{-1} BA also induced high (9.0) plantlet number (Table 1). The smallest number of plantlets (2.88–3) was obtained in the control medium and medium with 1.0 mg·L^{-1} BA only.

A maximum number of leaves (22 per explant) was established in the MS medium with 0.5 mg·L^{-1} NAA and 2 mg·L^{-1} BA. Many leaves (18 per explant) were also produced in the media containing 1 mg·L^{-1} NAA together with 2 mg·L^{-1} BA and 2 mg·L^{-1} NAA with 2 mg·L^{-1} BA. A minimum number of leaves (5.66 per explant) was obtained in the medium without PGRs (Table 1).

The analysis of variance (Table 2) revealed that in terms of all studied characteristics related to plantlet proliferation, there was a significant effect of the tested PGRs concentrations.

3.2. Callus Induction Percentage

The concentration of BA and interaction effect between BA and NAA greatly affected the in vitro induction of callus in *Z. zamiifolia* 'Dowon' (Table 2, Figure 1A,B). The highest callus production (78.33% per explant) was obtained on the medium augmented with 0.5 mg·L^{-1} BA together with 0.5 mg·L^{-1} NAA (Table 1). The share of explants forming callus cultured on the medium containing 0.5 mg·L^{-1} BA was also high (70%). The lowest rate of callus formation (32.33–33.33% of explants) was induced on the control object and medium with 2.0 mg·L^{-1} BA only (Table 1). Most of the calli were compact and firm.

Table 2. Analysis of variance of the effect of different concentrations of NAA and BA on the measured characteristics of *Zamioculcas zamiifolia* (Lodd.) Engl. 'Dowon' grown in vitro in the multiplication, rooting, and acclimatization steps.

Source of Variances	df	Plantlet Length	Plantlet Number	Leaf Number	Callus Percentage	Mean Root Length	Longest Root Length	Root Number	Acclimatization Efficiency
Block	2	119.73 *	0.11 **	0.13 **	119.00 ns	15.37 **	1.02 **	0.09 ns	74.70 ns
NAA	3	82.05 *	0.16 **	0.12 **	232.00 ns	24.94 **	2.49 **	2.72 **	735.00 **
BA	3	232.92 *	0.02 **	0.14 **	1839.00 **	2.72 **	0.53 **	0.19 ns	152.00 *
NAA × BA	9	44.11 *	0.12 **	0.11 **	994.00 *	10.43 **	0.70 **	1.10 **	298.00 **
Error	30	86.28	0.35	0.17	85.41	2.55	0.09	0.17	50.30
CV (%)	-	3.40	4.58	3.61	19.32	2.80	16.82	18.86	10.50

*, **: significant at the 0.05 and 0.01 probability level, respectively; ns: not significant; CV: coefficient of variation; df: degrees of freedom.

3.3. Microshoot Rooting

About 85% of the microshoots regenerated roots. The statistical analysis showed that there was a significant difference between various concentrations of NAA and BA (alone and in combination) regarding rooting effectiveness, i.e., mean root length and length of the longest root. The concentration of NAA alone and the interaction of NAA and BA had a significant effect on the root number (Table 2). The longest (2.66 cm) and shortest (0.26 cm) mean root lengths were found in the media enriched with 2 mg·L^{-1} NAA + 0.5 mg·L^{-1} BA (Figure 1F) and 0.5 mg·L^{-1} BA alone, respectively. The length of roots produced in the medium without PGRs (control) was low (0.6 cm) (Table 3).

Table 3. Effect of different concentrations of NAA and BA on the measured characteristics of *Zamioculcas zamiifolia* (Lodd.) Engl. 'Dowon' grown in vitro in the rooting and acclimatization steps.

NAA (mg·L^{-1})	BA (mg·L^{-1})	Mean Root Length (cm)	Longest Root Length (cm)	Root Number	Acclimatization (%)
0	0	0.60 ± 0.29 ^{e}	1.73 ± 0.61 ^{c,d}	1.97 ± 0.00 $^{b–d}$	53.30 ± 5.80 ^{e,f}
0	0.5	0.26 ± 0.27 ^{f}	1.47 ± 0.25 ^{d,e}	2.00 ± 0.12 $^{b–d}$	60.00 ± 10.00 $^{c–f}$
0	1	1.16 ± 0.33 ^{c}	1.32 ± 0.12 ^{f,g}	1.33 ± 0.57 ^{d}	60.00 ± 10.00 $^{c–f}$
0	2	1.13 ± 0.45 ^{c}	1.73 ± 0.06 ^{c,d}	1.33 ± 0.27 ^{d}	70.00 ± 0.00 ^{b,c}
0.5	0	1.46 ± 0.12 ^{b}	1.77 ± 0.31 ^{c,d}	2.01 ± 0.00 $^{b–d}$	64.90 ± 5.80 $^{c–e}$
0.5	0.5	1.20 ± 0.55 ^{b,c}	1.40 ± 0.36 ^{e,f}	2.00 ± 1.00 $^{b–d}$	70.00 ± 10.00 ^{b,c}
0.5	1	1.30 ± 0.35 ^{b,c}	1.63 ± 0.12 $^{c–e}$	2.00 ± 0.00 $^{b–d}$	80.00 ± 0.00 ^{b}
0.5	2	1.20 ± 0.31 ^{b,c}	1.63 ± 0.25 $^{c–e}$	1.67 ± 0.42 ^{c,d}	70.00 ± 0.00 ^{b,c}
1	0	1.13 ± 0.46 ^{c}	1.57 ± 0.32 $^{c–e}$	2.33 ± 0.74 ^{b,c}	93.30 ± 5.80 ^{a}
1	0.5	1.36 ± 0.23 ^{b,c}	1.27 ± 0.15 ^{f,g}	2.00 ± 0.00 $^{b–d}$	70.00 ± 10.00 ^{b,c}
1	1	1.20 ± 0.36 ^{b,c}	1.57 ± 0.15 $^{c–e}$	2.33 ± 0.66 ^{b,c}	63.30 ± 5.80 $^{c–e}$
1	2	0.80 ± 0.20 ^{d}	1.14 ± 0.26 ^{g}	2.66 ± 0.58 ^{a,b}	80.00 ± 0.00 ^{b}
2	0	1.76 ± 0.22 ^{a,b}	2.60 ± 0.36 ^{b}	3.33 ± 1.15 ^{a}	63.00 ± 10.00 $^{c–e}$
2	0.5	2.66 ± 0.35 ^{a}	3.60 ± 0.46 ^{a}	2.67 ± 0.57 ^{a,b}	66.60 ± 5.80 ^{c,d}
2	1	1.60 ± 0.12 ^{a,b}	1.60 ± 0.10 $^{c–e}$	2.34 ± 0.35 ^{b,c}	56.60 ± 5.80 $^{d–f}$
2	2	1.76 ± 0.53 ^{a,b}	1.90 ± 0.26 ^{c}	2.66 ± 0.19 ^{a,b}	70.00 ± 10.00 ^{b,c}

Means with different letters in the same column are significantly different ($p < 0.05$) based on Duncan's test.

The highest root number (3.33 per microshoot) was regenerated in a medium containing only 2 mg·L^{-1} NAA. All other experimental objects regenerated less than 3 roots per microshoot. The lowest number of roots (1.33 per microshoot) was produced in the media with 1 and 2 mg·L^{-1} BA, both without NAA (Table 3).

3.4. Acclimatization

There was a significant impact of NAA concentration, BA, and interaction effect of both PGRs on the acclimatization efficiency of ZZ plants (Table 2). An average of 68.19% of the plantlets survived acclimation to the greenhouse environmental conditions (Figure 1J–L). Table 3 shows that a maximum survival rate (93.33%) in the acclimatization process was observed in plantlets generated on a medium supplemented with 2 mg·L^{-1} NAA. Plantlets produced in the control medium showed the lowest survival rate (53.3%).

4. Discussion

In vitro propagation on several aroids plants (Araceae family) has previously been carried out [7], but there is no published information on the micropropagation of the black-leaved ZZ plant. In some species, in vitro cultured plant tissues begin to grow upon encountering a cytokinin or auxin alone, but the presence of both of them is required to continue growth or increases the number of shoots regenerated per explant [18,19]. Exogenous application of these PGRs can increase the endogenous concentration of phytohormones. A study by Chen and Yeh [20] on the micropropagation of *Aglaonema* sp. reported that the shoot number increased linearly with elevating concentration of BA. Our work confirms this finding, but the addition of NAA was essential for caulogenesis; the highest number of plantlets (11) and leaves (22) per explant (after 12 weeks of culture) were produced in the medium supplemented with 0.5 mg·L^{-1} NAA and 2.0 mg·L^{-1} BA. These results are very promising, as Ghoochani Khorasani et al. [21] obtained a very low (2.62) multiplication ratio of green-leaved ZZ in vitro.

Several species need a low level of auxin in combination with a high concentration of cytokinins to increase shoot multiplication [22]. For example, Ali et al. [23] showed that excellent multiple shoots were formed in *Caladium bicolor* when 0.25 mg·L^{-1} NAA was added to 1 mg·L^{-1} of BA. Islam et al. [24] reported that the addition of NAA at variable concentrations in a medium containing 1 mg·L^{-1} BA induced shoot induction in *Anthurium andreanum* cv. Nitta. Treatment using 3 mg·L^{-1} BA and 0.2 mg·L^{-1} NAA recorded the highest shoot proliferation rate (6.0) and shoot length (7.75 cm) in *Aglaonema widuri*. Likewise, 4 mg·L^{-1} BA together with 0.1 mg·L^{-1} NAA and 0.5 mg·L^{-1} TDZ produced the maximum number of nodes (13.25 per explant) in this species. The highest number of leaves (4.25 per explant), on the other hand, was produced in the medium containing 3.5 mg·L^{-1} BA and 0.2 mg·L^{-1} NAA [6]. The present results are in agreement with these findings. Cytokinins are involved in many aspects of plant development, including the regulation of cell proliferation. Studies imply that these PGRs may elevate cell division rates by the induction of CycD3 expression, which encodes a D-type cyclin involved in the G1→M transition of the cell cycle [25]. Auxins, on the other hand, act as a signaling molecule to regulate many developmental processes throughout all stages of plant ontogenesis, e.g., cell elongation [26]. Interestingly, in the present study, callus, shoot, and root production were also observed in the PGRs-free culture medium, which suggested that the explants have enough endogenous phytohormones to stimulate organogenesis in vitro, although at a low rate. A similar phenomenon was reported by some researchers on a few aroid members [27–30]. The main advantage of this phenomenon is that there is no need to use exogenous PGRs and, as a result, reduction in propagation costs and somaclonal variation occurrence during micropropagation.

A study by Fang et al. [19] on *Aglaonema* 'Lady Valentine' reported that the longest shoots were produced on a medium with 5 mg·L^{-1} BA. Likewise, the use of BA was effective for axillary bud outgrowth in *Dieffenbachia compacta* [31], also a member of Araceae. In our study, the longest shoots were produced in the presence of 2 mg·L^{-1} BA together with 1 mg·L^{-1} NAA, probably due to the impact of auxin on the elongation of cells [26]. The superiority of BA over other cytokinins such as kinetin, N6-(Δ^2-isopentenyl) adenine (2-iP), and TDZ in promoting shoot elongation has been found in several other ornamental Araceae [19,23,32].

The current study revealed that leaf explants cultured on the MS medium containing 0.5 mg·L^{-1} BA and 0.5 mg·L^{-1} NAA produced the greatest number of calli (response rate). Similar findings were reported in *Caladium bicolor* (Aiton) Vent. with 4 mg·L^{-1} BA along with 0.5 mg·L^{-1} NAA [33]. The auxin commonly used for callus induction is 2,4-D, but IAA and NAA are also used, alone or in combination with BA [34]. Yu et al. [35] achieved a compact callus mass from leaf and petiole explants in *Anthurium andreanum* Linden ex André by a combination of 2,4-D and BA, from which plant formation was obtained. Leaflet and petiole explants of *Z. zamiifolia* were cultured onto a callus induction medium composed of $\frac{1}{2}$ MS salts, 4 mg·L^{-1} 2,4-D, and 0.2 mg·L^{-1} BA by [15]. The cultures were transferred to a fresh medium biweekly and stored in the dark at a temperature of 25–27 °C. Callus was visible on the explants after about 4.5 weeks. Next, calli were transferred to a shoot induction medium containing $\frac{1}{2}$ MS enriched with 1 mg·L^{-1} BA. The development of adventitious buds was observed after 11 weeks of culture. All rooted plantlets transferred to the greenhouse developed normally [15]. Callus obtained from leaf laminas is a suitable source for the production of plantlets through indirect organogenesis and somatic embryogenesis. However, the genetic stability of such plants is uncertain, thus, the comparison of the PGRs-treated and control plantlets is necessary. In the present study, no phenotypical differences were observed among plants from various experimental objects, although further molecular analyses are recommended.

Our findings revealed the positive effect of NAA on root production in ZZ 'Dowon'. Microshoots cultured on a medium augmented with 2 mg·L^{-1} NAA produced the highest number of roots, whereas microshoots from the experimental objects containing 1 and 2 mg·L^{-1} BA without NAA regenerated the lowest number of roots. In a study on *Caladium* micropropagation, Ali et al. [23] reported that the best rooting was obtained on a medium augmented with 1.0 mg·L^{-1} NAA. Auxin type and concentration considerably affected rooting percentage and root length. In *C. bicolor*, the maximum root number was produced in a medium with 3 mg·L^{-1} BA and 0.5 mg·L^{-1} NAA [20]. In *A. widuri*, the best root initiation and development (14.25 per explant) was found on a medium containing 3 mg·L^{-1} BA and 0.2 mg·L^{-1} NAA. On the other hand, the combination of 3.5 mg·L^{-1} BA and 0.2 mg·L^{-1} NAA was most suitable for root elongation (8.25 cm per explant) [6]. In contrast, microcuttings produced from tissue culture of *Aglaonema* rooted when treated with IBA and NAA [20]. Auxins trigger complex growth and developmental processes. They facilitate fast switching between gene activation and transcriptional repression via the auxin-dependent degradation of transcriptional repressors. The nuclear auxin signaling pathway consists of a small number of core components, but each component is represented by a large gene family [36]. Exogenous application of synthetic auxins, such as NAA, is crucial for root initiation and development [37].

It was found that the in vitro obtained plantlets of *Z. zamiifolia* 'Dowon' acclimatized suitably through cultivation in pots filled with coco peat. Different substrates such as sand, vermicompost, vermiculite, perlite, and organic humus have been used for the acclimatization of in vitro-regenerated Araceae plantlets, with high survival ratios ranging from 60% to 98% [30,38]. The high survival level of the plantlets obtained in the present study (reaching even 93.33%) confirms the suitability of the present protocol in the commercial reproduction of ZZ plant 'Dowon'.

5. Conclusions

This is the first complete report on the in vitro propagation of black-leaved *Z. zamiifolia* 'Dowon', which is a starting point for mass-propagation programs of this valuable cultivar. Our procedure led to an increased production of healthy and disease-free propagules compared to the traditional in vivo propagation systems. A combination of BA and NAA in suitable proportion induced most shoots and roots. The maximum number of shoots (11 per explant) and leaves (22 per explant) was obtained in a medium enriched with 2 mg·L^{-1} BA and 0.5 mg·L^{-1} NAA. On the other hand, the highest number of roots (3.33 per plant) was achieved in a medium with 2 mg·L^{-1} NAA. The survival of the plantlets during the acclimatization process was high (even 93.33%) and they were phenotypically similar to the mother plant. The application of easily accessible PGRs allows for the effortless implementation of our protocol in horticultural practice, but also opens the possibility to further study the combined effect of BA and NAA with less conventional PGRs.

Author Contributions: All authors made a significant contribution to the work presented. All authors have read and agreed to the published version of the manuscript.

Funding: This research received no external funding.

Data Availability Statement: Data are available by e-mail upon reasonable request.

Conflicts of Interest: The authors declare no conflict of interest.

References

1. Sriprapat, W.; Thiravetyan, P. Phytoremediation of BTEX from indoor air by *Zamioculcas zamiifolia*. *Water Air Soil Pollut.* **2013**, *224*, 1482. [CrossRef]
2. Chen, J.; Henny, R.J.; Henny, P.S. ZZ: A unique tropical ornamental foliage plant. *HortTechnol.* **2003**, *13*, 458–462. [CrossRef]
3. Lopez, R.G.; Blanchard, M.G.; Runkle, E.S. Propagation and production of *Zamioculcas zamiifolia*. *Acta Hortic.* **2009**, *813*, 559–564. [CrossRef]
4. Le Moullec, A.; Juvik, O.J.; Fossen, T. First identification of natural products from the African medicinal plant *Zamioculcas zamiifolia*—A drought resistant survivor through millions of years. *Fitoterapia* **2015**, *106*, 280–285. [CrossRef]
5. Prathibha, B.R.; Nirmala, K.S.; Satyanarayana, B.N.; Anithe, P.; Chinnaswamy, K.P. Induction of multiple shoots in *Zamioculcas zamiifolia* Engl. under in vivo conditions. *Int. J. Chem. Studies* **2018**, *6*, 667–671.
6. Kaviani, B.; Sedaghathoor, S.; Motlagh, M.R.S.; Rouhi, S. Influence of plant growth regulators (BA, TDZ, 2-iP and NAA) on micropropagation of *Aglaonema widuri*. *Iran. J. Plant Physiol.* **2019**, *9*, 2703–2718.
7. Chen, J.; Wei, X. Thidiazuron in micropropagation of aroid plants. In *Thidiazuron: From Urea Derivative to Plant Growth Regulator*; Faisal, M., Ahmad, N., Eds.; Springer: Berlin/Heidelberg, Germany, 2018; pp. 95–113.
8. Murashige, T.; Skoog, F. A revised medium for rapid growth and bioassays with tobacco tissue culture'. *Plant Physiol.* **1962**, *15*, 473–497. [CrossRef]
9. Chen, J.; Henny, R.J.; Devanand, P.S.; Chao, C.T. AFLP analysis of nephthytis (*Syngonium podophyllum* Schott) selected from somaclonal variants. *Plant Cell Rep.* **2006**, *24*, 743–749. [CrossRef]
10. Chen, F.; Wang, C.; Fang, J. Micropropagation of self-heading philodendron via direct shoot regeneration. *Sci. Hortic.* **2012**, *141*, 23–29. [CrossRef]
11. Stanly, C.; Bhatt, A.; Sulaiman, B.; Keng, C.L. Micropropagation of *Homalomena pineodora* Sulaiman & Boyce (Araceae): A new species from Malaysia. *Hortic. Bras.* **2012**, *30*, 39–43.
12. Heping, S.; Peng, L. Plantlet regeneration from leaf explants of *Zamioculcas zamiifolia*. *Acta Hort. Sin.* **2003**, *5*, 621–622.
13. Papafotiou, M.; Martini, A. Effect of position and orientation of leaflet explants with respect to plant growth regulators on micropropagation of *Zamioculcas zamiifolia* Engl. (ZZ). *Sci. Hortic.* **2009**, *120*, 115–120. [CrossRef]
14. Sayadi Nejad, M.; Sadeghi, S. Optimization of callus production and regeneration of zamiifolia (*Zamioculcas zamiifolia*). *J. Hortic. Sci.* **2019**, *33*, 405–415. [CrossRef]
15. Vanize-Canton, S.D.; Leonhardt, K.W. In vitro callus induction and plantlet regeneration protocol developed for the oryzalin treatment of *Zamioculcas zamiifolia* (Lodd.) Engl. (Araceae). *Acta Hortic.* **2009**, *813*, 201–208. [CrossRef]
16. Lee, H.J. *Zamioculcas* Plant Named 'Dowon'. US Patent USPP30035P3, 25 December 2018.
17. Zamioculcas zamiifolia 'Dowon' Common Name(s): Jet-Black Raven, Raven™ ZZ. Available online: https://plants.ces.ncsu.edu/plants/zamioculcas-zamiifolia-dowon/ (accessed on 5 January 2023).
18. Dewir, Y.D.; Chakrabarty, E.; Hahn, A.; Paek, K. A simple method for mass propagation of *Spathiphyllum cannifolium* using an airlift bioreactor'. *Vitro Cell. Dev. Biol. Plant* **2006**, *42*, 291–297. [CrossRef]
19. Fang, J.Y.; Hsul, Y.R.; Chen, F.C. Development of an efficient micropropagation procedure for *Aglaonema* 'Lady Valentine' through adventitious shoot induction and proliferation. *Plant Biotechnol.* **2013**, *30*, 423–431. [CrossRef]
20. Chen, W.L.; Yeh, D.M. Elimination of in vitro contamination, shoot multiplication, and *ex vitro* rooting of *Aglaonema*. *HortSci* **2007**, *42*, 629–632. [CrossRef]
21. Ghoochani Khorasani, A.; Roohollahi, I.; Golkar, P. Effect of different concentrations of nitrogen and 2,4-D on callus and plantlet production of *Zamioculcas zamiifolia* Engl. under in vitro condition. *J. Hortic. Sci.* **2021**, *35*, 55–62.
22. Kaviani, B.; Hashemabadi, D.; Khodabakhsh, H.; Onsinejad, R.; Ansari, M.H.; Haghighat, N. Micropropagation of *Begonia rex* Putz. by 6-benzyladenine (BA) and α-naphthalene acetic acid (NAA). *Int. J. Bioscie.* **2015**, *6*, 8–15.
23. Ali, A.; Munawar, A.; Naz, S. An in vitro study on micropropagation of *Caladium bicolor*. *Int. J. Agric. Biol.* **2007**, *9*, 731–735.
24. Islam, S.A.; Dewan, M.H.R.; Mukul, M.A.; Hossain, M.A.; Khatun, F. In vitro regeneration of *Anthurium andreanum* cv. Nitta. *Bangladesh J. Agril. Res.* **2010**, *35*, 217–226. [CrossRef]

25. Fathy, M.; Saad Eldin, S.M.; Naseem, M.; Dandekar, T.; Othman, E.M. Cytokinins: Wide-spread signaling hormones from plants to humans with high medical potential. *Nutrients* **2022**, *14*, 1495. [CrossRef]
26. Zhang, Y.; Yu, J.; Xu, X.; Wang, R.; Liu, Y.; Huang, S.; Wei, H.; Wei, Z. Molecular mechanisms of diverse auxin responses during plant growth and development. *Int. J. Mol. Sci.* **2022**, *23*, 12495. [CrossRef]
27. Viegas, J.; Rocha, M.T.R.; Ferreira-Moura, I.; Rosa, J.D.L.; Souza, A.; Correa, M.G.S.; Silva, J.A.T. *Anthurium andraeanum* (Linden ex Andre) culture: In vitro and ex vitro. *Floric. Ornam. Biotech.* **2007**, *1*, 61–65.
28. Liendo, M.; Mogollon, N. Multiplicacion clonal in vitro del anturio (*Anthurium andreanum* Lind. cv. Nicoya). *Bioagro* **2009**, *21*, 179–182.
29. Farsi, M.M.E.; Taghavizadeh, Y.; Qasemiomran, V. Micropropagation of *Anthurium andreanum* cv. Terra. *Afr. J. Biotechnol.* **2012**, *11*, 13162–13166.
30. Murillo-Gómez, P.A.; Naranjo, E.; Callejas, R.; Atehortúa, L.; Urrea, A. Micropropagation of the native species *Anthurium antioquiense* Engl. for conservation purposes. *Agronomia Colombiana* **2014**, *32*, 334–340. [CrossRef]
31. Azza, M.; El-Sheikh, K.; Mutasim, M.K. In vitro shoot micropropagation and plant establishment of an ornamental plant dumb cane (*Dieffenbachia compacta*). *Int. J. Curr. Res.* **2010**, *6*, 27–32.
32. Kozak, D.; Stelmaszczuk, M. The effect of benzyladenine on shoot regeneration in vitro of *Zantedeschia aethiopica* 'Green Goddess'. *Ann. UMCS Hortic.* **2009**, *19*, 14–18. [CrossRef]
33. Seydi, S.; Negahdar, N.; Andevari, R.T.; Ansari, M.H.; Kaviani, B. Effect of BAP and NAA on Micropropagation of *Caladium bicolor* (Aiton) Vent., an Ornamental Plant. *J. Ornament. Plants* **2016**, *6*, 59–66.
34. Kaviani, B. Some useful information about micropropagation. *J. Ornam. Plants* **2015**, *5*, 29–40.
35. Yu, Y.X.; Liu, L.; Liu, J.X.; Wang, J. Plant regeneration by callus-mediated protocorm-like body induction of *Anthurium andraeanum*. *Hort. Agric. Sci. China* **2009**, *8*, 572–577. [CrossRef]
36. Lavy, M.; Estelle, M. Mechanisms of auxin signaling. *Development* **2016**, *143*, 3226–3229. [CrossRef]
37. Matsuo, S.; Kikuchi, K.; Nagasuga, K.; Ueno, H.; Imanishi, S. Transcriptional regulation of auxin metabolic-enzyme genes during tomato fruit development. *Sci. Hortic.* **2018**, *241*, 329–338. [CrossRef]
38. Atak, C.; Celik, O. Micropropagation of *Anthurium andraeanum* from leaf explants. *Pakistan J. Bot.* **2009**, *41*, 1155–1161.

horticulturae

MDPI

Article

Ectopic Expression of *PgF3′5′H* in Commercial *Gypsophila paniculata* Cultivar through Optimized Agrobacterium-Mediated Transformation

Chunlian Jin †, Dan Sun †, Lulin Ma, Xijun Mo *, Chunmei Yang * and Fan Li *

Key Laboratory for Flower Breeding of Yunnan Province, National Engineering Research Center for Ornamental Horticulture, Floriculture Research Institute, Yunnan Academy of Agricultural Sciences, Kunming 650205, China
* Correspondence: moxj427@163.com (X.M.); yangchunmei@yaas.org.cn (C.Y.); lifan@yaas.org.cn (F.L.)
† These authors contributed equally to this work.

Abstract: *Gypsophila paniculata* is one of the most popular cut flowers in the world whose major cultivars are blooming white. As is well known, blue flowers could be generated via the overexpression of the gene encoding flavonoid 3′5′-hydroxylase (F3′5′H) in species that naturally lack it. In this study, we established the regeneration and a genetic transformation system for the commercial cultivar 'YX4' of *G. paniculata* and introduced the *F3′5′H* of *Platycodon grandiflorus* (*PgF3′5′H*) successfully into 'YX4' using the established protocol. A total of 281 hygromycin (Hyg)-resistant plantlets were obtained, and 38 of them were polymerase chain reaction (PCR) positive, indicating a 13.5% transformation efficiency. Shoot apex without meristem was more suitable for explant due to its high regeneration capacity, and the supplement of thidiazuron (TDZ) provided the most efficient promotion of adventitious bud induction, whereas the supplement of 6-Benzyladenine (6-BA) and 1-naphthaleneacetic acid (NAA) did not affect much. Additionally, the combination of 1 day (d) pre-culture, 5 d co-culture, 10 min infection, 30 mg·L^{-1} additional acetosyringone (AS) supplement, and 10 mg·L^{-1} Hyg selection formed the optimized system for 'YX4' transformation. This reliable and efficient agrobacterium-mediated transformation of the valuable commercial cultivar 'YX4' will contribute not only to the creation and improvement of *G. paniculata* cultivars, but also to the function research of genes associated with important ornamental traits.

Keywords: *Gypsophila paniculata*; regeneration; genetic transformation; commercial cultivar; *PgF3′5′H*

Citation: Jin, C.; Sun, D.; Ma, L.; Mo, X.; Yang, C.; Li, F. Ectopic Expression of *PgF3′5′H* in Commercial *Gypsophila paniculata* Cultivar through Optimized Agrobacterium-Mediated Transformation. *Horticulturae* **2023**, *9*, 321. https://doi.org/10.3390/horticulturae9030321

Academic Editors: Alicja Tymoszuk and Dariusz Kulus

Received: 28 January 2023
Revised: 23 February 2023
Accepted: 24 February 2023
Published: 1 March 2023

1. Introduction

Gypsophila paniculata, also known as baby's breath, is a perennial herbaceous flower of the Caryophyllaceae family [1]. Since the clouds of tiny white or pink flowers cover the bunches of the branching stems after blooming, *G. paniculata* is commonly used as fresh or dried filler in flower arrangements and bouquets [2]. As the only species in the genus *Gypsophila* used as a cut flower, *G. paniculata* is one of the top ten best-selling cut flowers in the world [3]. To meet the large demand for seedlings in production, tissue culture, which can rapidly and massively provide standard and uniform seedlings in plant factories, is widely used in the propagation of *G. paniculata* [4].

In contrast to the systematic and mature propagation industry, the cultivar innovation of *G. paniculata* remains slow. There were only 78 *G. paniculata* cultivars registered in the EU until 2017, while hundreds of commercial cultivars of other ornamental crops have been released every year [5]. Difficulties in crosses and low seed formation rates due to the double flower phenotype might explain the slowness of the *G. paniculata* cultivar's release. To overcome this slowness, researchers have tried to establish a transgenic system for *G. paniculata* since 1990s. Regenerated shoots were obtained from leaf or segmented stem explants for several cultivars [6,7]. Later, an *Agrobacterium tumefaciens* (*A. tumefaciens*)-mediated transformation system assisted by gibberellic acid was established for three cultivars [8]. However, neither a gene function study nor the traits modified using genetic transformation have been reported in *G. paniculata* since then.

Flower colour is one of the most critical ornamental traits and is the primary breeding target of many floricultural species, such as rose, carnation, lisianthus, etc. Most of the natural *G. paniculata* cultivars bloom white except 'Flamengo', with pale-pink flowers [9]. Nevertheless, there is a huge market demand for the colourful *G. paniculata*. Thus, dyeing technology has been developed to change the flower colour into carmine, tartrazine, blue, and so on. However, the artificial variegation fails to stain the petals evenly, therefore the dyed flowers usually display white spots, and some of them even wither. It is also environmentally risky due to the potential pollution to the river system.

Over decades, the genetic and molecular networks regulating the formation of flower colour have been studied in various ornamental species [10–14]. The water-soluble anthocyanins belonging to flavonoids are responsible for the flower colour, which ranges from orange/red to violet/blue [15]. Moreover, three well-known flavonoids, pelargonidin, cyanidin, and delphinidin, contribute to the development of the red, purple, and blue colours of the flowers, respectively [16]. An anthocyanin biosynthetic pathway is conserved in most plant species, which involves various enzymes. 4-coumaroyl-CoA and 3-malonyl-CoA are the first substrates in the synthetic pathway, which are then catalyzed by chalcone synthase (CHS) and form naringenin chalcone. This is then converted to naringenin flavanone under the action of chalcone isomerase (CHI). Flavanone 3-hydroxylase (F3H) turns naringenin into dihydrokaempferol. To display blue, a precursor chemical, dihydromyricetin, is necessary, and flavonoid 3′,5′-hydroxylase, encoded by *F3′5′H*, is considered a critical enzyme to introduce hydroxylation of the flavonoid B ring of dihydrokaempferol. The dihydromyricetin is then catalyzed by dihydroflavonol 4-reductase (DFR), anthocyanidin synthase (ANS), and flavonoid 3-O-glucosyltransferase (UFGT) step by step and is finally converted into delphinidin 3-glucoside [17]. Lacking F3′5′H, the key enzyme that promotes delphinidin synthesis, in species such as rose, chrysanthemum, and carnation, are unable to bloom blue through traditional hybridization or mutation [7,18], which might be fixed by introducing exogenous *F3′5′H* into the non-blue cultivars [11,19–21]. The previous RNA-seq data [22] of 'YX4' from three independent stages (flower bud stage, flower semi-open stage, and flower fully open stage) revealed that *G. paniculata* harbouring *F3′5′H* did not express somehow (Figure 1), resulting in the formation of white flowers.

It was reported that *F3′5′H* of different species exhibit diversity in catalytic efficiency, and that the *Kal* from *Campanula medium* displayed the highest catalytic efficiency with a unique sequence structure of nine amino acids (SKLDSSASA) [23,24]. The *F3′5′H* of *Platycodon grandifloras* (*PgF3′5′H*) was the second gene reported to possess this special sequence, whereas its catalytic function required further exploration [25]. In this study, we established the regeneration and genetic transformation system for the commercial cultivar 'YX4' of *G. paniculata* and introduced *PgF3′5′H* into this cultivar, providing a basal tool for further gene function analyses and gene editing.

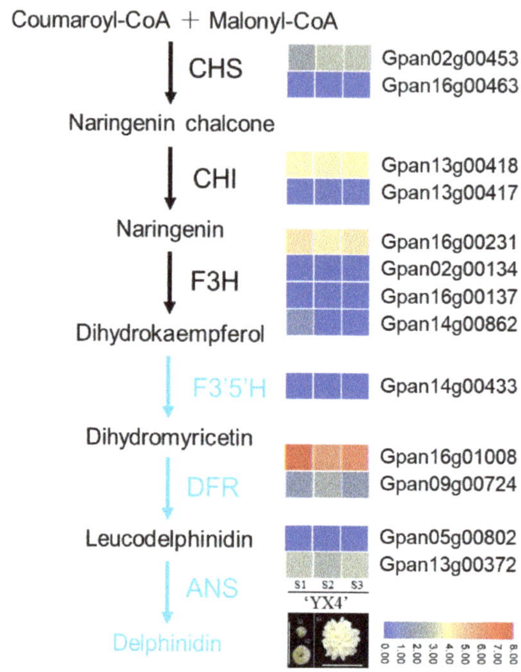

Figure 1. The delphinidin biosynthesis pathway in *G. paniculata*. Heat map of the expression of delphinidin biosynthesis pathway genes was constructed using TBtools. The S1, S2, and S3 are flower bud stage (S1), semi-open stage (S2), and fully open stage (S3), respectively. The scale bar is 1 cm. Abbreviations include CHS: chalcone synthase; CHI: chalcone isomerase; F3H: flavanone 3-hydroxylase; F3′5′H: flavonoid 3′,5′-hydroxylase; DFR: dihydroflavonol 4-reductase; ANS: anthocyanidin synthase.

2. Materials and Methods

2.1. Plant Materials and Culture Conditions

Aseptic plantlets of *G. paniculata* cultivar 'YX4' were obtained from Yuxi Yunxing Biological Technology Co., Ltd. (Yuxi, Yunnan Province, China). The aseptic plantlets were propagated every month with Murashige–Skoog (MS, Duchefa, Haarlem, The Netherlands) [26] medium supplemented with 1 $mg{\cdot}L^{-1}$ 6-Benzyladenine (6-BA, Sigma–Aldrich, St. Louis, MO, USA), 0.1 $mg{\cdot}L^{-1}$ 1-naphthaleneacetic acid (NAA, Sigma–Aldrich), 30 $g{\cdot}L^{-1}$ sucrose, and 8 $g{\cdot}L^{-1}$ agar. The pH of the propagation medium was adjusted to 5.95. The plantlets were cultured at 23 ± 2 °C with 1500 lx light intensity (16 h light/8 h dark, GreenPower LED with red and blue light in 4500 K, Philips, Amsterdam, The Netherlands).

2.2. Establishment of Regeneration System of *G. paniculata*

One-month-old robust plantlets were used in this study. First, the shoots were cut, and all the leaves were removed. Then, the stem apexes were cut and sliced up into 2–3 pieces. The segments were inoculated on MS-based adventitious bud induction medium supplemented with different concentrations of cytokinin (6-BA or thidiazuron, TDZ, Sigma–Aldrich) combined with 0.1 $mg{\cdot}L^{-1}$ NAA (Table 1). MS1 medium containing only 0.1 $mg{\cdot}L^{-1}$ NAA was taken as control. All the medium contained 30 $g{\cdot}L^{-1}$ sucrose and 8 $g{\cdot}L^{-1}$ agar and were adjusted to pH = 5.95. Thirty explants were included in each treatment, which was repeated three times. The explants were cultured in a growth room with the conditions mentioned above and relative data were collected one month later.

Table 1. Composition of mediums used to establish regeneration system.

NO.	PGRs' Combination	NO.	PGRs' Combination
T1	0.5 mg·L^{-1} TDZ + 0.1 mg·L^{-1} NAA	B1	0.5 mg·L^{-1} 6-BA + 0.1 mg·L^{-1} NAA
T2	1.0 mg·L^{-1} TDZ + 0.1 mg·L^{-1} NAA	B2	1.0 mg·L^{-1} 6-BA + 0.1 mg·L^{-1} NAA
T3	1.5 mg·L^{-1} TDZ + 0.1 mg·L^{-1} NAA	B3	1.5 mg·L^{-1} 6-BA + 0.1 mg·L^{-1} NAA
T4	2.0 mg·L^{-1} TDZ + 0.1 mg·L^{-1} NAA	B4	2.0 mg·L^{-1} 6-BA + 0.1 mg·L^{-1} NAA
T5	2.5 mg·L^{-1} TDZ + 0.1 mg·L^{-1} NAA	B5	2.5 mg·L^{-1} 6-BA + 0.1 mg·L^{-1} NAA
T6	3.0 mg·L^{-1} TDZ + 0.1 mg·L^{-1} NAA	B6	3.0 mg·L^{-1} 6-BA + 0.1 mg·L^{-1} NAA

PGRs, plant growth regulators, TDZ, thidiazuron; NAA, 1-naphthaleneacetic acid; 6-BA, 6-Benzyladenine.

2.3. Determination of Hygromycin (Hyg) Selection Pressure

Twenty explants were cultured with MS medium containing Hyg at different concentrations (0, 5, 10, 15, 20 mg·L^{-1}), respectively. The explants were cultured in the dark for 5 d and then transferred to a regular growth room. The number of surviving adventitious buds was counted after one month.

2.4. A. tumefaciens-Based Transformation

The *A. tumefaciens* strain EHA 105 containing pCAMBIA1301 with 35S-*GUS* and 35S-*PgF3′5′H* cassettes (pCAMBIA1301-*GUS*-*PgF3′5′H*, provided by Dr. Lulin Ma) was activated on the solid Luria–Bertani (LB) medium containing 50 mg·L^{-1} kanamycin (Kan) and rifampicin (Rif), followed by incubation at 28 °C for 16 h. The positive single colony was then inoculated into the liquid LB medium containing 50 mg·L^{-1} Kan and shaken cultured overnight (28 °C, 180 rpm) until the OD_{600} value reached 0.5–0.6. Stem apexes pre-cultured for 1 d were soaked in the *A. tumefaciens* solution and shaken cultured at 200 rpm for 10 min. The explants were blotted on sterile filter paper and grown on the M3 (for co-culture) medium in the dark for 5 d. They were then transferred to the M4 (for first selection) medium and cultured for two weeks. The adventitious buds were cut and grown on the M5 (for second selection) medium when they reached 1.5 cm. Two weeks later, the selected adventitious buds were transferred to the M6 medium and recovered for one month. Regeneration plantlets with 6–7 pairs of leaves were rooting on the M7 medium. The compositions of the medium mentioned above were listed in Table 2.

Table 2. Mediums used for *G. paniculata* genetic transformation.

Medium Number	Composition
Pre-culture: M2	MS + 2.5 mg·L^{-1} TDZ + 0.1 mg·L^{-1} NAA
Co-culture: M3	MS + 2.5 mg·L^{-1} TDZ + 0.1 mg·L^{-1} NAA + 10 mg·L^{-1} AS
First selection: M4	MS + 2.5 mg·L^{-1} TDZ + 0.1 mg·L^{-1} NAA + 200 mg·L^{-1} Cef + 10 mg·L^{-1} Hyg
Second selection: M5	MS + 1.0 mg·L^{-1} 6-BA + 0.1 mg·L^{-1} NAA + 200 mg·L^{-1} Cef + 10 mg·L^{-1} Hyg
Recover: M6	MS + 1.0 mg·L^{-1} 6-BA + 0.1 mg·L^{-1} NAA + 200 mg·L^{-1} Cef
Rooting: M7	MS + 0.2 mg·L^{-1} IAA + 0.3 mg·L^{-1} NAA

AS, acetosyringone; Cef, cefotaxime sodium; Hyg, hygromycin; IAA, Indole-3-Acetic Acid.

2.5. Effects of Different Factors on Transformation

The effects of the pre-culture period (1, 2, and 3 d), co-culture period (3, 4, and 5 d), infection period (10, 20, and 30 min) and AS concentration (10, 20 and 30 mg·L^{-1}) on the *A. tumefaciens* transformation efficiency were tested using an orthogonal experiment. All the treatments had three replicates, and each replicate was conducted using 50 explants. The number of adventitious buds was counted one month later.

2.6. β-glucuronidase Test

β-glucuronidase (GUS, Sigma–Aldrich) staining was performed according to [27]. Callus induced from explants co-cultured with *A. tumefaciens* containing pCAMBIA1301-*GUS-PgF3′5′H* were used for the histochemical analysis of GUS expression. The calli were soaked in x-gluc staining solution (Solarbio, Beijing, China) at 37 °C for 12 h and decolourised with 70% (v/v) ethanol for 2 h to remove chlorophyll.

2.7. Verification of Transgenic Plantlets Using PCR

The total DNA of plantlets which were resistant to Hyg and non-transformed plants (used as negative control) was extracted using the hexadecyltrimethylammonium bromi (CTAB) method [28]. The presence of the *Hyg* and *PgF3′5′H* genes was detected by using polymerase chain reaction (PCR) with pCAMBIA1301-*GUS-PgF3′5′H* as the positive control. The primers used were Hyg-F: GTTTCCACTATCGGCGAGTA, Hyg-R: GAGCCTGACCTATTGCATCTC; *PgF3′5′H*-F: TTCCTCCTCATCGTCCTC, *PgF3′5′H*-R: TGGCTAGGCAGTGTAAGC. The PCR was performed with the following reaction system: 94 °C for 3 min; 35 cycles of 94 °C for 1 min, 58 °C for 30 s, 72 °C for 1 min; 72 °C for 5 min. The amplified products were separated via electrophoresis using 1% agarose gel.

2.8. Statistical Analasis

Images of callus in the GUS staining experiment were taken using a stereomicroscope. To determine the effects of plant growth regulators on adventitious bud regeneration, one-way analysis of variance (ANOVA) was used, and means were compared using a Tukey's Honest Significant Difference (HSD) test ($p > 0.05$). The statistic calculation was performed using the Data Processing System (v15.10, Hangzhou Ruifeng Information Technology Co., Ltd., Hangzhou, China).

3. Results and Discussion

3.1. Establishing the Regeneration System of G. paniculata Cultivar 'YX4'

Thirteen mediums with different combinations of plant growth regulators were tested for 'YX4' regeneration using segments of shoot apex without meristem as explants. All the mediums showed 100% callus induction capacity while the adventitious bud induction rates and induction co-efficiency varied. The induction rate of adventitious bud was up to 90% in our test, and it was 67% when using leaves as explants [7]. The supplement of TDZ increased the number of adventitious buds induced significantly whereas 6-BA did not contribute to the induction of adventitious shoot much, correlating with the former report [7]. It was also observed that the higher concentration (2.5 $mg{\cdot}L^{-1}$) of TDZ reduced the number of explants that formed the adventitious shoot but increased the induction coefficient to 4.14 ± 0.22 (Table 3). Thus, we consider MS + 2.5 $mg{\cdot}L^{-1}$ TDZ + 0.01 $mg{\cdot}L^{-1}$ NAA as the optimized regeneration medium for further study.

3.2. Determination of Hyg Screening Pressure

To screen the Hyg-resistant concentration of *G. paniculata* shoot apical segments, we cultured the explants on M2 mediums containing 0, 5, 10, 15, or 20 $mg{\cdot}L^{-1}$ Hyg. The results showed that the survival ratio of the regenerated buds dropped as the concentration of Hyg increased. The survival ratio dropped to 25% when supplemented with 10 $mg{\cdot}L^{-1}$ Hyg, and no bud survived when the concentration was 20 $mg{\cdot}L^{-1}$ (Table 4). We also noticed that, when adding 10 $mg{\cdot}L^{-1}$ Hyg to the medium, the meristem of the regenerated bud was bleached although the leaves remained green, indicating the death of buds. Thus, 10 $mg{\cdot}L^{-1}$ was chosen as the selection concentration for 'YX4' transformation.

Table 3. The effects of plant growth regulators on adventitious bud regeneration.

Medium Number	Number of Explants	Number of Explants Forming Callus [1]	Number of Explants with Adventitious Buds	Number of Adventitious Buds [2]	Bud Induction Rates [3]	Induction Coefficient [4]
MS1	30	30	27.33 ± 0.58 ab	37.33 ± 1.53 f	0.91 ± 0.02 ab	1.37 ± 0.07 fg
T1	30	30	27.00 ± 2.00 ab	80.67 ± 4.51 b	0.90 ± 0.07 ab	3.00 ± 0.33 b
T2	30	30	27.33 ± 1.15 ab	66.00 ± 3.61 cd	0.91 ± 0.04 ab	2.42 ± 0.22 cd
T3	30	30	26.33 ± 1.53 ab	55.33 ± 2.52 de	0.88 ± 0.05 ab	2.10 ± 0.03 de
T4	30	30	25.00 ± 1.73 abc	69.00 ± 18.68 bc	0.83 ± 0.06 abc	2.74 ± 0.56 bc
T5	30	30	23.00 ± 3.46 c	95.00 ± 13.57 a	0.77 ± 0.12 c	4.14 ± 0.22 a
T6	30	30	27.67 ± 1.53 a	80.00 ± 3.00 b	0.92 ± 0.05 a	2.89 ± 0.12 b
B1	30	30	25.33 ± 2.08 abc	44.33 ± 8.14 ef	0.84 ± 0.07 abc	1.74 ± 0.19 ef
B2	30	30	24.33 ± 1.53 bc	43.00 ± 5.57 ef	0.81 ± 0.05 bc	1.76 ± 0.12 ef
B3	30	30	26.67 ± 2.52 ab	45.67 ± 7.37 ef	0.89 ± 0.08 ab	1.72 ± 0.26 efg
B4	30	30	27.33 ± 0.58 ab	35.67 ± 8.02 f	0.91 ± 0.02 ab	1.31 ± 0.31 g
B5	30	30	26.33 ± 2.08 ab	43.33 ± 1.53 ef	0.88 ± 0.07 ab	1.65 ± 0.09 fg
B6	30	30	26.67 ± 0.58 ab	46.67 ± 1.53 ef	0.89 ± 0.02 ab	1.75 ± 0.07 ef

Statistical analysis was performed using Tukey's HSD analysis ($p > 0.05$), different lowercase letters (a–g) indicate significant differences. [1] explants display visible callus; [2] bud regenerated from calli with visible leaves; [3] number of explants with adventitious buds/number of explants forming callus; [4] number of adventitious buds/number of explants with adventitious buds.

Table 4. The screen of Hyg selection pressure for commercial cultivar 'YX4'.

Concentration of Hyg (mg·L^{-1})	Number of Explants	Number of Explants Forming Callus	Number of Explants Forming Survived Adventitious Buds	Survival Ratio (%)
0	20	20	20	100
5	20	20	13	65.0
10	20	20	5	25.0
15	20	20	2	10.0
20	20	20	0	0

3.3. Optimization of Transformation Conditions and Exogenous Gene Transformation

The only reported *G. paniculata*- and *A. tumefaciens*-based transformation was performed using stem segments as explants, and the inoculation in *A. tumefaciens* was conducted directly without pre-culture [8]. To obtain the best procedure for the genetic transformation of 'YX4' using shoot apexes without meristem, we conducted the orthogonal experiment, including four factors, pre-culture period, co-culture period, infection period, and AS concentration, with three levels. *GUS* promoted by 35S promoter was introduced into the explants in this experiment. As shown in Figure 2, *GUS* expression was detected in the transgenic calli but exhibited blue pigmentation with varying degrees compared with the non-transgenic calli, indicating the successful transformation of pCAMBIA1301-*GUS*-*PgF3′5′H*. The observation of regenerated buds indicated that the infection time was the most influential factor for 'YX4' transformation, and the co-culture period ranked second, followed by the pre-culture period and AS concentration (Tables 5 and 6). The shorter pre-culture period and longer co-culture duration provided a higher differential ratio while the elongation of the infection duration affected the regeneration capacity obviously. To summarize, the combination of 1d pre-culture, 5d co-culture, 10 min infection, and 30 mg·L^{-1} additional AS forms the optimized system for 'YX4' transformation.

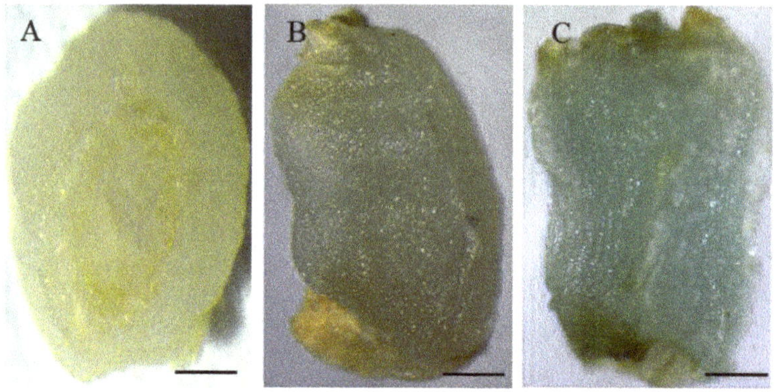

Figure 2. Histochemical staining of GUS transgenic calli. (**A**), wild type callus (**B**), transgenic calli. (**C**), Scale bar = 500 μm.

Table 5. The differentiation ratio of each orthogonal test group.

Test Number	Number of Explants	Pre-culture Period (d)	Co-culture Period (d)	Infection Period (min)	AS Concentration (mg·L^{-1})	Number of Explants Forming Buds	Differentiation Ratio
1	60	1	3	10	10	50	0.83
2	60	1	4	20	20	41	0.68
3	60	1	5	30	30	47	0.78
4	60	2	3	20	30	30	0.50
5	60	2	4	30	10	35	0.58
6	60	2	5	10	20	53	0.88
7	60	3	3	30	20	33	0.55
8	60	3	4	10	30	55	0.92
9	60	3	5	20	10	36	0.60

Table 6. Statistics analysis of the intersubjective effect.

T Value	Pre-culture Period	Co-culture Period	Infection Period	AS Concentration
T1	138	113	158	121
T2	118	131	107	127
T3	124	136	115	132
t1	46.00	37.67	52.67	40.33
t2	39.33	43.67	35.67	42.33
t3	41.33	45.33	38.33	44.00
R	6.67	7.66	14.34	3.67

Note: T1, T2, and T3 represent the sum of the number of the differentiation explants under different levels of the same factor, respectively; t1, t2, and t3 represent the average number of differentiation explants under different levels of the same factor; R is the range value.

3.4. Construction of Transgenic Blue G. paniculata

To test whether exogenous *F3′5′H* would promote the synthesis of delphinidin and facilitate the creation of blue *G. paniculata*, we introduced *PgF3′5′H* into 'YX4' through the optimized *A. tumefaciens*-mediated genetic transformation. The explants were cultured on the M2 medium for 1 d, followed by a 10 min infection of *A. tumefaciens* and a subsequent 5 d co-culture on the M3 medium. The induction of transgenic adventitious buds was then controlled by two rounds of Hyg selection (Figure 3). The regenerated Hyg-resistant plantlets were then verified using PCR. To increase the reliability of PCR verification, we detected both *Hyg* and *PgF3′5′H* in the Hyg-resistant plants. A total of 38 out of 281 resistant plantlets which possessed both *Hyg* and *PgF3′5′H* in the genomic DNA were obtained, indicating that the transformation efficiency of 'YX4' reached 13.5%. The PCR-positive plantlets were then rooted and transferred to the soil for growth. Unfortunately, we did not observe blue flowers in any of them, which indicated that the function of exogenous *F3′5′H* was also silenced in 'YX4'. This could be caused by chimeric adventitious bud formation, defects in exogenous gene transcription, or translation and silence in gene function. However, the underlying mechanisms remain to be further explored.

Figure 3. Construction of *PgF3′5′H* transgenic *G. paniculata*. (**A**), the schematic diagram of explants' acquisition; (**B**), the freshly cut explants; (**C**), regenerated shoots under first selection; (**D**), surviving regenerated shoots after second selection; (**E**), rooted seedlings on soil; (**F**), grown-up transgenic 'YX4' plant and flowers; (**G**), represented gel of PCR verification (*p*, plasmid; N, Non-transgenic plant; W, water; M, DNA ladder; 1–12, Hyg-resistant plantlets), Scale bar = 1 cm.

4. Conclusions

G. paniculata is the only species in the genus *Gypsophila* that has been used as cut flowers, and there is massive demand for new cultivars in the floricultural market. In contrast to its rising status, the breeding of *G. paniculata* has stagnated, and new cultivars with novel or improved traits are badly in need. Molecular genetics has become an important tool in floricultural breeding, in which novel flower cultivars have been created through genetic transformation, such as the blue rose. In this study, we established the regeneration and genetic transformation system for the commercial cultivar 'YX4' of *G. paniculata* and introduced the *PgF3′5′H* successfully into 'YX4' using a shoot apex without meristem as an explant. This reliable and efficient agrobacterium-mediated transformation of the valuable commercial cultivar 'YX4' will facilitate genetic improvement, as well as the application of gene-editing technology in this species. Nevertheless, one may notice that the expected phenotype was not obtained in this study, as well as in many other attempts when the traits of ornamental crops were modified. In addition to the efficient transformation system, genetic study of the gene regulatory network controlling key traits also plays a critical role in the breeding procedure, which requires more attention and should be involved.

Author Contributions: C.J. and D.S. conceived and designed the research; L.M. provided the *PgF3′5′H* plasmid and cultivated the transgenic plants. C.J., D.S. and F.L. wrote the manuscript; F.L., X.M. and C.Y. supervised the project. All authors have read and agreed to the published version of the manuscript.

Funding: This research was funded by Yunnan Fundamental Research Projects (202101AT070147), the National Natural Science Foundation of China (No. 31960608, 31960612), the Open Funding from the National Engineering Research Center for Ornamental Horticulture and Key Laboratory for Flower Breeding of Yunnan Province (FKL-202001, FKL-202002), High-level Talent Introduction Program of Yunnan Province -Industrial Talent Special Project (YNQR-CYRC-2020-004), and the Green Food Brand -Build a special project (floriculture) supported by science and technology (530000210000000013742).

Data Availability Statement: Not applicable.

Conflicts of Interest: The authors declare no conflict of interest.

References

1. Jin, C.; Sun, D.; Wei, C.; Guo, Z.; Yang, C.; Li, F. Gas Chromatography-Mass Spectrometry Analysis of Natural Products in *Gypsophila paniculata*. *Hortscience* **2021**, *56*, 1195–1198. [CrossRef]
2. Li, F.; Wang, G.; Yu, R.; Wu, M.; Shan, Q.; Wu, L.; Ruan, J.; Yang, C. Effects of Seasonal Variation and Gibberellic Acid Treatment on the Growth and Development of *Gypsophila paniculata*. *Hortscience* **2019**, *54*, 1370–1374. [CrossRef]
3. Li, F.; Mo, X.; Wu, L.; Yang, C. A Novel Double-flowered Cultivar of *Gypsophila paniculata* Mutagenized by 60Co γ-Ray. *Hortscience* **2020**, *55*, 1531–1532. [CrossRef]
4. Kozai, T. *Smart Plant Factory: The Next Generation Indoor Vertical Farms*, 1st ed.; Springer Nature Singapore Pte Ltd.: Singapore, 2018. [CrossRef]
5. Huang, L.; Wu, L.; He, F.; Yang, C.; Yu, R.; Shan, Q.; Ruan, J.; Wang, G.; Qu, Y. Optimization of chromosome sectioning technique and polyploidy induction in *Gypsophila paniculata*. *Southwest China J. Agricult. Sci.* **2017**, *30*, 2327–2333.
6. Ahroni, A.; Zuker, A.; Rozen, Y.; Shejtman, H.; Vainstein, A. An efficient method for adventitious shoot regeneration from stem-segment explants of gypsophila. *Plant Cell Tissue Organ Cult.* **1997**, *49*, 101–106. [CrossRef]
7. Zuker, A.; Ahroni, A.; Shejtman, H.; Vainstein, A. Adventitious shoot regeneration from leaf explants of *Gypsophila paniculata* L. *Plant Cell Rep.* **1997**, *16*, 775–778. [CrossRef] [PubMed]
8. Ben Zvi, M.M.; Zuker, A.; Ovadis, M.; Shklarman, E.; Ben-Meir, H.; Zenvirt, S.; Vainstein, A. Agrobacterium-mediated transformation of gypsophila (*Gypsophila paniculata* L.). *Mol. Breed.* **2008**, *22*, 543–553. [CrossRef]
9. Ayeh, K.O.; Selliah, R.; Hvoslef-Eide, A.K. Effect of Media on Some in vitro and ex-vitro Parameters in Micropropagation of *Gypsophila paniculata* L. *Acad. J. Plant Sci.* **2009**, *2*, 260–266.
10. Nakatsuka, T.; Mishiba, K.-I.; Kubota, A.; Abe, Y.; Yamamura, S.; Nakamura, N.; Tanaka, Y.; Nishihara, M. Genetic engineering of novel flower colour by suppression of anthocyanin modification genes in gentian. *J. Plant Physiol.* **2010**, *167*, 231–237. [CrossRef]
11. Noda, N.; Yoshioka, S.; Kishimoto, S.; Nakayama, M.; Douzono, M.; Tanaka, Y.; Aida, R. Generation of blue chrysanthemums by anthocyanin B-ring hydroxylation and glucosylation and its coloration mechanism. *Sci. Adv.* **2017**, *3*, e1602785. [CrossRef]
12. Zhao, D.; Tao, J. Recent advances on the development and regulation of flower color in ornamental plants. *Front. Plant Sci.* **2015**, *6*, 261. [CrossRef]
13. Urs, A.N.N.; Hu, Y.; Li, P.; Yuchi, Z.; Chen, Y.; Zhang, Y. Cloning and Expression of a Nonribosomal Peptide Synthetase to Generate Blue Rose. *ACS Synth. Biol.* **2018**, *8*, 1698–1704. [CrossRef]
14. Van Huylenbroeck, J. *Ornamental Crops*; Springer: Berlin/Heidelberg, Germany, 2018; Volume 11.
15. Tanaka, Y.; Brugliera, F.; Chandler, S. Recent Progress of Flower Colour Modification by Biotechnology. *Int. J. Mol. Sci.* **2009**, *10*, 5350–5369. [CrossRef] [PubMed]

16. Liu, Y.; Qian, J.; Li, J.; Xing, M.; Grierson, D.; Sun, C.; Xu, C.; Li, X.; Chen, K. Hydroxylation decoration patterns of flavonoids in horticultural crops: Chemistry, bioactivity, and biosynthesis. *Hortic. Res.* **2022**, *9*, uhab068. [CrossRef] [PubMed]
17. Sunil, L.; Shetty, N.P. Biosynthesis and regulation of anthocyanin pathway genes. *Appl. Microbiol. Biotechnol.* **2022**, *106*, 1783–1798. [CrossRef] [PubMed]
18. Noda, N. Recent advances in the research and development of blue flowers. *Breed. Sci.* **2018**, *68*, 79–87. [CrossRef]
19. Bai, L.; Jia, J.; Sun, J.; Li, P.; Zhao, M.; Pan, A. Cloning, expression and immunological identification of F3′5′H gene in transgenic carnation. *Plant Physiol. Commun.* **2009**, *46*, 205–209.
20. Katsumoto, Y.; Fukuchi-Mizutani, M.; Fukui, Y.; Brugliera, F.; Holton, T.; Karan, M.; Nakamura, N.; Yonekura-Sakakibara, K.; Togami, J.; Pigeaire, A.; et al. Engineering of the Rose Flavonoid Biosynthetic Pathway Successfully Generated Blue-Hued Flowers Accumulating Delphinidin. *Plant Cell Physiol.* **2007**, *48*, 1589–1600. [CrossRef]
21. Brugliera, F.; Tao, G.-Q.; Tems, U.; Kalc, G.; Mouradova, E.; Price, K.; Stevenson, K.; Nakamura, N.; Stacey, I.; Katsumoto, Y.; et al. Violet/Blue Chrysanthemums—Metabolic Engineering of the Anthocyanin Biosynthetic Pathway Results in Novel Petal Colors. *Plant Cell Physiol.* **2013**, *54*, 1696–1710. [CrossRef]
22. Li, F.; Gao, Y.; Jin, C.; Wen, X.; Geng, H.; Cheng, Y.; Qu, H.; Liu, X.; Feng, S.; Zhang, F.; et al. The chromosome-level genome of *Gypsophila paniculata* reveals the molecular mechanism of floral development and ethylene insensitivity. *Hortic. Res.* **2022**, *9*, uhac176. [CrossRef]
23. Okinaka, Y.; Shimada, Y.; Nakano-Shimada, R.; Ohbayashi, M.; Kiyokawa, S.; Kikuchi, Y. Selective accumulation of delphinidin derivatives in tobacco using a putative flavonoid 3′, 5′-hydroxylase cDNA from Campanula medium. *Biosci. Biotech. Bioch.* **2003**, *67*, 161–165. [CrossRef] [PubMed]
24. Ma, L.; Jia, W.; Duan, Q.; Wang, X.; Jiang, Y.; Guo, Y.; Li, J.; Wang, J. Clone and expression analysis of PgF3′5′H gene in *Platycodon grandiflorus. Acta Bot. Boreali Occident. Sin.* **2014**, *34*, 40–46.
25. Ma, L.; Duan, Q.; Du, W.; Cui, G.; Jia, W.; Wang, X.; Wang, J. Differential Expression Analysis of *Platycodon grandiflorus* PgF3′5′H and Its Mutant Gene in *Pichia pasteris. Mol. Plant Breed.* **2021**, *19*, 2256–2264.
26. Classic Murashige, T.; Skoog, F. A revised medium for rapid growth and bioassays with tobacco tissue cultures. *Physiol. Plant* **1962**, *15*, 473–497. [CrossRef]
27. Janssen, B.-J.; Gardner, R.C. Localized transient expression of GUS in leaf discs following cocultivation with Agrobacterium. *Plant Mol. Biol.* **1990**, *14*, 61–72. [CrossRef] [PubMed]
28. Li, F.; Cheng, Y.; Zhao, X.; Yu, R.; Li, H.; Wang, L.; Li, S.; Shan, Q. Haploid induction via unpollinated ovule culture in Gerbera hybrida. *Sci. Rep.* **2020**, *10*, 1702. [CrossRef] [PubMed]

horticulturae

MDPI

Article

Thidiazuron Induced In Vitro Plant Regeneration, Phenolic Contents, Antioxidant Potential, GC-MS Profiles and Nuclear Genome Stability of *Plectranthus amboinicus* (Lour.) Spreng

Mohammad Faisal *, Ahmed A. Qahtan and Abdulrahman A. Alatar

Department of Botany and Microbiology, College of Science, King Saud University, P.O. Box 2455, Riyadh 11451, Saudi Arabia
* Correspondence: faisalm15@yahoo.com; Tel.: +966-(011)-4675877

Abstract: *Plectranthus amboinicus* (Lour.) Spreng is a perennial plant from the mint family with aromatic, succulent leaves and several health benefits. Multiple shoot regeneration was accomplished in vitro using nodal segments (NS) explants of *P. amboinicus* pretreated with 0, 0.5, 5, 25, 50, and 100 μM thidiazuron (TDZ) for 4 h, then transferred to a growth regulator-free media. After 8 weeks of growth, NS explants pre-treated with 25 μM TDZ for 4 h and then transferred to TDZ-free Murashige and Skoog (MS) media produced the greatest number of shoots (27.3 per NS) with the longest average shoot length (4.9 mm) in 97.2% of cultures. On the same medium, regeneration of roots in most of the *P. amboinicus* shoots occurred spontaneously. The in vitro-regenerated *P. amboinicus* plantlets were adequately hardened off and adapted to the ex-vitro environment with a 90% survival rate. Total phenolic, tannin, and flavonoid contents, as well as 2,2-diphenyl-1-picrylhydrazyl (DPPH) radical-scavenging/antioxidant activity, were significantly higher in in vitro-regenerated plants than in ex vitro-plants. Flow cytometry (FCM) analysis validated the nuclear genome stability of the in vitro generated plants, which assessed their nuclear DNA content and found it to be comparable in genome size to that of the field-grown plants. The study found a quick and efficient method for in vitro multiplication of *P. amboinicus* which can aid to increased availability and accessibility of this plant species for various purposes. The genetic and phytochemical analysis of the in vitro propagated plants can also provide valuable insight into the plant's properties and potential applications, which can further assist in its preservation and sustainable usage.

Keywords: plant biotechnology; in vitro regeneration; medicinal plant; tissue culture; nuclear DNA content; phytochemicals

Citation: Faisal, M.; Qahtan, A.A.; Alatar, A.A. Thidiazuron Induced In Vitro Plant Regeneration, Phenolic Contents, Antioxidant Potential, GC-MS Profiles and Nuclear Genome Stability of *Plectranthus amboinicus* (Lour.) Spreng. *Horticulturae* **2023**, *9*, 277. https://doi.org/10.3390/horticulturae9020277

Academic Editors: Alicja Tymoszuk and Dariusz Kulus

Received: 31 January 2023
Revised: 16 February 2023
Accepted: 16 February 2023
Published: 18 February 2023

1. Introduction

Plectranthus amboinicus (Lour.) Spreng is a species of plant in the mint family, 'Lamiaceae', frequently known as 'Country Borage', 'Mexican mint' or 'Spanish thyme', and native to eastern and southern Africa, as well as naturally widespread throughout the tropical and warm regions of the globe [1]. In addition to its ornamental value, this plant has long been valued for its medicinal properties, particularly in the treatment of such conditions as a runny nose, cough, sore throat, bronchitis, diarrhea, and insect bites [2,3]. It has several useful pharmacological effects, including antimicrobial [4,5], antitumor, anti-inflammatory [6,7], anti-rheumatoid arthritis [8], anti-cancer [9], antioxidant [10–12], antithrombotic [3], antihyperglycemic and antihyperlipidemic [13] properties. The plethora of bioactive chemicals, including terpenoids, alkaloids, phenols, amino acids, glycosides, flavonoids, tannins, and sesquiterpenoids, in this plant is responsible for its numerous beneficial effects [2,3,14]. The Gas chromatography–mass spectrometry (GC-MS) analysis showed that the extract had 25 bioactive compounds. Total phenol (48 mg GAE/g) and total flavonoid (25 mg CE/g) content were found to be rather high in hexane leaf extracts of *P. amboinicus* in an analysis reported by Ashaari et al. [2]. In addition, phenolic content

(146.77 g GAE/kg extract) and antioxidant activity (0.491, 0.396, and 0.643 mol TE/kg extract in 2,2-diphenyl-1-picrylhydrazyl (DPPH), ferric reducing antioxidant power (FRAP), and 2,2-azinobis (3-ethyl-benzothiazoline-6-sulfonic acid) (ABTS) tests, respectively) were detected in *P. amboinicus* methanolic leaf extract [12].

The practice of cultivating plant cells, tissues, and organs in vitro plays a crucial role in plant biotechnology and crop improvement, and it has proven to be a highly effective method in producing a wide array of rare, endangered, medicinal, and economically valuable plants, thereby ensuring a steady supply of vital raw materials. Clonal multiplication in vitro offers scalable options for large-scale plant propagation, conservation, and sustainable utilization [15,16]. Multiple investigations on the in vitro micropropagation of several species of *Plectranthus*, including *P. amboinicus*, have been reported by a number of researchers in the past [17–27]. To this day, there are no reports on the use of thidiazuron (TDZ) for in vitro proliferation and plant regeneration of *P. amboinicus*. TDZ (N-phenyl N-1,2,3-thiazol-5-yl urea) was initially developed as a cotton defoliant; it is a diphenylurea synthetic herbicide and plant growth regulator with strong cytokinin-like activities that has been found to be effective in promoting shoot formation, improving shoot quality, and increasing the number of shoots produced from explants. [28–30]. The significant effect of TDZ on tissue culture regeneration is related to endogenous cytokinin production, changes in cell membranes, nutritional absorption, metabolism, and better nutrient assimilation in plant tissues, continuing to make it distinguish out from other plant growth regulators (PGRs) [28,31,32]. The biological effect of TDZ is very variable and is determined by factors such as the concentration of the compound, the amount of time it is exposed for, the kind of explant that is utilized, and the genotype of the plant [33,34]. During the process of tissue culture regeneration, the impacts of numerous elements, including as light, elicitors, and PGRs, not only have an influence on the process of shoot regeneration and the production of biomass, but they also have an impact on the synthesis of a variety of bioactive compounds [35]. Plants are subjected to a variety of stresses during the processes of in vitro culture and acclimatization and, in order for the plant's cells to cope with these stresses and maintain growth and development, they produce phytochemical compounds such as phenols, alkaloids, flavonoids, terpenoids, tannins, and many others. It has been shown that TDZ is effective for in vitro regeneration and synthesis of phytochemicals in a broad range of medicinal plants, including *Merwilla plumbea* [36], *Dendrobium nobile* [37], *Crocus sativus* [38], *Tecoma stans* [28], *Linum usitatissimum* [35,39] and *Lagerstroemia speciosa* [29].

Long-term exposure to TDZ has been found to have a negative impact on plant growth and development leading to the appearance of undesired shoot abnormalities such hyperhydricity, fasciation, shortening, and thickening. These abnormalities can then cause poor elongation and weak rooting capacity, ultimately affecting the overall health and productivity of the plant [40–42]. Regeneration of *Arachis hypogaea* [43] and *Dianthus caryophyllus* [44] in a continuous culture that contained media with TDZ resulted in the formation of unusual bud primordia that were unable to mature into fully developed plantlets. Kadota and Niimi [45] reported that TDZ significantly increased the occurrence of hyperhydric shoots in *Pyrus pyrifolia* by approximately ten times, compared benzyl adenine or kinetin. Ivanova and van Staden [46] suggested that TDZ should be used with caution in plant tissue culture, as its prolonged exposure can lead to the development of flattened shoot anomalies in *Aloe polyphylla* that ultimately failed to develop into plantlets. Furthermore, prolonged exposure to TDZ has been shown to negatively impact the rooting capacity of *Solanum melongena*, leading to weak and poorly developed root systems [47]. The possibility of TDZ to induce genetic changes in the regenerated plants is a major concern for its use in tissue culture and regeneration. The incidence of DNA polymorphism was found to increase in *Phalaenopsis bellina* plants grown in vitro after being treated with TDZ [48]. In this study, we develop an effective in vitro multiple shoot induction and user-friendly plant regeneration system for *P. amboinicus* using nodal explants exposed with TDZ for a very short period of time before being transfer to MS basal medium. In addition, for the first time, the nuclear DNA content, genome size, phenolic contents, antioxidant potential and GC-MS profiles of in vitro-regenerated plants were evaluated and compared to the ex-vitro donor-plant.

2. Materials and Methods

2.1. Starting Materials and Aseptic Cultures

Young proliferating shoots were collected from a *Plectranthus amboinicus* (Lour.) Spreng plants growing in the growth room at the Botany & Microbiology Department, College of Science, King Saud University. The collected shoots underwent a thorough washing using the laboratory's running tap water, after which they were subjected to a Tween-20 (0.1%, *v*/*v*), treatment for five minutes, and a subsequent rinsing in ultra-pure water (UPW) to remove any traces of dust and debris. The shoot samples, now cleaned, were further sterilized with 0.1% (*w*/*v*) aqueous mercury chloride ($HgCl_2$, Riedel-de Haan AG, Seelze, Germany) for 3 min, and then thoroughly washed with sterile UPW five times to remove to the traces of mercury. To prepare the starting materials (nodal segments, NS) for the subsequent in vitro culture studies, the sterilized shoots were cut into pieces that were 0.5–0.7 cm in length containing axillary node.

2.2. Thidiazuron (TDZ) Treatment and Shoot Regeneration

In the experiments TDZ (CAS number 51707-55-2, Duchefa Biochemie B.V., Haarlem, The Netherlands) were used for the pulse treatment of the sterilized NS explants. The NS explants were pre-soaked in a solution of Murashige and Skoog (MS) medium [49] containing micro and macro salts including vitamins (plant cell culture tested, MDL number MFCD00240976, Sigma-Aldrich, St. Louis, MO, USA) supplied with 0.5, 5, 25, 50 and 100 µM of TDZ for 4 h while gently shaking on a rotatory shaker (Incu-Shaker™ Mini, Benchmark Scientific, Sayreville NJ, USA) at 100 rpm. The treated NS explants were blotted dry on sterile filter papers and then transferred to agar (0.7%, *w*/*v*, plant cell culture tested, CAS number 9002-18-0, Sigma-Aldrich, St. Louis, MO, USA) solidified MS basal medium for further growth and regeneration. The untreated NS explants cultured on MS basal medium serve as control. The pH of the culture medium was adjusted to 5.7 using 1 N aqueous NaOH (96.0%, WINLAB, Middlesex, UK) solution before autoclaving at 121 °C (1.05 kg cm^{-2}) for 20 min. Cultures were initiated on Petri dishes and further maintained in 100 mL culture bottles at 23 ± 2 °C under 50 µmol m^{-2} s^{-1} cool white fluorescent light (Philips 39-Watt T5 linear fluorescent tubes, Philips, Pila, Poland) with a 16/8 h day/night photoperiod.

2.3. Effect of Basal Media on Shoot Proliferation

Different basal media was also evaluated, after the determination of the most suitable TDZ pre-soaking of NS for induction and proliferation of shoots. In addition to MS, the four other types of media that were evaluated were: Gamborg B5 [50] medium (B5, Product number G0209), Linsmaier and Skoog [51] medium (LS, Product number L0230) Nitsch [52] medium (N, Product number N0224); Schenk and Hildebrandt [53] medium (SH, Product number S0225), all of which were purchased from Duchefa Biochemie B.V. (Haarlem, The Netherlands). Once every three weeks, the explants were moved to new culture vials with fresh medium. The number of regenerated shoots per treated explant and the frequency with which this occurred were determined after 8 weeks of culture on the medium.

2.4. Rooting and Acclimatization

The majority of the *P. amboinicus* shoots developed roots concurrently on the same in vitro culture medium, and the shoots that were unable to do so were switched to $\frac{1}{2}$MS media with 0.5 µM indole 3-butyric acid (IBA, CAS number 133-32-4, Duchefa Biochemie B.V., Haarlem, The Netherlands). After washing with sterile water to remove any remnants of the basal nutrient media, the TDZ-induced shoots that had already developed a root system were carefully removed from the culture vials and transplanted into 12 cm diameter pots filled with a soil mixture (Planta-Guard Germany). The plantlets were grown in a controlled environment (Plant Growth Chamber, Conviron Adaptis, Manitoba, Canada) at a temperature of 24 ± 2 °C under 16-h day and 8-h night photoperiod with light intensity of 50 μmol m^{-2} s^{-1}. The plants were watered twice a week with a nutrient solution consisting of $\frac{1}{2}$MS basal media, which only included micro and macro salts, applied through sprinkling from above. After 8 weeks, the survival rate of the plants was recorded, and the acclimated in vitro *P. amboinicus* plants were transferred to the growth room for continued growth and development.

2.5. Flow Cytometry (FCM) and Nuclear DNA Content

FCM analysis were carried out to determine the nuclear DNA (nDNA) content and genome size of *P. amboinicus* donor plants and in vitro regenerated ones using a Beckman Coulter XL-MCL flow cytometer system (Beckman Coulter, Inc. USA). From 100 mg of fresh leaf tissues, nuclei were extracted by chopping them using a scalpel and suspending the pieces in 1.0 mL of filtered Galbraith buffer [54] (pH 7.0) containing 20 mM MOPS (CAS number 1132-61-2), 45 mM $MgCl_2$ (CAS number 7791-18-6, Loba Chemie, Mumbai India), 30 mM sodium citrate (CAS number 6132-04-3, Loba Chemie, Mumbai India) and 0.1% (*v/v*) Triton X-100 (CAS number 9036-19-5, Sigma-Aldrich, St. Louis, MO, USA). The leaf-homogenates were filtered through a 22 μm membrane using micro syringe after being chopped, and then they were mixed for 15 min with a 50 g/mL solution of propidium iodide (PI, CAS number 25535-16-4, Sigma-Aldrich, St. Louis, MO, USA). To avoid the staining of RNA, RNase (50 μ g/mL, DNase and protease-free, Catalog number EN0531, Thermo Scientific™, Massachusetts, USA) was also included in the samples. The relative nuclear DNA (nDNA, 2C) content of each leaf sample was determined by analyzing it three times using flow cytometry (FCM) with a green laser (532 nm). The G_1 peak mean values were used to compute the nDNA content, as described by Doležel and Bartoš [55], and the values were expressed in picograms (pg).

2C nDNA content of sample = standard 2C value × (sample 2C mean peak value/ standard 2C mean peak value)

The genome size of *P. amboinicus* samples was determined by applying the formula as follows: genome size (bps) = $(0.978 \times 10^9) \times$ 1C DNA content. This calculation relied on the previously reported information that 1 pg of DNA is equivalent to 978 million base pairs, as stated by Doležel et al. [56].

2.6. Preparation of Plant Extracts

Leaf samples from both in vitro-established and ex vitro-grown *P. amboinicus* donor plants were oven-dried at 50 °C for 48 h. Using a laboratory grinder, the dried leaf samples were processed to a fine powder. After grinding the materials to a fine powder, they were extracted with 50 mL of methanol for 24 h while being shaken gradually at room temperature. Following this step, the methanolic extracts were filtered using filter paper (Whatman filter paper, Grade 1). The solvent was evaporated with the help of a rotary evaporator (IKA Werke GmbH & Co. KG, Staufen im Breisgau, Germany).

2.6.1. Determination of Total Phenolics

The total phenolic content of *P. amboinicus* leaf extracts from both in vitro and ex vitro-grown plants was determined using a modified version of the technique reported by Singleton et al. [57]. The 10% (*v/v*) Folin-Ciocalteu reagent was added to a test tube containing 3.2 mL of the plant extract, and the mixture was mixed thoroughly before being let to stand for 5–10 min. Following the addition of 600 μL of a 20% (*w/v*) sodium carbonate solution to the mixture, it was left to incubate at room temperature for one hour. The absorbance of the mixtures was measured using a UV-Visible spectrophotometer at a wavelength of 765 nm, and the total phenolic content in both the leaf samples was calculated from the calibration curve ($y = 0.0021x + 0.0245$) using gallic acid as the reference standard. The results were expressed in terms of mg of gallic acid equivalents (GAE) per g dry weight of extract.

2.6.2. Determination of Total Tannins

The total tannin content of in vitro and ex vitro grown plants of *P. amboinicus* leaf extracts was measured using a technique reported by Chandran and Indira [58], with minor changes. Following the addition of 200 µL of plant extract, 3.2 mL of 10% (*v*/*v*) Folin-Ciocalteu reagent containing and 600 µL of sodium carbonate solution containing 35% were mixed together in the test tube. The materials were well mixed and then incubated in the dark for a period of thirty minutes so that the reaction could take place. The absorbance of the mixture samples was measured using a UV-Visible spectrophotometer at a wavelength of 700 nm. The total tannin content in both the leaf extracts was determined by using tannic acid as a reference standard and the calibration curve ($y = 0.0052x + 0.0171$) to calculate the total tannin content. The findings were represented as mg of tannic acid equivalents (TAE) per g dry weight of extract.

2.6.3. Determination of Total Flavonoid

The total flavonoid content of *P. amboinicus* leaf extracts from both in vitro and ex vitro grown plants was determined using a method reported by Ordoñez et al. [59]. It was determined by mixing 2 mL of the extract with the same volume of 2% aluminum chloride solution in a test tube. After waiting 30 min for the reaction to take place at room temperature, the absorbance was taken at a wavelength of 420 nm using a UV-Visible spectrophotometer. Using quercetin as a standard, the total flavonoid content of both the leaf extracts was determined from the calibration curve ($y = 0.0172x - 0.0657$) and represented as mg of quercetin equivalents (QE) per g dry weight (DW) of extract.

2.6.4. Antioxidant and 2,2-Diphenyl-1-picrylhydrazyl (DPPH) Assay

P. amboinicus extracts from both in vitro and ex vitro grown plants were tested for their capacity to scavenge free radicals using the DPPH radical scavenging assay [60]. A total of 0.5 mL of extract at several concentrations (100–500 g/mL) was mixed with 135 µM of DPPH solution in a test tube. After thoroughly mixing, the solution was permitted to rest in the dark for half an hour. Using a spectrophotometer, we measured the absorbance of the samples at 517 nm. By using the following formula: Antioxidant capacity = [(Absorbance of control solution—Absorbance of sample solution)/Absorbance of control solution] × 100, we were able to determine the sample's antioxidant capacity.

2.6.5. Gas Chromatography-Mass Spectrometry (GC-MS)

The phytochemical analysis of the leaf extract of both in vitro and ex vitro grown plants of *P. amboinicus* was performed by using a GC-MS (Turbomass, PerkinElmer Inc., Waltham, MA, USA). Individually, the samples of the extracts were injected into the Elite-5MS column, which was 30 m long, 0.25 m thick, and 0.25 m in diameter. The carrier gas was helium gas, which flowed at a rate of 1 mL/min. Starting at 40 °C, the temperature was held for 2 min before gradually rising to 200 °C (at a rate of 5 °C per minute) and then being held there for 2 min. The heat was turned up to 300 °C at a rate of 5 °C per minute, and it stayed there for the next two minutes. The resulting mass spectra were cross-referenced to those in the Adams and Wiley databases.

2.7. Statistical Analysis

The data on the shoot regeneration frequency and number of shoots from NS explants of *P. amboinicus* were evaluated using one-way analysis of variance (ANOVA) in IBM SPSS software version 26 (SPSS Inc., Chicago, IL, USA), and the graphs were made using the Microsoft Excel for Macintosh (Microsoft Office suite 2019, Microsoft Corporation, Washington, DC, USA). The data were given as means and standard errors (M ± SE) of the sum total of the three repeated experiments, with a total of 10 explants used in each experiment. Tukey's HSD significant difference test in the BM SPSS software was used to determine significant differences between the means using a p-value of 0.05.

3. Results and Discussion

3.1. Effect of TDZ Pre-Treatment and Shoot Regeneration

In compared to purine-based cytokinins, TDZ is the most efficient synthetic cytokinin for eliminating apical dominance in tissue culture, and it also fulfills the cytokinin and auxin needs of regeneration responses in several plants. In the present study, we evaluated the TDZ pulse treatment of *P. amboinicus* NS explants to minimize the deleterious effect of extended TDZ exposure on the growth of in vitro shoots. Pretreatment of explants with TDZ prior to culture initiation can improve the responsiveness of the explants to subsequent regeneration procedures, which may overcome the inhibitory effects of endogenous growth inhibitors, such as phenolic compounds, and can also help to break the dormancy of explants, allowing for more efficient regeneration. In this study, the NS explants pre-soaked with different concentrations (0, 0.5, 5, 25, 50 and 100 μM) of TDZ exhibited initial bud break after one week of transfer on MS basal medium (Figure 1A). Explants pre-treated with varying dosages of TDZ and subsequently transferred to MS media showed statistically significant differences in the percentage of explants that produced shoots (range: 23–97%) as well as the number of shoots (range: 7–27 shoot/explants) that were developed (Table 1).

Figure 1. Shoot induction and plant regeneration from nodal segment explants of *Plectranthus amboinicus* pretreated with thidiazuron for 4 h. (**A**) Shoot induction after 2 weeks of culture on Murashige and Skoog basal medium from nodal segment explants pretreated with 25 μM TDZ. (**B**) Proliferated shoot with roots after 8 weeks of culture on MS medium. (**C**) An acclimated plants regenerated from nodal segment explants.

Table 1. Effect of different concentrations of thidiazuron pretreatment on shoot regeneration from nodal segment explants of *Plectranthus amboinicus* after 8 weeks of transfer on MS basal medium.

Thidiazuron (μM)	% Regeneration	Number of Shoots Per Explant	Shoot Length (cm)
0.0	23.25 ± 1.25 e	1.50 ± 0.57 e	2.87 ± 0.42 cd
0.5	46.30 ± 2.50 d	6.57 ± 0.95 d	3.55 ± 0.33 bc
5.0	68.25 ± 2.36 c	12.25 ± 2.06 c	4.02 ± 0.23 ab
25.0	97.15 ± 2.90 a	27.30 ± 1.91 a	4.93 ± 0.28 a
50.0	87.50 ± 2.80 b	18.25 ± 1.70 b	3.17 ± 0.19 bcd
100.0	65.95 ± 3.55 c	11.00 ± 1.29 c	2.50 ± 0.17 d

Data presented in the table are mean and standard error (M ± SE) of three repeated experiments recorded after 8 weeks of transfer on MS basal medium. Tukey's HSD test with a p-value less than 0.05 indicates that distinct letters in each column indicate significant differences between treatments.

Only 1-2 shoots were produced by the untreated NS explants cultured on MS basal media, whereas the number and growth of the shoots were inhibited when the NS explants were exposed to greater concentrations of TDZ (100 μM). After 8 weeks of growth on MS basal media, the NS explants that were pre-treated with 25 μM TDZ produced the highest number of 27.3 shoots per explant, with the longest average shoot length (4.9 mm) in 97.2% of cultures (Figure 1B). Similarly, TDZ pretreatment improves in vitro shoot induction and plant regeneration in a number of plant species including *Rauvolfia tetraphylla* [41] *Curculigo orchioides* [61], *Begonia semperflorens* and *Begonia* spp. [62], *Nyctanthes arbor-tristis* [63], *Lepidium campestre* [64], *Mentha arvensis* [65], *Bacopa monnieri* [66], *Rhododendron mucronulatum* [67]. It is suggested that TDZ may be needed as a trigger for initiating the proliferation of shoot meristems and further incubation on TDZ free MS medium led the explants to further development [68]. The potential of TDZ to sustain activity even after plant materials have been transferred to a new growth medium may account for the extended activity resulting from a brief TDZ treatment [69].

3.2. Effect of Basal Media on Shoot Regeneration

The successful in vitro growth and differentiation of an explant is a complex process that is influenced by many factors, with the selection of a suitable nutrient-rich basal medium being a critical aspect. The right basal medium can provide the necessary nutrients and growth factors for the explant to thrive and develop into a mature plant. In this study, we attempted to find the most optimal culture medium for the in vitro growth of *P. amboinicus* explants. We compared the effectiveness of five different media, including MS, B5, LS, N and SH (as shown in Figure 2) in inducing in vitro shoot growth and plant regeneration from *P. amboinicus* explants that were pre-treated with 25 μM TDZ for 4 h. The results of this comparison will help to identify the most appropriate culture medium for future in vitro growth experiments with *P. amboinicus* and other medicinal plants. Number of shoots produced per explant varies significantly depending on the type of media used, with 27.3 shoots/explants on MS medium and 12.2 shoots/explants on LS medium, respectively (Figure 2). As a result of the comparison of various culture media, we found that MS medium was the most suitable for shoot regeneration from *P. amboinicus* NS explants that had been pre-treated with 25 μM TDZ. The use of MS medium, along with regular sub-culturing, was crucial in achieving successful shoot regeneration from *P. amboinicus* explants. The MS basal media contained a balanced mix of micro and macro salts, as well as essential vitamins, to provide the necessary nutrients for the continued proliferation of the shoots. This process ensured that the shoots received a consistent supply of nutrients, promoting their growth and development over time.

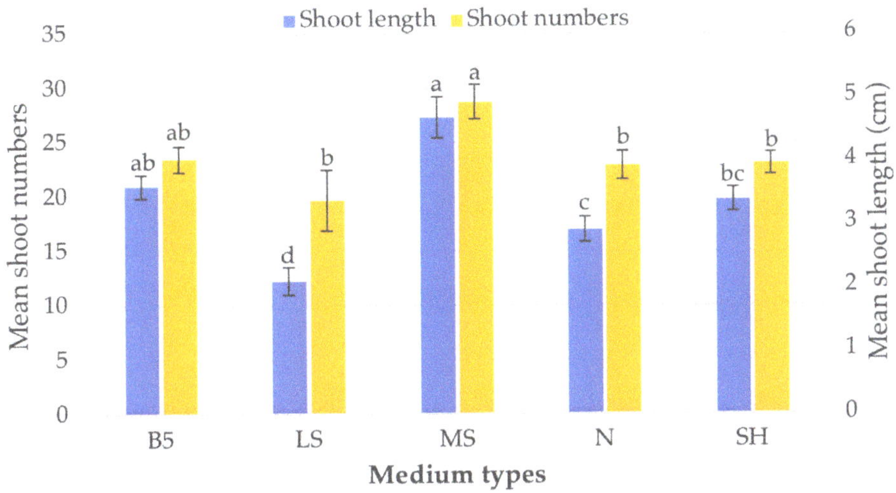

Figure 2. Data presented in the form of bar-graph are mean and standard error (M ± SE) of three repeated experiments recorded after 8 weeks of transfer on different basal medium. Tukey's HSD test with a *p*-value less than 0.05 indicates that bar with distinct letters indicate significant differences between medium.

3.3. Rooting and Acclimatization

In the present study, shoots that were regenerated from NS explants that had been pretreated with TDZ developed roots on their own in the same MS basal media, resulted in a reduction in the extra time and expenses required for the in vitro rooting of a microshoot. It has been estimated that the spontaneous rooting of microshoots may save costs from around 35 to 75 percent of the overall budget required for an in vitro regeneration [70]. The shoots that were not yet have roots were rooted in $\frac{1}{2}$MS medium that contained 0.5 µM IBA. Eight weeks after being transferred to MS basal media, shoots that had been regenerated from NS explants that were pre-treated with optimal TDZ concentration (25 µM) produced an average of 4.3 roots which was 4.7 cm in length. The development of spontaneous roots system together with shoot on the same culture medium may be related to the endogenous amount of auxin as well as some other intrinsic rooting factors [71]. After 4 weeks of acclimatization in a growth chamber, the regenerated *P. amboinicus* micropropagated plants with robust shoot and root systems were successfully transplanted into an ex-vitro environment (Figure 1C), where they sustained a 90% survival rate.

3.4. Flow Cytometry (FCM) and Nuclear DNA Content

FCM is a fast, high-throughput technique for assessing a large number of cells in a short period of time, and it may be used to examine the genetic stability of in vitro regenerated plants by determining the DNA content and chromosome number of individual plant cells. The flow cytometer counts the number of chromosomes and the ploidy level by sorting cells based on the fluorescence intensity of DNA-specific dyes. This can help in figuring out whether there were any alterations to the DNA content throughout the in vitro regeneration process, which might have an effect on the stability and growth of the regenerated plants. In the present study, the acclimated plants that were regenerated from NS explants after being pre-treated with 25 µM TDZ were examined by FCM and compared with donor plants of *P. amboinicus* in order to make sure that the genetic stability of the plants was maintained. The histograms derived by the FCM analysis of both in vitro and ex vitro plants showed no significant variations in mean PI fluorescence peak, indicating that there was no difference nDNA content and ploidy level of the plants (Figure 3). Both the relative 2C nDNA content and the genome size of the *P. amboinicus* plants that were grown in vitro were determined to be 5.63 pg and 275.8 Mbs respectively (Figure 4). There were no discernible differences observed in either the amount of nDNA or the size of the genome among the donor plants and in vitro plants that were regenerated from NS explants pretreated with 25 μM TDZ. Similarly, the stability of nDNA content and genome size were also reported in the in vitro plants of *Viola uliginosa* [72], *Bacopa monnieri* [66] and *Rhododendron mucronulatum* [67].

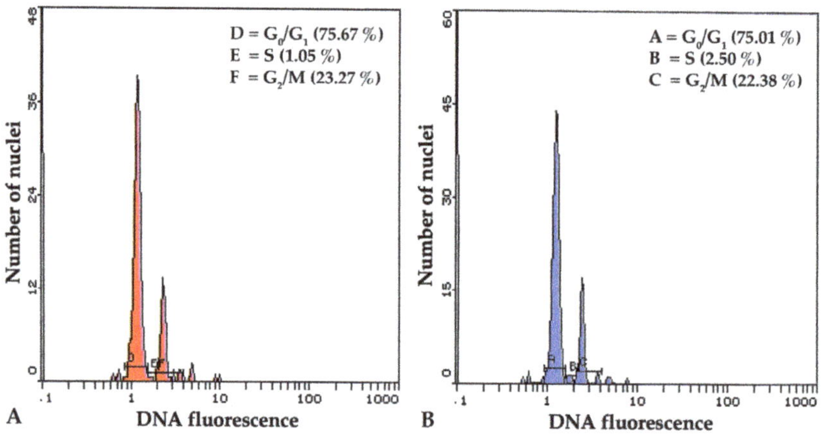

Figure 3. FCM profiles of *Plectranthus amboinicus* showing the relative DNA fluorescence intensity). (**A**). In vitro plants; (**B**). Ex vitro donor plants. Letter A–F represents the phases of cell cycle.

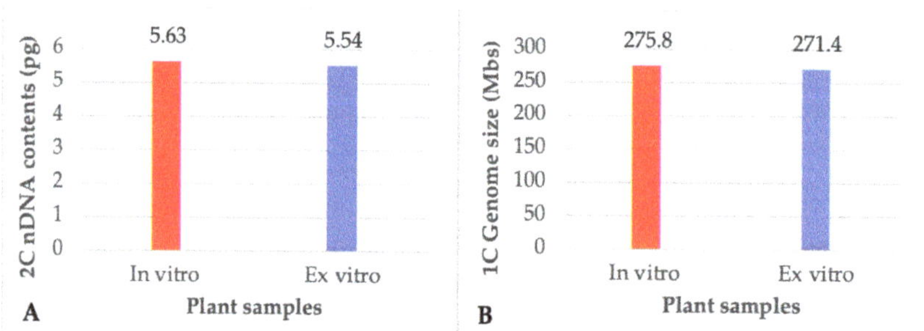

Figure 4. Relative 2C nDNA contents (**A**) and genome size (**B**) of the in vitro regenerated and ex-vitro donor plants of *Plectranthus amboinicus*. pg = picogram; Mbs = Megabase pairs.

3.5. Phytochemical Analysis

Total phenolics, flavonoids, and tannin contents in leaf extracts of in vitro plants regenerated from NS explants pre-treated with 25 μM of TDZ were assessed and compared to those of ex vitro grown donor plants of *P. amboinicus* (Table 2). Phenolic compounds have emerged as potential targets for in vitro culture, which not only creates the optimal circumstances for their production but also encourages increased secondary metabolite accumulation and excretion by biosynthetic cells or tissues [73]. In the present study, higher concentrations of total phenolics and total tannins were detected in plant extracts grown in vitro compared to those of the ex-vitro grown donor plants (Table 2). While no significant difference was observed in the contents of total flavonoids between the donor plants and the in vitro plants that were regenerated from NS explants pre-treated with 25 μM of TDZ for 4 h (Table 2). Total phenolics, tannins, and flavonoid were measured to be 81.2, 55.6, and 42.7 mg GAE, TAE, and QE/g DW respectively, in the leaf extracts from in vitro regenerated *P. amboinicus* plants. It has been suggested that the discrepancy in the concentration of secondary metabolites in plants grown in vitro is regulated by the composition of the media and growth regulators in which the plants are grown [36,74,75].

Table 2. Total phenolic compound, tannin and flavonoid contents of *Plectranthus amboinicus* plants grown in vitro and ex vitro.

Phytoconstituents	In Vitro Plants	Ex Vitro Plants
Total phenolic (mg GAE/g DW)	81.23 ± 0.72 a	56.64 ± 0.31 b
Total Tannin (mg TAE/g DW)	55.68 ± 0.48 a	23.41 ± 52 b
Total Flavonoid (mg QE/g DW)	42.77 ± 0.43 a	40.27 ± 0.39 a

Data presented in the table are mean and standard error (M ± SE) of three repeated experiments recorded after 8 weeks of transfer on MS basal medium. Tukey's HSD test with a *p*-value less than 0.05 indicates that distinct letters in each column indicate significant differences between treatments.

3.6. Antioxidant and DPPH Radical Scavenging Assay

The antioxidant potential of methanol extracts of leaves of the ex vitro donor plants and the plants that were regenerated in vitro from NS explants *P. amboinicus* pretreated with 25 μM of TDZ was determined using DPPH free radicals scavenging assay. It was determined that there was a concentration-dependent antioxidant activity against DPPH, and that the activity was at its highest at 300 μg/mL. The strongest levels of antioxidant activity were found in in vitro regenerated plants of *P. amboinicus* (67.0%), compared to ex vitro donor plants (59.3%) (Figure 5). The results of the study revealed that the antioxidant capability of regenerated plants was regulated directly by the quantities of total phenolics and total tannins. Tannins are secondary antioxidants that have the capacity to bind metal ions like Fe(II), interfere with one of the reaction stages in the Fenton reaction, and slow down the oxidation process [76]. By binding to metal ions and inhibiting the Fenton reaction, tannins in regenerated plants effectively reduce the concentration of ROS and prevent the damage caused by in vitro-induced oxidative stress. In addition to their antioxidant properties, tannins have also been shown to possess anti-inflammatory, anti-tumor, and anti-microbial properties, making them a highly valuable component of plant defense mechanisms. Similarly, high antioxidant activity was reported in in vitro regenerated plants of *Ceropegia santapaui* [77], *Coleonema pulchellum* [75], *Cucumis anguria* [78], *Angelica glauca* [79] and *Ruta chalepensis* [80] all of which are in agreement with our findings.

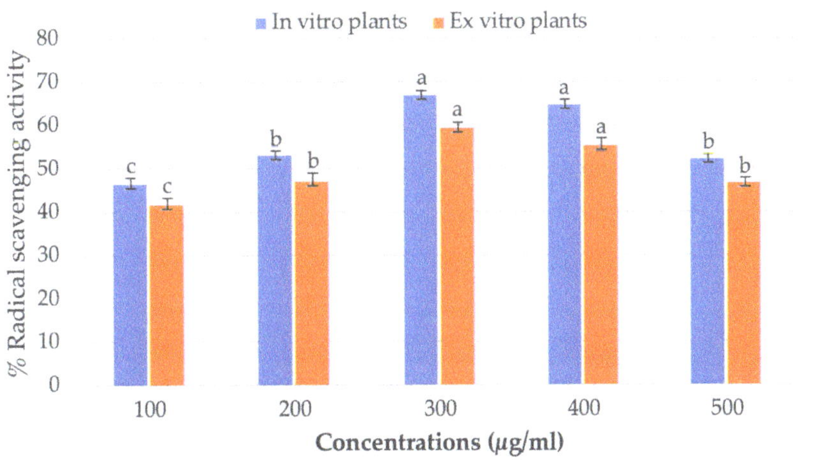

Figure 5. Free radical-scavenging activity of *Plectranthus amboinicus* plants grown in vitro and ex vitro by the DPPH method. Tukey's HSD test with a *p*-value less than 0.05 indicates that bar with distinct letters indicate significant differences between medium.

3.7. Gas Chromatography-Mass Spectrometry (GC-MS) Analysis

GC-MS is a widely used analytical technique for identifying and analyzing the chemical composition of plants, and it may offer useful information about the potential therapeutic effects of various chemicals such as alkaloids, flavonoids, amino acids, organic acids etc. The analysis can also be used to identify any changes in the metabolic profile of the in vitro regenerated plants compared to their wild-type counterparts, providing insight into the genetic and environmental factors that influence plant growth and development. In the present study, the chromatographic examination made it possible to identify 19 and 21, respectively, of the bioactive compounds that were present in ex vitro and in vitro *P. amboinicus* plants (Table 3).

Table 3. Gas chromatography-mass spectrometry (GC-MS) analysis from leaf extracts of *Plectranthus amboinicus* plants grown in vitro and ex vitro.

Name	Retention Time (min)		Area %		Formula	Molecular Weight (g/mol)
	Ex Vitro	In Vitro	Ex Vitro	In Vitro		
Beta-myrcene	5.03	5.05	0.08	0.04	$C_{10}H_{16}$	136.23
1-methyl-2-(1-methylethyl)-benzene	5.38	5.39	1.54	2.38	$C_{10}H_{14}$	134.22
Gamma-terpinene	5.66	5.67	2.65	4.75	$C_{10}H_{16}$	136.23
3-Hydroxy-2-methyl-4H-pyran-4-one	-	6.11	-	0.03	$C_6H_6O_3$	126.11
Cyclononanone	6.24	6.24	0.1	0.13	$C_9H_{16}O$	140.22
Terpinen-4-ol	6.7	6.69	0.11	0.12	$C_{10}H_{18}O$	154.25
5-(Hydroxymethyl)-2-furancarboxaldehyde	-	6.96	-	0.25	$C_6H_6O_3$	126.11
2-Furanmethanol	6.98	-	0.31	-	$C_5H_6O_2$	98.1
2,5-Cyclohexadiene-1,4-dione	7.2	7.18	0.07	0.05	$C_6H_4O_2$	108.09
5-Methyl-2-(1-methylethyl)-phenol	7.5	7.45	91.38	86.69	$C_{10}H_{14}O$	150.22
Alpha-copaene	8.2	8.14	0.12	0.18	$C_{15}H_{24}$	204.36
Trans-caryophyllene	8.54	8.47	1.88	2.2	$C_{15}H_{24}$	204.35
Alpha-humulene	-	8.7	-	0.22	$C_{15}H_{24}$	204.35
Trans ocimene	8.8	-	0.17	-	$C_{10}H_{16}$	136.23
Aromadendrene	-	8.91	-	0.34	$C_{15}H_{24}$	204.35
Delta-cadinene	9.22	9.04	0.09	0.11	$C_{15}H_{24}$	204.35
(-)-Curcuhydroquinone	9.49	-	0.23	-	$C_{15}H_{22}O_2$	234.16
Caryophyllene oxide	-	9.51	-	0.19	$C_{15}H_{24}O$	220.35
(-)-Caryophyllene oxide	9.74	-	0.26	-	$C_{15}H_{24}O$	220.35
Hybridalactone	10.14	-	0.08	-	$C_{21}H_{30}O_3$	330.5
cis-(Z)-alpha-Bisabolene epoxide	10.28	-	0.04	-	$C_{15}H_{24}O$	220.35
trans-Z- alpha-Bisabolene epoxide	10.4	-	0.03	-	$C_{15}H_{24}O$	220.35
BIS[(1S,2S,3S,5R)-(+)-isopinocamphyl] phosphorochloridate	-	11.15	-	0.16	$C_{10}H_{18}O$	154.25
Hexadecanoic acid	-	11.4	-	0.51	$C_{16}H_{32}O_2$	256.42
3,7,11,15-Tetramethyl 2-hexadecen-1-ol	-	11.59	-	0.02	$C_{20}H_{40}O$	296.53
10-Heneicosene	-	12.24	-	0.32	$C_{21}H_{42}$	294.56
9,12,15-Octadecatrienoic acid	-	12.4	-	0.59	$C_{18}H_{30}O_2$	278.43
Octadecanoic acid	-	12.52	-	0.1	$C_{18}H_{36}O_2$	284.48
5-Methyl-2-(1-methylethylidene) cyclohexanone	12.56	-	0.16	-	$C_{10}H_{16}O$	152.23
Tridecanoic Acid	15.12	-	0.1	-	$C_{13}H_{26}O_2$	214.34

The heatmap in Figure 6 depicts a cluster analysis based on the levels of phytochemicals in leaf extracts from in vitro and donor plants, as measured by GC-MS. The study found that both ex vitro and in vitro plants contained eleven chemicals that were chemically identical to one another. In both the ex vitro and in vitro leaf extracts of *P. amboinicus* plants, it was found that 5-methyl-2-(1-methylethyl)-phenol was the main chemical constituent, accounting for 91.38 and 86.69 percent of the total, respectively. Eleven phytocompounds were found in in vitro plants, but they were not present in ex vitro donor plants, some of which have biological activities such as octadecanoic acid (0.1%), caryophyllene oxide (0.19%), Alpha-humulene (0.22%), 5-(Hydroxymethyl)-2-furancarboxaldehyde (0.25%), 10-heneicosene (0.32%), aromadendrene (0.34%), hexadecanoic acid (0.51%), 9,12,15-octadecatrienoic acid (0.59%). The possible reasons for the increased number of phytochemical substances in in vitro propagated plantlets compared to the field grown plants could be influenced by various factors, including the tissue culture conditions, the supply of the nutrient media, the presence of growth regulators, genetic modifications and chemical and abiotic elicitors [80–82].

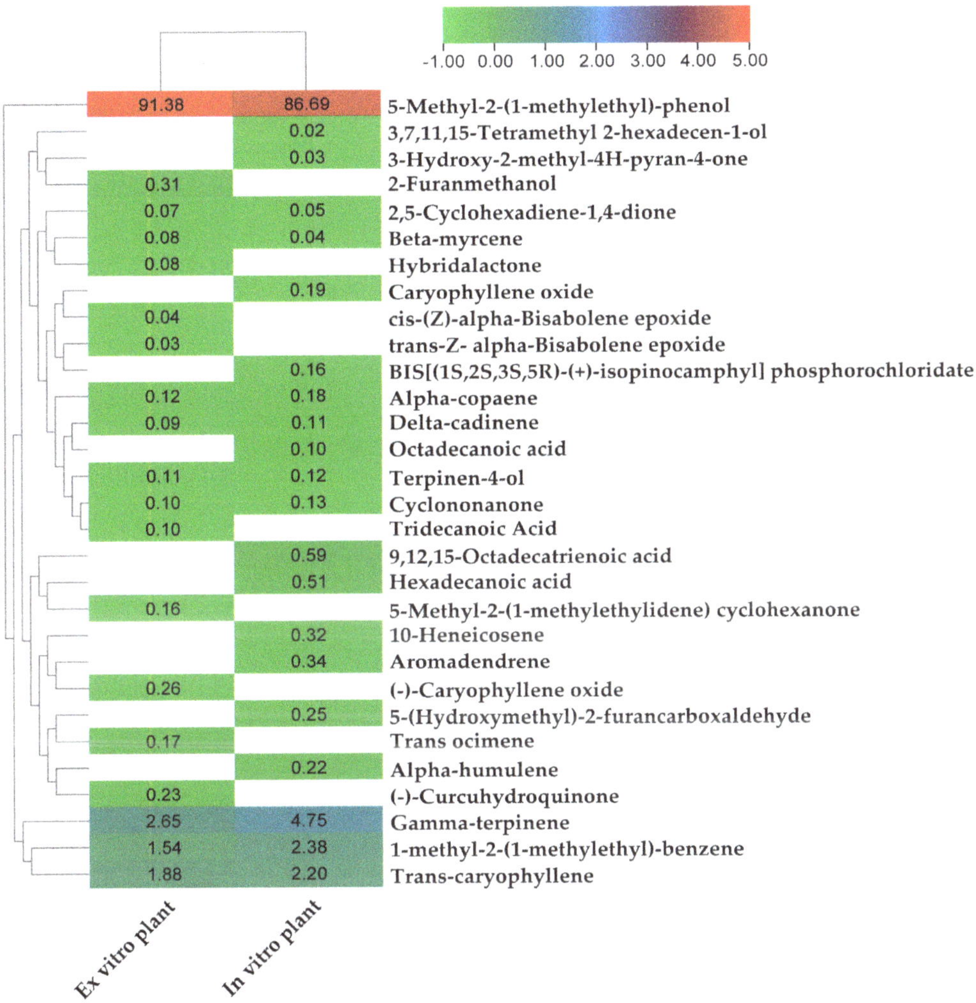

Figure 6. Relative abundance of phytochemicals measured by GC-MS in leaf extracts of *Plectranthus amboinicus* plants grown in vitro and ex vitro.

4. Conclusions

This is the first report describing the TDZ induced plant regeneration, as well as total phenolics, total tannin, total flavonoid contents, DPPH radical-scavenging activity and GC-MS profiles of leaf extracts from in vitro and ex vitro growing plants of *P. amboinicus*. In this study a simple and efficient protocol for plant regeneration was established from NS explants that had been pretreated with 25 μM of TDZ followed by their transfer onto the MS basal media which offer an alternative approach for rapid in vitro propagation and mass multiplication of *P. amboinicus*. The application of TDZ pulse treatment for a short duration of 4h allows the production of healthy shoots with no abnormalities, as well as the development of spontaneous roots on the same medium, saving money and time in comparison to the technique of prolonged exposure to TDZ and other growth regulators. Moreover, the characteristics of the in vitro-regenerated plants were analyzed and compared to the ex-vitro donor plants with regards to the nuclear DNA content, genome size, phenolic content, antioxidant capacity, and GC-MS profiles. The results of the nuclear genome stability and biochemical analyses of the in vitro propagated *P. amboinicus* plants could lead to the development of conservation and sustainable use strategies for this valuable natural resource.

Author Contributions: Conceptualization, M.F.; methodology, M.F. and A.A.Q.; validation, M.F. and A.A.A.; formal analysis, M.F., A.A.Q. and A.A.A.; investigation, M.F. and A.A.A.; resources, M.F. and A.A.A.; data curation, M.F.; writing—original draft preparation, M.F.; writing—review and editing, M.F., A.A.Q. and A.A.A.; project administration, M.F. and A.A.A.; funding acquisition, M.F. and A.A.A. All authors have read and agreed to the published version of the manuscript.

Funding: The work was financially supported by the Researchers Supporting Project Number (RSP-2023R86), King Saud University, Riyadh, Saudi Arabia.

Institutional Review Board Statement: Not applicable.

Informed Consent Statement: Not applicable.

Data Availability Statement: Not applicable.

Conflicts of Interest: The authors declare no conflict of interest.

References

1. Arumugam, G.; Swamy, M.K.; Sinniah, U.R. *Plectranthus amboinicus* (Lour.) Spreng: Botanical, Phytochemical, Pharmacological and Nutritional Significance. *Molecules* **2016**, *21*, 369. [CrossRef]
2. Ashaari, N.S.; Mohamad, N.E.; Afzinizam, A.H.; Ab. Rahim, M.-H.; Lai, K.S.; Ong Abdullah, J. Chemical Composition of Hexane-Extracted *Plectranthus amboinicus* Leaf Essential Oil: Maximizing Contents on Harvested Plant Materials. *Appl. Sci.* **2021**, *11*, 10838. [CrossRef]
3. Bhatt, P.; Joseph, G.S.; Negi, P.S.; Varadaraj, M.C. Chemical Composition and Nutraceutical Potential of Indian Borage (*Plectranthus amboinicus*) Stem Extract. *J. Chem.* **2013**, *2013*, 320329. [CrossRef]
4. Vasconcelos, S.E.C.B.; Melo, H.M.; Cavalcante, T.T.A.; Júnior, F.E.A.C.; de Carvalho, M.G.; Menezes, F.G.R.; de Sousa, O.V.; Costa, R.A. *Plectranthus amboinicus* essential oil and carvacrol bioactive against planktonic and biofilm of oxacillin- and vancomycin-resistant Staphylococcus aureus. *BMC Complement. Altern. Med.* **2017**, *17*, 462. [CrossRef] [PubMed]
5. Sindhu, M.; Poonkothai, M. Phytochemical and antimicrobial analyses of *Plectranthus amboinicus* leaf extracts. *Res. J. Pharm. Technol.* **2021**, *14*, 6379–6384. [CrossRef]
6. Gurgel, A.P.A.D.; da Silva, J.G.; Grangeiro, A.R.S.; Oliveira, D.C.; Lima, C.M.P.; da Silva, A.C.P.; Oliveira, R.A.G.; Souza, I.A. In vivo study of the anti-inflammatory and antitumor activities of leaves from *Plectranthus amboinicus* (Lour.) Spreng (Lamiaceae). *J. Ethnopharmacol.* **2009**, *125*, 361–363. [CrossRef]
7. Akinbo, D.B.; Onyeaghala, A.A.; Emomidue, J.O.; Ogbhemhe, S.O.; Okpoli, H.C. Phytochemical and anti-inflammatory activities of aqueous leaf extract of Indian borage (oregano) on rats induced with inflammation. *Cancer Biomark.* **2018**, *22*, 257–265. [CrossRef] [PubMed]
8. Chang, J.-M.; Cheng, C.-M.; Hung, L.-M.; Chung, Y.-S.; Wu, R.-Y. Potential Use of *Plectranthus amboinicus* in the Treatment of Rheumatoid Arthritis. *Evid.-Based Complement. Altern. Med.* **2010**, *7*, 174726. [CrossRef]
9. Manjamalai, A.; Grace, V.M.B. The Chemotherapeutic Effect of Essential Oil of *Plectranthus amboinicus* (Lour) on Lung Metastasis Developed by B16F-10 Cell Line in C57BL/6 Mice. *Cancer Investig.* **2013**, *31*, 74–82. [CrossRef] [PubMed]
10. Kumaran, A.; Joel Karunakaran, R. Antioxidant and free radical scavenging activity of an aqueous extract of *Coleus aromaticus*. *Food Chem.* **2006**, *97*, 109–114. [CrossRef]
11. Bhatt, P.; Negi, P.S. Antioxidant and antibacterial activities in the leaf extracts of Indian borage (*Plectranthus amboinicus*). *Food Nutr. Sci.* **2012**, *2012*, 146–152.
12. Kozłowska, M.; Ścibisz, I.; Przybył, J.; Ziarno, M.; Żbikowska, A.; Majewska, E. Phenolic Contents and Antioxidant Activity of Extracts of Selected Fresh and Dried Herbal Materials. *Pol. J. Food Nutr. Sci.* **2021**, *71*, 269–278. [CrossRef]
13. Viswanathaswamy, A.; Koti, B.; Gore, A.; Thippeswamy, A.; Kulkarni, R. Antihyperglycemic and antihyperlipidemic activity of *Plectranthus amboinicus* on normal and alloxan-induced diabetic rats. *Indian J. Pharm. Sci.* **2011**, *73*, 139. [PubMed]
14. Patel, R.D.; Mahobia, N.K.; Singh, M.P.; Singh, A.; Sheikh, N.W.; Alam, G.; Singh, S.K. Antioxidant potential of leaves of *Plectranthus amboinicus* (Lour) Spreng. *Pharm. Lett.* **2010**, *2*, 240–245.
15. Varshney, A.; Anis, M. *Trees: Propagation and Conservation: Biotechnological Approaches for Propagation of a Multipurpose Tree, Balanites Aegyptiaca Del*; Springer Science & Business Media: New Delhi, India, 2014.
16. Ahmad, N.; Javed, S.B.; Khan, M.I.; Anis, M. Rapid plant regeneration and analysis of genetic fidelity in micropropagated plants of *Vitex trifolia*: An important medicinal plant. *Acta Physiol. Plant.* **2013**, *35*, 2493–2500. [CrossRef]
17. Ab Rahman, Z.; Noor, E.S.M.; Ali, M.S.M.; Mirad, R.; Othman, A.N. In vitro micropropagation of a valuable medicinal plant, *Plectranthus amboinicus*. *Am. J. Plant Sci.* **2015**, *6*, 1091–1097. [CrossRef]

18. Arumugam, G.; Sinniah, U.R.; Swamy, M.K.; Lynch, P.T. Micropropagation and essential oil characterization of *Plectranthus amboinicus* (Lour.) Sprengel, an aromatic medicinal plant. *In Vitro Cell. Dev. Biol.-Plant* **2020**, *56*, 491–503. [CrossRef]
19. Faisal, M.; Alatar, A.A. Establishment of an Efficient In Vitro Propagation Method for a Sustainable Supply of *Plectranthus amboinicus* (Lour.) and Genetic Homogeneity Using Flow Cytometry and SPAR Markers. *Horticulturae* **2022**, *8*, 693. [CrossRef]
20. Sreedevi, E.; Anuradha, M.; Pullaiah, T. Plant regeneration from leaf-derived callus in *Plectranthus barbatus* Andr. [Syn.: *Coleus forskohlii* (Wild.) Briq.]. *Afr. J. Biotechnol.* **2013**, *12*, 2441–2448.
21. Tsegaw, M.; Feyissa, T. Micropropagation of *Plectranthus edulis* (Vatke) Agnew from meristem cultur. *Afr. J. Biotechnol.* **2014**, *13*, 3682–3688. [CrossRef]
22. Kebede, B.; Abera, B. Micropropagation of *Plectranthus edulis* (Vatke) Agnew from shoot tip and nodal explants. *Afr. J. Agric. Res.* **2015**, *10*, 6–13. [CrossRef]
23. Thaniarasu, R.; Kumar, T.S.; Rao, M. In vitro Propagation of *Plectranthus bourneae* Gamble-An Endemic Red Listed Plant. *Plant Tissue Cult. Biotechnol.* **2015**, *25*, 273–284. [CrossRef]
24. Thaniarasu, R.; Senthil Kumar, T.; Rao, M.V. Mass propagation of *Plectranthus bourneae* Gamble through indirect organogenesis from leaf and internode explants. *Physiol. Mol. Biol. Plants* **2016**, *22*, 143–151. [CrossRef]
25. Fonseka, D.; Wickramaarachchi, W.; Situge, C. Mass production of *Plectranthus zeylanicus*-A valuable medicinal and aromatic plant with a future value. *Int. J. Minor Fruits Med. Aromat. Plants* **2019**, *5*, 15–20.
26. Mahmoud, D.S.; Sayed, L.M.; Diab, M.; Fahmy, E.M. In vitro propagation of *Plectranthus barbatus* andrews as important medicinal plant. *Arab. Univ. J. Agric. Sci.* **2019**, *27*, 511–517. [CrossRef]
27. Kujeke, G.T.; Chitendera, T.C.; Masekesa, R.T.; Mazarura, U.; Ngadze, E.; Rugare, J.T.; Matikiti, A. Micropropagation of Livingstone Potato (*Plectranthus esculentus* N.E.Br). *Adv. Agric.* **2020**, *2020*, 8364153. [CrossRef]
28. Hussain, S.A.; Ahmad, N.; Anis, M.; Hakeem, K.R. Development of an efficient micropropagation system for *Tecoma stans* (L.) Juss. ex Kunth using thidiazuron and effects on phytochemical constitution. *In Vitro Cell. Dev. Biol.-Plant* **2019**, *55*, 442–453. [CrossRef]
29. Ahmad, N.; Faisal, M.; Ahmad, A.; Alatar, A.A.; Qahtan, A.A.; Alok, A. Thidiazuron Induced In Vitro Clonal Propagation of *Lagerstroemia speciosa* (L.) Pers.—An Important Avenue Tree. *Horticulturae* **2022**, *8*, 359. [CrossRef]
30. Gharari, Z.; Bagheri, K.; Sharafi, A.; Danafar, H. Thidiazuron induced efficient in vitro organogenesis and regeneration of *Scutellaria bornmuelleri*: An important medicinal plant. *In Vitro Cell. Dev. Biol.-Plant* **2019**, *55*, 133–138. [CrossRef]
31. Mok, M.; Mok, D.; Armstrong, D.; Shudo, K.; Isogai, Y.; Okamoto, T. Cytokinin activity of N-phenyl-N′-1, 2, 3-thiadiazol-5-ylurea (thidiazuron). *Phytochemistry* **1982**, *21*, 1509–1511. [CrossRef]
32. Murthy, B.N.S.; Murch, S.J.; Saxena, P.K. Thidiazuron: A potent regulator of in vitro plant morphogenesis. *In Vitro Cell. Dev. Biol.-Plant* **1998**, *34*, 267. [CrossRef]
33. Ćosić, T.; Motyka, V.; Raspor, M.; Savić, J.; Cingel, A.; Vinterhalter, B.; Vinterhalter, D.; Trávníčková, A.; Dobrev, P.I.; Bohanec, B.; et al. In vitro shoot organogenesis and comparative analysis of endogenous phytohormones in kohlrabi (*Brassica oleracea* var. *gongylodes*): Effects of genotype, explant type and applied cytokinins. *Plant Cell Tissue Organ Cult.* **2015**, *121*, 741–760. [CrossRef]
34. Guo, B.; Abbasi, B.H.; Zeb, A.; Xu, L.; Wei, Y. Thidiazuron: A multi-dimensional plant growth regulator. *Afr. J. Biotechnol.* **2011**, *10*, 8984–9000. [CrossRef]
35. Khan, I.; Khan, M.A.; Shehzad, M.A.; Ali, A.; Mohammad, S.; Ali, H.; Alyemeni, M.N.; Ahmad, P. Micropropagation and Production of Health Promoting Lignans in *Linum usitatissimum*. *Plants* **2020**, *9*, 728. [PubMed]
36. Baskaran, P.; Ncube, B.; Van Staden, J. In vitro propagation and secondary product production by *Merwilla plumbea* (Lindl.) Speta. *Plant Growth Regul.* **2012**, *67*, 235–245. [CrossRef]
37. Bhattacharyya, P.; Kumaria, S.; Diengdoh, R.; Tandon, P. Genetic stability and phytochemical analysis of the in vitro regenerated plants of *Dendrobium nobile* Lindl., an endangered medicinal orchid. *Meta Gene* **2014**, *2*, 489–504. [CrossRef]
38. Moradi, A.; Zarinkamar, F.; Caretto, S.; Azadi, P. Influence of thidiazuron on callus induction and crocin production in corm and style explants of *Crocus sativus* L. *Acta Physiol. Plant.* **2018**, *40*, 185. [CrossRef]
39. Anjum, S.; Abbasi, B.H. Thidiazuron-enhanced biosynthesis and antimicrobial efficacy of silver nanoparticles via improving phytochemical reducing potential in callus culture of *Linum usitatissimum* L. *Int. J. Nanomed.* **2016**, *11*, 715. [CrossRef]
40. Huetteman, C.A.; Preece, J.E. Thidiazuron: A potent cytokinin for woody plant tissue culture. *Plant Cell Tissue Organ Cult.* **1993**, *33*, 105–119. [CrossRef]
41. Faisal, M.; Ahmad, N.; Anis, M. Shoot multiplication in *Rauvolfia tetraphylla* L. using thidiazuron. *Plant Cell Tissue Organ Cult.* **2005**, *80*, 187–190. [CrossRef]
42. Dewir, Y.H.; Nurmansyah; Naidoo, Y.; Teixeira da Silva, J.A. Thidiazuron-induced abnormalities in plant tissue cultures. *Plant Cell Rep.* **2018**, *37*, 1451–1470. [CrossRef]
43. Akasaka, Y.; Daimon, H.; Mii, M. Improved plant regeneration from cultured leaf segments in peanut (*Arachis hypogaea* L.) by limited exposure to thidiazuron. *Plant Sci.* **2000**, *156*, 169–175. [CrossRef]
44. Ahmad, N.; Srivastava, R.; Anis, M. Improvement in carnation shoot multiplication using thidiazuron in vitro. *Propag. Ornam. Plants* **2006**, *6*, 109–113.
45. Kadota, M.; Niimi, Y. Effects of cytokinin types and their concentrations on shoot proliferation and hyperhydricity in in vitro pear cultivar shoots. *Plant Cell Tissue Organ Cult.* **2003**, *72*, 261–265. [CrossRef]
46. Ivanova, M.; van Staden, J. Effect of ammonium ions and cytokinins on hyperhydricity and multiplication rate of in vitro regenerated shoots of *Aloe polyphylla*. *Plant Cell Tissue Organ Cult.* **2008**, *92*, 227–231. [CrossRef]

47. Magioli, C.; Rocha, A.P.M.; de Oliveira, D.E.; Mansur, E. Efficient shoot organogenesis of eggplant (*Solanum melongena* L.) induced by thidiazuron. *Plant Cell Rep.* **1998**, *17*, 661–663. [CrossRef] [PubMed]
48. Khoddamzadeh, A.A.; Sinniah, U.R.; Kadir, M.A.; Kadzimin, S.; Mahmood, M.; Sreeramanan, S. Detection of somaclonal variation by random amplified polymorphic DNA analysis during micropropagation of *Phalaenopsis bellina* (Rchb. f.) Christenson. *Afr. J. Biotechnol.* **2010**, *9*, 6632–6639.
49. Murashige, T.; Skoog, F. A revised medium for rapid growth and bio assays with tobacco tissue cultures. *Physiol. Plant.* **1962**, *15*, 473–497. [CrossRef]
50. Gamborg, O.L.; Miller, R.A.; Ojima, K. Nutrient requirements of suspension cultures of soybean root cells. *Exp. Cell Res.* **1968**, *50*, 151–158.
51. Linsmaier, E.M.; Skoog, F. Organic growth factor requirements of tobacco tissue cultures. *Physiol. Plant.* **1965**, *18*, 100–127.
52. Nitsch, J.P.; Nitsch, C. Haploid Plants from Pollen Grains. *Science* **1969**, *163*, 85–87. [CrossRef]
53. Schenk, R.U.; Hildebrandt, A. Medium and techniques for induction and growth of monocotyledonous and dicotyledonous plant cell cultures. *Can. J. Bot.* **1972**, *50*, 199–204. [CrossRef]
54. Galbraith, D.W.; Harkins, K.R.; Maddox, J.M.; Ayres, N.M.; Sharma, D.P.; Firoozabady, E. Rapid flow cytometric analysis of the cell cycle in intact plant tissues. *Science* **1983**, *220*, 1049–1051. [CrossRef]
55. Doležel, J.; Bartoš, J. Plant DNA flow cytometry and estimation of nuclear genome size. *Ann. Bot.* **2005**, *95*, 99–110. [CrossRef] [PubMed]
56. Doležel, J.; Bartoš, J.; Voglmayr, H.; Greilhuber, J. Nuclear DNA Content and Genome Size of Trout and Human. *Cytom. Part A* **2003**, *51*, 127–128. [CrossRef] [PubMed]
57. Singleton, V.L.; Orthofer, R.; Lamuela-Raventós, R.M. Analysis of total phenols and other oxidation substrates and antioxidants by means of folin-ciocalteu reagent. In *Methods in Enzymology*; Academic Press: Cambridge, MA, USA, 1999; Volume 299, pp. 152–178.
58. Chandran, K.C.; Indira, G. Quantitative estimation of total phenolic, flavonoids, tannin and chlorophyll content of leaves of *Strobilanthes Kunthiana* (Neelakurinji). *J. Med. Plants Stud.* **2016**, *4*, 282–286.
59. Ordoñez, A.A.L.; Gomez, J.D.; Vattuone, M.A.; lsla, M.I. Antioxidant activities of *Sechium edule* (Jacq.) Swartz extracts. *Food Chem.* **2006**, *97*, 452–458. [CrossRef]
60. Liyana-Pathirana, C.M.; Shahidi, F. Antioxidant activity of commercial soft and hard wheat (*Triticum aestivum* L.) as affected by gastric pH conditions. *J. Agric. Food Chem.* **2005**, *53*, 2433–2440. [PubMed]
61. Thomas, T.D. Pretreatment in thidiazuron improves the in vitro shoot induction from leaves in *Curculigo orchioides* Gaertn., an endangered medicinal plant. *Acta Physiol. Plant.* **2007**, *29*, 455–461. [CrossRef]
62. Kereša, S.; Mihovilović, A.; Barić, M.; Jerčić, I.H. Efficient plant regeneration of *Begonia semperflorens* and *Begonia* spp. from petiole and leaf explants. *J. Food Agric. Environ.* **2011**, *9*, 240–244.
63. Jahan, A.A.; Anis, M.; Aref, I.M. Preconditioning of Axillary Buds in Thidiazuron-Supplemented Liquid Media Improves In Vitro Shoot Multiplication in *Nyctanthes arbor-tristis* L. *Appl. Biochem. Biotechnol.* **2011**, *163*, 851–859. [CrossRef] [PubMed]
64. Ivarson, E.; Ahlman, A.; Li, X.; Zhu, L.-H. Development of an efficient regeneration and transformation method for the new potential oilseed crop *Lepidium campestre*. *BMC Plant Biol.* **2013**, *13*, 115. [CrossRef] [PubMed]
65. Faisal, M.; Alatar, A.A.; Hegazy, A.K.; Alharbi, S.A.; El-Sheikh, M.; Okla, M.K. Thidiazuron induced in vitro multiplication of *Mentha arvensis* and evaluation of genetic stability by flow cytometry and molecular markers. *Ind. Crops Prod.* **2014**, *62*, 100–106. [CrossRef]
66. Faisal, M.; Alatar, A.A.; El-Sheikh, M.A.; Abdel-Salam, E.M.; Qahtan, A.A. Thidiazuron induced in vitro morphogenesis for sustainable supply of genetically true quality plantlets of Brahmi. *Ind. Crops Prod.* **2018**, *118*, 173–179. [CrossRef]
67. Novikova, T.I.; Asbaganov, S.V.; Ambros, E.V.; Zaytseva, Y.G. TDZ-induced axillary shoot proliferation of *Rhododendron mucronulatum* Turcz and assessment of clonal fidelity using DNA-based markers and flow cytometry. *In Vitro Cell. Dev. Biol.-Plant* **2020**, *56*, 307–317. [CrossRef]
68. Khalafalla, M.M.; Hattori, K. A combination of thidiazuron and benzyladenine promotes multiple shoot production from cotyledonary node explants of faba bean (*Vicia faba* L.). *Plant Growth Regul.* **1999**, *27*, 145–148. [CrossRef]
69. Capelle, S.C.; Mok, D.W.; Kirchner, S.C.; Mok, M.C. Effects of thidiazuron on cytokinin autonomy and the metabolism of N 6-(Δ2-isopentenyl)[8-14C] adenosine in callus tissues of *Phaseolus lunatus* L. *Plant Physiol.* **1983**, *73*, 796–802. [CrossRef]
70. Debergh, P.C.; Maene, L.J. A scheme for commercial propagation of ornamental plants by tissue culture. *Sci. Hortic.* **1981**, *14*, 335–345. [CrossRef]
71. Agretious, T.K.; Martin, K.; Hariharan, M. In vitro clonal multiplication of *Alpinia calcarata* Rosc. *Phytomorphology* **1996**, *46*, 133–138.
72. Slazak, B.; Sliwinska, E.; Saługa, M.; Ronikier, M.; Bujak, J.; Słomka, A.; Göransson, U.; Kuta, E. Micropropagation of *Viola uliginosa* (Violaceae) for endangered species conservation and for somaclonal variation-enhanced cyclotide biosynthesis. *Plant Cell Tissue Organ Cult.* **2015**, *120*, 179–190. [CrossRef]
73. Matkowski, A. Plant in vitro culture for the production of antioxidants—A review. *Biotechnol. Adv.* **2008**, *26*, 548–560. [CrossRef]
74. Abbasi, B.H.; Khan, M.A.; Mahmood, T.; Ahmad, M.; Chaudhary, M.F.; Khan, M.A. Shoot regeneration and free-radical scavenging activity in *Silybum marianum* L. *Plant Cell Tissue Organ Cult.* **2010**, *101*, 371–376. [CrossRef]

75. Baskaran, P.; Moyo, M.; Van Staden, J. In vitro plant regeneration, phenolic compound production and pharmacological activities of *Coleonema pulchellum*. *S. Afr. J. Bot.* **2014**, *90*, 74–79. [CrossRef]
76. Karamać, M. Chelation of Cu(II), Zn(II), and Fe(II) by Tannin Constituents of Selected Edible Nuts. *Int. J. Mol. Sci.* **2009**, *10*, 5485–5497. [CrossRef] [PubMed]
77. Chavan, J.J.; Gaikwad, N.B.; Umdale, S.D.; Kshirsagar, P.R.; Bhat, K.V.; Yadav, S.R. Efficiency of direct and indirect shoot organogenesis, molecular profiling, secondary metabolite production and antioxidant activity of micropropagated *Ceropegia santapaui*. *Plant Growth Regul.* **2014**, *72*, 1–15. [CrossRef]
78. Thiruvengadam, M.; Chung, I.-M. Phenolic compound production and biological activities from *in vitro* regenerated plants of gherkin (*Cucumis anguria* L.). *Electron. J. Biotechnol.* **2015**, *18*, 295–301. [CrossRef]
79. Rawat, J.M.; Bhandari, A.; Mishra, S.; Rawat, B.; Dhakad, A.K.; Thakur, A.; Chandra, A. Genetic stability and phytochemical profiling of the in vitro regenerated plants of *Angelica glauca* Edgew.: An endangered medicinal plant of Himalaya. *Plant Cell Tissue Organ Cult.* **2018**, *135*, 111–118. [CrossRef]
80. Qahtan, A.A.; Faisal, M.; Alatar, A.A.; Abdel-Salam, E.M. Callus-Mediated High-Frequency Plant Regeneration, Phytochemical Profiling, Antioxidant Activity and Genetic Stability in *Ruta chalepensis* L. *Plants* **2022**, *11*, 1614. [CrossRef]
81. López-Laredo, A.R.; Ramírez-Flores, F.D.; Sepúlveda-Jiménez, G.; Trejo-Tapia, G. Comparison of metabolite levels in callus of *Tecoma stans* (L.) Juss. ex Kunth. cultured in photoperiod and darkness. *In Vitro Cell. Dev. Biol. Plant* **2009**, *45*, 550–558. [CrossRef]
82. Upadhyay, A.; Shahzad, A.; Ahmad, Z.; Alatar, A.A.; Guerriero, G.; Faisal, M. High frequency direct organogenesis, genetic homogeneity, chemical characterization and leaf ultra-structural study of regenerants in *Diplocyclos palmatus* (L.) C. Jeffrey. *Agronomy* **2021**, *11*, 2164. [CrossRef]

horticulturae

MDPI

Article

Investigation of the Effects of the Explant Type and Different Plant Growth Regulators on Micropropagation of Five Mediterranean *Salvia* spp. Native to Greece

Maria Papafotiou *, Georgia Vlachou and Aikaterini N. Martini

Laboratory of Floriculture and Landscape Architecture, Department of Crop Science, School of Plant Sciences, Agricultural University of Athens, Iera Odos 75, 11855 Athens, Greece
* Correspondence: mpapaf@aua.gr; Tel.: +30-210-5294-555

Abstract: Sages are medicinal and aromatic plants that constitute a large pool from which active compounds of great pharmaceutical potential can be derived, while at the same time, they also have ornamental value. The purpose of this study was to develop the micropropagation protocols of *Salvia fruticosa*, *S. officinalis*, *S. ringens*, *S. tomentosa*, and *S. pomifera* ssp. *pomifera* to facilitate their exploitation in the pharmaceutical and floriculture industries. In vitro cultures of *S. ringens* and *S. pomifera* ssp. *pomifera* was studied for the first time. Shoot tips and single node explants from in vitro seedlings were initially cultured on hormone free (Hf)-MS medium, followed by subcultures on MS medium supplemented with 6-benzyladenine (BA) for all species, as well as with zeatin (ZEA), kinetin (KIN), 6-(γ,γ-dimethylallyamino) purine (2iP), or meta-topolin (mT) for *S. fruticosa* and *S. officinalis*, at concentrations 0.0 to 3.2 mg L^{-1}, in combination with 0.01 mg L^{-1} 1-naphthaleneacetic acid (NAA). *S. officinalis* was the most efficient in shoot multiplication of all the studied species. The highest multiplication indices were found using 0.8 mg L^{-1} BA for *S. fruticosa*, 0.4 mg L^{-1} BA, or mT for *S. officinalis*, and lower than 0.8 mg L^{-1} BA for the other three species. Hyperhydricity was a problem at the multiplication stage, and was most pronounced in single node explants, increasing in proportion to cytokinin concentration. Microshoots rooted at high percentages (75–85%) on half-strength MS medium with 0.0 or 0.5 mg L^{-1} Indole-3-butyric acid (IBA), except for those of *S. ringens*, which rooted best at 1.0–2.0 mg L^{-1} IBA. Ex vitro acclimatization was highly successful (80–95%) on peat–perlite substrate (1:1 *v*/*v*). Thus, the present study resulted in efficient micropropagation protocols for five Mediterranean sage species native to Greece, which will facilitate breeding programs and the promotion of these species in the floriculture and pharmaceutical industries.

Keywords: cytokinin; explant type; hyperhydricity; in vitro; sage; *Salvia fruticosa*; *Salvia officinalis*; *Salvia pomifera* ssp. *pomifera*; *Salvia ringens*; *Salvia tomentosa*

Citation: Papafotiou, M.; Vlachou, G.; Martini, A.N. Investigation of the Effects of the Explant Type and Different Plant Growth Regulators on Micropropagation of Five Mediterranean *Salvia* spp. Native to Greece. *Horticulturae* **2023**, *9*, 96. https://doi.org/10.3390/horticulturae9010096

Academic Editor: Alicja Tymoszuk

Received: 3 December 2022
Revised: 6 January 2023
Accepted: 7 January 2023
Published: 11 January 2023

1. Introduction

The genus *Salvia* (tribe Mentheae) is the largest in the Lamiaceae family, with 986 accepted species names [1] displaying a remarkable range of variation; it is not a monophyletic genus [2]. It has undergone marked species radiations in three regions of the world, Mediterranean and Central Asia, Eastern Asia, and Central and South America [2]. *Salvia* is distinguished from other Mentheae by its staminal structure, having two stamens instead of the four found in most members of the tribe Menthae, and a distinct stamen morphology [3]. The name *Salvia* comes from the Latin salvus (healthy), and a plant called "*Salvia*" by the Romans, likely the type species for the genus *Salvia*, *S. officinalis*, was described for the first time by Pliny the Elder. The common name in English, sage, is attributed to different species of the genus *Salvia*, which are widely used as ornamental or medicinal plants [4].

Sages are medicinal and aromatic plants from which active compounds of great pharmaceutical potential can be derived. The leaves of sages are rich in essential oils and secondary metabolites such as phenolics and terpenoids, and there is widespread knowledge and use of many native *Salvia* spp. in various countries for medicinal and perfumery purposes, as well as culinary plants, supported by a huge number of scientific publications [5–12]. In addition, numerous cultivars, varieties, and hybrids of *Salvia* are widely used worldwide as ornamental plants for their flower interest and fragrant foliage. Further, *Salvia* spp. are valuable along with other aromatic herbs for healing gardens and therapeutic landscapes, which are gaining ground internationally [13].

The climate crisis has reinforced the need for the ecological and sustainable design of green projects. Therefore, *Salvia* spp. native to eastern Mediterranean regions are worth promoting for use in these landscape regeneration contexts, due to their resistance to xerothermic conditions, low maintenance needs, and their contribution to the rescue of insect pollinators [14–18]. In addition, the floriculture industry is constantly looking for new plant species to compete in international markets, and the Mediterranean flora is an inexhaustible reservoir.

In Greece are found 30 taxa (species and subspecies) of the genus *Salvia* [19]. Two of these are widely known commercially as medicinal and ornamental plants, *S. officinalis* and *S. fruticosa* (*S. triloba*), and there are hundreds of research papers on their properties [9,15–17,20–24]. However, a number of the other native sages of Greece have also been studied for their medicinal, cosmetic, and floricultural value [16,17,23,25–33].

The research project SALVIA-BREED-GR (https://www.salvia-breed-gr.com/el/) was carried out for four years (2019–2022) in order to explore the native sages of Greece, with the aim of promoting to the ornamental plant industry new sage products of high ornamental value suitable for xeriscaping. Four species, *S. fruticosa* Mill., *S. officinalis* L., *S. ringens* Sibth., and Sm., *S. tomentosa* Mill.; and one subspecies, *S. pomifera* L. ssp. *pomifera*, members of macchia vegetation, were selected for this project, and among other parameters, their seed and clonal propagation were studied.

S. fruticosa (Greek sage Figure 1a), is a strongly aromatic perennial evergreen shrub (120 cm tall), bearing lilac or pink flowers (Figure 1b) in early spring. It is endemic to the Mediterranean coastal zone, with a wider distribution from Sicily to Israel. In Greece, it is found in Central Greece, the Peloponnese, and the Aegean islands [34]. It is widely used for the preparation of an herbal tea (faskomilo).

Figure 1. Native to Greece *Salvia* spp.; *S. fruticosa* plant (**a**) and inflorescence (**b**), *S. officinalis* plant (**c**) and inflorescence (**d**), *S. ringens* plant (**e**) and inflorescence (**f**), *S. tomentosa* plant (**g**) and inflorescence (**h**), *S. pomifera* ssp. *pomifera* plant (**i**) and inflorescence (**j**).

S. officinalis (Dalmatian sage, common sage, sage, (Figure 1c) is a strongly aromatic perennial evergreen shrub (60 cm tall), bearing violet-blue flowers (Figure 1d) in May–July. It is naturally widespread on the Apennines and the eastern Adriatic coast, but naturalized in many places throughout the world [35]. In Greece, it is found in the north and the east, and in the Ionian islands. Since antiquity, it has been used as a medicinal plant, while nowadays is cultivated in many varieties as a medicinal and ornamental plant, being one of the most important *Salvia* species worldwide.

S. ringens (Figure 1e) is a perennial evergreen plant with a slightly woody base and low vegetation (30 cm). Its leaves have a light aroma. It bears a few violet-blue flowers on branching long (60 cm) stocks (Figure 1f) during late spring through summer. It is found in the southern and eastern parts of the Balkan Peninsula. In Greece, it spreads north, and in the highlands of Macedonia, Epirus, and Central Greece [35].

S. tomentosa (Figure 1g) is a strongly aromatic perennial evergreen plant (80 cm tall), bearing violet flowers with reddish-brown calyces (Figure 1h) in late spring or early summer. Its geographical distribution extends from South-Eastern Europe to Transcaucasia [36]. In Greece, it is found in the central and north-eastern areas, and in Eastern Aegean Islands [19].

S. pomifera ssp. *pomifera* (Figure 1i) is an endemic shrub of Greece found in Crete and the Peloponnese [34]. It is a strongly aromatic perennial evergreen plant (80 cm tall), bearing pink or violet flowers (Figure 1j) in spring to early summer on slightly curving flower stocks.

All the above sages have been found to be suitable for use as ornamental plants under limited water supply [14–17], which makes them valuable plants for xeriscaping and landscaping in arid and semi-arid regions. Research on their propagation using stem cuttings has shown that this is a suitable method for their commercial production [37], although its effectiveness may depend on the season or climatic conditions. In contrast, the seed propagation of these sage species was found to be unstable in terms of germination rates, and in most species, especially in *S. ringens* and *S. tomentosa*, the germination rate was extremely low [33,38,39], verifying that seed germination is a global problem of the *Salvia* spp. [40,41]. Concerning the micropropagation of these Mediterranean *Salvia* spp., there are many works published on the in vitro culture of *S. officinalis* [42–51], two on *S. fruticosa* [42,52], and one on *S. tomentosa* [53] while there were no reports found on the micropropagation of *S. ringens* and *S. pomifera* ssp. *pomifera*.

The commercial production of many ornamental, medicinal, and aromatic plants is based on the use of tissue culture and micropropagation at some stage of their development; micropropagation technique is important for developing, selecting, multiplying and conserving the critical genotypes of these plants [54–56]. In medicinal plants, tissue culture is used to produce active compounds for pharmaceutical and herbal industries. In addition, in vitro culture techniques are applied for the conservation of the genetic material of many threatened medicinal plants [55].

Shoot proliferation in most micropropagation protocols proposed for *S. officinalis* was low, as in most cases, up to three shoots per explant were produced using shoot tip or nodal explants on MS medium [57] supplemented with rather low concentrations (0.5 to 1.5 mg L^{-1}) of 6-benzyladenine (BA), alone or in combination with a low concentration (0.1 mg L^{-1}) of 1-naphthaleneacetic acid (NAA) [33,45,46,50,58]. Even lower BA concentrations (0.2–0.5 mg L^{-1}) were found to be appropriate for the shoot proliferation of *S. fruticosa* [52], inducing lower proliferation indices compared to *S. officinalis*. In *S. officinalis*, a liquid culture in MS with BA and a low concentration of auxin resulted in similarly low proliferation, as in solid medium [48], while the proliferation was doubled when triacontanol at 20 mg L^{-I} was added into the liquid medium [48]. Similarly increased proliferation was achieved when adventitious shoots were induced from nodal or leaf explants excised from in vitro seedlings on MS medium with 1.5 mg L^{-1} thidiazuron (TDZ), combined with 0.1 mg L^{-1} indole-3-acetic acid (IAA) [51].

The nature and origin of explants has been shown to significantly influence the in vitro response [59–62], as the ability for totipotency differs in plant cells. In the micropropagation of *S. officinalis*, both shoot tip explants [46,48,49] and nodal explants [43,47,50,58] have been used, and more often, shoot tips showed higher response rates [49], as has also been shown in other medicinal Lamiaceae herbs [63,64]. In *S. fruticosa* [52] and other Mediterranean sages [65,66], as well as in other medicinal Lamiaceae herbs, nodal explants have been shown to produce more shoots than the shoot tip explants [67–71]. Further, the juvenile or adult origins of the explants may affect their response to micropropagation differently. Explants originating from seedlings often present high shoot proliferation, as has been shown for some native Mediterranean species [72,73]. The use of seedlings as mother plants is recommended when native species are reintroduced to the landscape, because a higher genetic diversity is promoted. In addition, seedling mother material contributes to the selection of specific genotypes that present special characteristics as high medicinal value.

The in vitro rooting of medicinal Lamiaceae herbs, including *Salvia* spp., presented no difficulties [38,50,52,53,65–69,71,72,74–76], and in a number of these species, auxin supplementation was not necessary for rooting [38,45,65,72,75]. In contrast, *S. fruticosa* microshoots rooted only in the presence of auxin [52]. Half-strength MS medium supplemented with IBA at 0.5–1.5 mg L^{-1} and occasionally up to 4.0 mg L^{-1} has been the most appropriate protocol, with a few exceptions in which full-strength MS medium was used [52,66].

The aim of this work was to develop protocols for the micropropagation of *S. fruticosa*, *S. officinalis*, *S. ringens*, *S. tomentosa*, and *S. pomifera* ssp. *Pomifera*, in order to facilitate their introduction into the pharmaceutical and floriculture industries, by providing high-value plant material for cultivation without seasonal constraints and the selection and conservation of critical genotypes. The micropropagation of *S. ringens* and *S. pomifera* ssp. *pomifera* were studied for the first time to our knowledge. The in vitro micropropagation studies of *S. fruticosa* and *S. officinalis* aimed to further improve the protocols due to the commercial importance of these species. Thus, five types of cytokinins at various concentrations were tested as for their efficacy in shoot proliferation from shoot tip and singe node explants of seedling origin. In vitro rooting and ex vitro acclimatization were also studied, and the responses of the five species to all stages of micropropagation were compared and discussed.

2. Materials and Methods

2.1. In Vitro Culture Establishment Stage

The establishment of in vitro cultures was performed on hormone free (Hf)-MS medium [57] with 30 g L^{-1} sucrose, under the culture conditions described below. Shoot tip and single node explants of the 1st, 2nd, or 3rd visible node below the shoot tip, about 0.6 cm long, excised from 3-month-old in vitro grown seedlings of *S. fruticosa*, *S. officinalis*, *S. ringens*, *S. tomentosa*, and *S. pomifera* ssp. *pomifera* were used as explants. The seedlings were germinated in vitro, following the method described by Vlachou et al. [77], and grown on Hf-MS medium with 30 g L^{-1} sucrose.

2.2. Shoot Multiplication Stage

Aiming to increase the plant material to be used in experiments on shoot proliferation, up to six subcultures (the number depending on *Salvia* spp.) on MS medium supplemented with 0.4 mg L^{-1} 6-benzyladenine (BA) and 0.01 mg L^{-1} 1-naphthaleneacetic acid (NAA) were performed for each *Salvia* species.

The media used in the shoot multiplication stage were based on preliminary work, and on previous reports on Mediterranean sages and other Lamiaceae [33,38,45,50,58].

2.2.1. Effect of Explant Type and BA Concentration on the Proliferation of the Five *Salvia* spp.

The aim of this experiment was to investigate the effect of BA concentration in relation to explant type (shoot tip and single node) and *Salvia* species, on shoot proliferation. Shoot tip and single node explants of the five *Salvia* spp. were cultured, either on Hf-MS medium or on MS medium supplemented with BA at four concentrations, i.e., 0.4, 0.8, 1.6, and 3.2 mg L^{-1}, in combination with 0.01 mg L^{-1} NAA.

The explants were 0.6 cm long and they were excised from microshoots grown on MS medium supplemented with 0.4 mg L^{-1} BA and 0.01 mg L^{-1} NAA.

2.2.2. Effect of Explant Type, and Cytokinin Type and Concentration on the Proliferation of *S. fruticosa* and *S. officinalis*

Shoot tip and single node explants, 0.6 cm long, were excised from the microshoots of in vitro cultures (on MS medium with 0.4 mg L^{-1} BA and 0.01 mg L^{-1} NAA) of *S. fruticosa* and *S. officinalis*, and cultured on MS medium with 30 g L^{-1} sucrose. The medium was either Hf (control) or supplemented with 0.01 mg L^{-1} 1-naphthaleneacetic acid (NAA) in combination with a cytokinin of five different types, i.e., BA, zeatin (ZEA), kinetin (KIN), 6-(γ,γ-dimethylallylamino) purine (2iP), or meta-topolin (mT), at four concentrations, i.e., 0.4, 0.8, 1.6, and 3.2 mg·L^{-1}.

2.2.3. Explant Number and Data Collection at the Shoot Multiplication Stage

The number of explants used in each experiment at the multiplication stage is presented at the base of each relevant data table.

Data were collected after 30 d of culture. The percentage of explants that responded to form shoots, the percentage of those that formed normal shoots or normal, together with hyperhydrated shoots and the percentage of explants that formed only hyperhydrated shoots (Figure 2), were recorded. The number of normal and hyperhydrated shoots produced, shoot length, and the number of nodes per shoot of the normal shoots were also recorded. The "multiplication index" of each culture was calculated by multiplying the percentage of explants that produced normal shoots (i.e., explants with all shoots hyperhydrated were not included) by the mean number of normal shoots per responding explant, and by the mean node number per normal shoot.

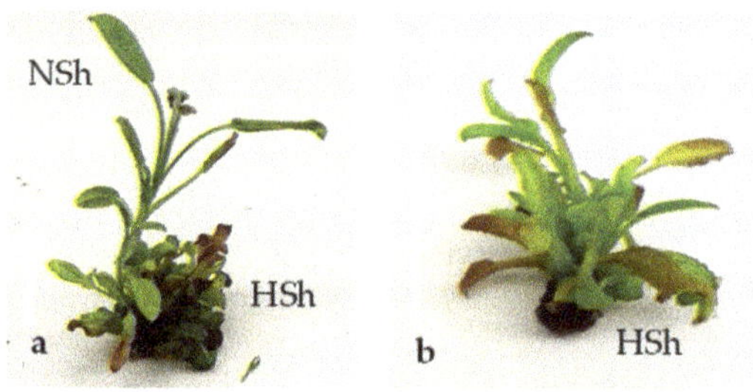

Figure 2. Responses of single node explants of *S. officinalis* during the multiplication stage to produce normal and hyperhydrated shoots, (**a**) or hyperhydrated shoots only (**b**). NSh = normal shoot, HSh = hyperhydrated shoot. Size bars= 1.0 cm.

2.3. *In Vitro Rooting*

For the experiment on in vitro root induction and development, microshoots produced on MS medium supplemented with 0.4 or 0.8 mg L^{-1} BA and 0.01 mg L^{-1} NAA, approximately 2.0 cm long, were used. Microshoots were cultured on half-strength MS medium with 20 g L^{-1} sucrose supplemented with various concentrations of IBA, i.e., 0.0, 0.5, 1.0, 2.0, or 4.0 mg L^{-1}.

Three replicates of 10 microshoots were used for each treatment. Data were collected after 30 d of culture. The percentage of explants that formed roots, and the number and length of roots were recorded.

2.4. *Ex Vitro Acclimatization*

For ex vitro acclimatization, the rooted microshoots of all rooting media after being rinsed thoroughly with running tap water to remove growth medium were transferred to

500 mL containers (eight plantlets per container), on peat (high-moor with adjusted pH of up to 5.5–6.5, Klasmann-Delimann Gmbh, Geeste, Germany) and perlite (particles diameter 1–5 mm, Perloflor, ISOCON S.A., Athens, Greece) substrate 1:1 (*v*/*v*), were transferred ex vitro into trays (eight plantlets per 500 mL volume tray) with a mixture of peat: perlite (1:1, *v*/*v*). The trays were covered with plastic wrap (SANITAS; Sarantis S.A., Greece) and placed in a growth chamber (20 °C and 16 h cool white fluorescent light, 37.5 mmol·m^{-2}·s^{-1}) for 1 week. Then, the plastic wrap was removed and transferred to a heated glasshouse (37°58′58.0″ N, 23°42′19.2″ E) for an additional 4 weeks.

Six to 10 replicates of eight rooted microshoots were used, and their survival was estimated at 30 d after transfer to the greenhouse.

2.5. In Vitro Culture Conditions

All media were solidified with 8 g L^{-1} agar, and their pH was adjusted to 5.7 before agar addition and autoclaving (121 °C for 20 min). Initial cultures from in vitro seedlings, subcultures, and rooting experiments took place in 145 mL glass vessels with 25 mL medium (four explants or microshoots per vessel), covered with a magenta plastic cap. The cultures were maintained at 25 ± 2 °C with a 16 h photoperiod at 37.5 µmol m^{-2} s^{-1} fluorescent light, provided by cool-white fluorescent lamps.

2.6. Statistical Analysis

A completely randomized design was used. The significance of the results was tested through either one-, two-, or three-way analysis of variance (ANOVA), and the means of the treatments were compared via Student's *t* test at $p \leq 0.05$ (JMP 13.0 software, SAS Institute Inc., Cary, NC, USA).

3. Results

3.1. Establishment Stage

For the establishment of in vitro cultures, seedlings that germinated and grew in vitro were used as mother plants. The germination rates of *S. ringens*, *S. pomifera* ssp. *pomifera*, and *S. tomentosa* were very low, and especially in *S. tomentosa*, it was lower than 10% [77]. Thus, for these sage species, the mother material was limited. At the culture establishment stage, which took place on Hf-MS medium, shoot tip explants of all five *Salvia* spp. responded at higher percentages to produce normal shoots compared to single node explants (26–93% vs. 10–75%, depending on the species), as the latter showed higher hyperhydricity, which in the case of *S. pomifera* ssp. *pomifera* and *S. tomentosa* reached 40–50%. Apart from hyperhydricity problems, *S. tomentosa* also showed much a lower explant response to producing shoots, compared to all other species (Table 1). In all of the sage species, single node explants, which bear two buds due to the phyllotaxis of *Salvia* spp., produced more shoots compared to shoot tip explants, but shoots on single node explants were shorter and with fewer nodes (Table 1, Figure 3) and thus, the shoot tip explants presented higher multiplication indices compared to single node explants (Table 1), with the exception of *S. officinalis*.

Table 1. Effect of *Salvia* species and explant type on axillary shoot production of explants excised from in vitro seedlings of Greek sage species, at the establishment stage of in vitro cultures on a hormone-free MS medium.

Salvia sp.	Explant Type	Shoot-Producing Explants ½ (%)	Mean Number of NSh ₸/HSh ₸₸	Mean Shoot Length (cm)	Mean NSh Node Number	Multi-Plication Index ₸
S. fruticosa	Shoot tip	67.0 c ^{z}/17 de	1.2 bc/0.2 cd	1.6 cd	3.2 bc	2.6 c
(*n* = 18)	Single node	56.0 d/22 cd	1.9 a/0.7 b	0.8 d	1.1 d	1.2 de
S. officinalis	Shoot tip	93.3 a/7 f	1.1 c/0.1 d	3.6 a	4.7 a	4.8 b
(*n* = 30)	Single node	75.0 b/23 c	2.1 a/0.8 ab	3.0 b	3.7 b	5.8 a
S. pomifera ssp.	Shoot tip	92.0 a/8 f	1.0 c/0.1 d	1.6 d	2.1 cd	1.9 cd
pomifera (*n* = 12)	Single node	46.0 e/50 a	1.8 a/1.0 a	1.0 d	1.6 d	1.3 cde
S. ringens	Shoot tip	87.0 a/0 g	1.0 c/0.0 d	3.0 ab	5.2 a	4.5 b
(*n* = 40)	Single node	27.1 f/18 cd	1.4 b/0.4 c	2.4 bc	4.5 a	1.6 cde

Table 1. *Cont.*

Salvia sp.	Explant Type	Shoot-Producing Explants [1]/[2] (%)	Mean Number of NSh Ŧ/HSh ŦŦ	Mean Shoot Length (cm)	Mean NSh Node Number	Multi-Plication Index Ŧ
S. tomentosa (n = 7)	Shoot tip	26.0 f/10 ef	1.0 c/0.2 cd	1.9 bcd	2.4 cd	0.6 de
	Single node	10.0 g/40 b	2.0 a/0.9 ab	1.4 cd	2.0 cd	0.4 e
$F_{Salvia\ sp.}$		-/-	-/-	***	***	-
$F_{explant\ type}$		-/-	-/-	**	***	-
$F_{Salvia\ sp. \times expl.\ type}$		***/***	**/*	NS	NS	***
$F_{one\text{-}way\ ANOVA}$		***/***	***/***	***	***	***

[z] Mean separation in columns using Student's t, $p \leq 0.05$; means followed by the same letter are not significantly different at $p \leq 0.05$. NS: nonsignificant or *, **, *** significant at $p \leq 0.05$ or $p \leq 0.01$ or $p \leq 0.001$, respectively, n = 7–58. Multiplication index = shoot-producing explants [1] (%) × mean shoot number Ŧ × mean node number Ŧ. [1] The explants produced normal and hyperhydrated shoots. [2] The explants produced hyperhydrated shoots only. Ŧ NSh = normal shoot. ŦŦ HSh = hyperhydrated shoot.

Figure 3. Typical responses of shoot tip and single node explants excised from in vitro seedlings of *Salvia* spp., during the in vitro establishment stage on hormone-free MS medium. Size bars = 1.0 cm.

3.2. Shoot Multiplication on BA Media

3.2.1. *S. fruticosa*

In *S. fruticosa*, both type of explants, shoot tip, and single node, responded at high percentages (70–100%) to produce shoots in all media tested, with single node explants showing a decrease in response with increasing BA concentration, which was significantly lower compared to the shoot tip explants in all BA media (Figure 4a). Some of the explants produced only hyperhydrated shoots, while others produced both normal and hyperhydrated shoots. An increase in BA concentration resulted in an increase in hyperhydrated explants that was more pronounced in single node ones, which at the two highest BA concentrations, only 20–30% produced normal shoots and 40–50% produced only hyperhydrated shoots (Figure 4b). In contrast, at the highest BA concentration, shoot tip explants produced normal shoots in 68%, and only hyperhydrated shoots in 32% (Figure 4b).

The number of shoots per responding explant was higher in single node explants when normal and hyperhydrated shoots were counted together (Figures 5a and 6a), but as single node explants formed more hyperhydrated shoots compared to the tip explants, the production of normal shoots was not statistically significantly different between the two types of explants (Figure 5b). At 0.8 mg L^{-1} BA; however, single node explants produced significantly more normal shoots compared to the tip ones (Figure 5b).

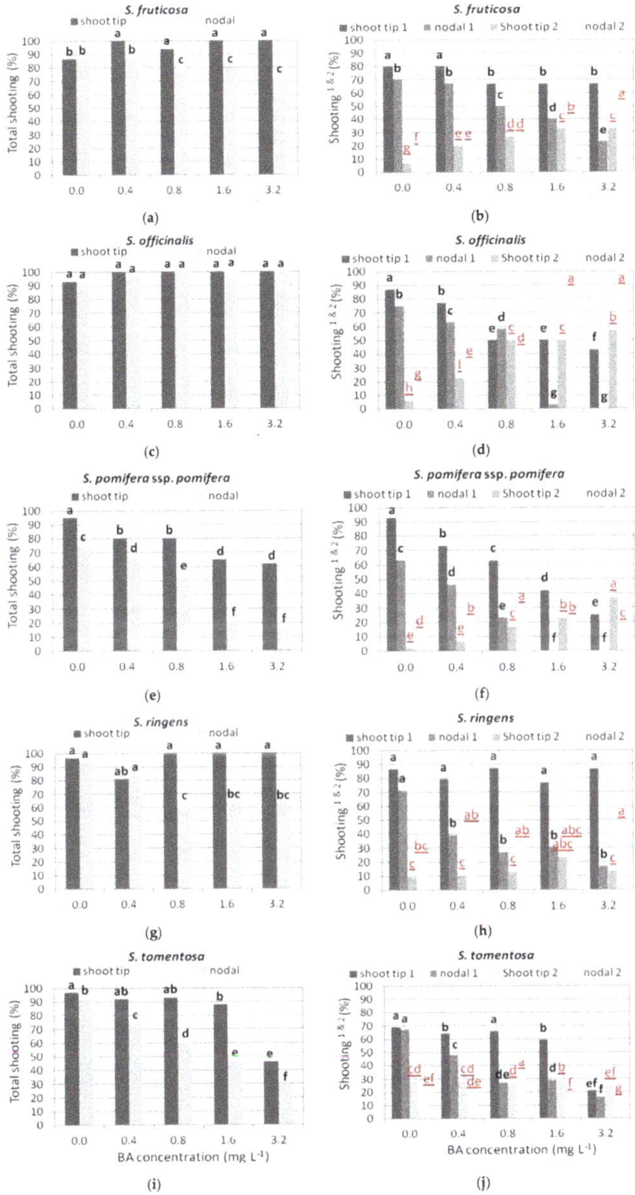

Figure 4. Effect of 6-benzyladenine (BA) concentration on explant response percentage for shoot production when shoot tip or single node (nodal) explants of each *Salvia* species shown were cultured on MS medium, either hormone free or supplemented with 0.01 mg L^{-1} 1-naphthaleneacetic acid (NAA) and BA concentration marked. Total shooting (%): percentage of explants that responded for shoot production. Explants formed both normal and hyperhydrated shoots. Explants formed only hyperhydrated shoots. Response of explants (black bold letters) was statistically analyzed separately from the respective explants (red underlined letters). Mean separation using Student's t, $p \leq 0.05$; means followed by the same letter are not significantly different at $p \leq 0.05$, $n = 30$.

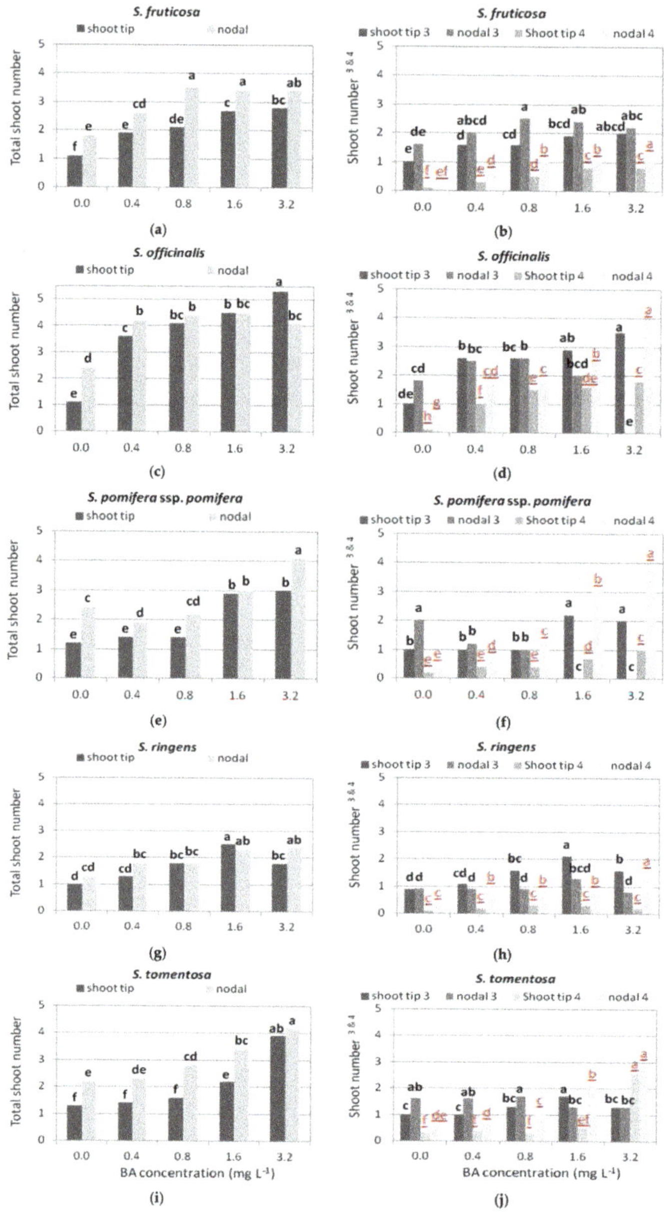

Figure 5. Effect of BA concentration on normal, hyperhydrated, and total number of shoots produced per explant of *Salvia* spp. shown when shoot tip or single node (nodal) explants were cultured on MS medium, either hormone free or supplemented with 0.01 mg L^{-1} NAA and BA concentration marked. Response for normal and hyperhydrated shoot production was statistically analyzed separately (black bold letters and red underlined letters, respectively). Mean separation via Student's t, $p \leq 0.05$; means followed by the same letter are not significantly different at $p \leq 0.05$, $n = 30$. Normal shoots per explant. Hyperhydrated shoots per explant.

Figure 6. Characteristic response of shoot tip and single node explants of *Salvia fruticosa* (**a**), *S. officinalis* (**b**), *S. pomifera* spp. *pomifera* (**c**), *S. ringens* (**d**) and *S. tomentosa* (**e**) during the multiplication stage, when cultured on MS medium either hormone free (Hf) or supplemented with 0.01 mg L^{-1} NAA and BA concentration (mg L^{-1}) marked. Size bar s = 1.0 cm.

Shoot length and number of nodes per shoot were the highest in Hf medium, while all media with BA reduced to a similar extent the shoot length and number of nodes in the generated shoots, to half of that of Hf medium (Figure 7a,b).

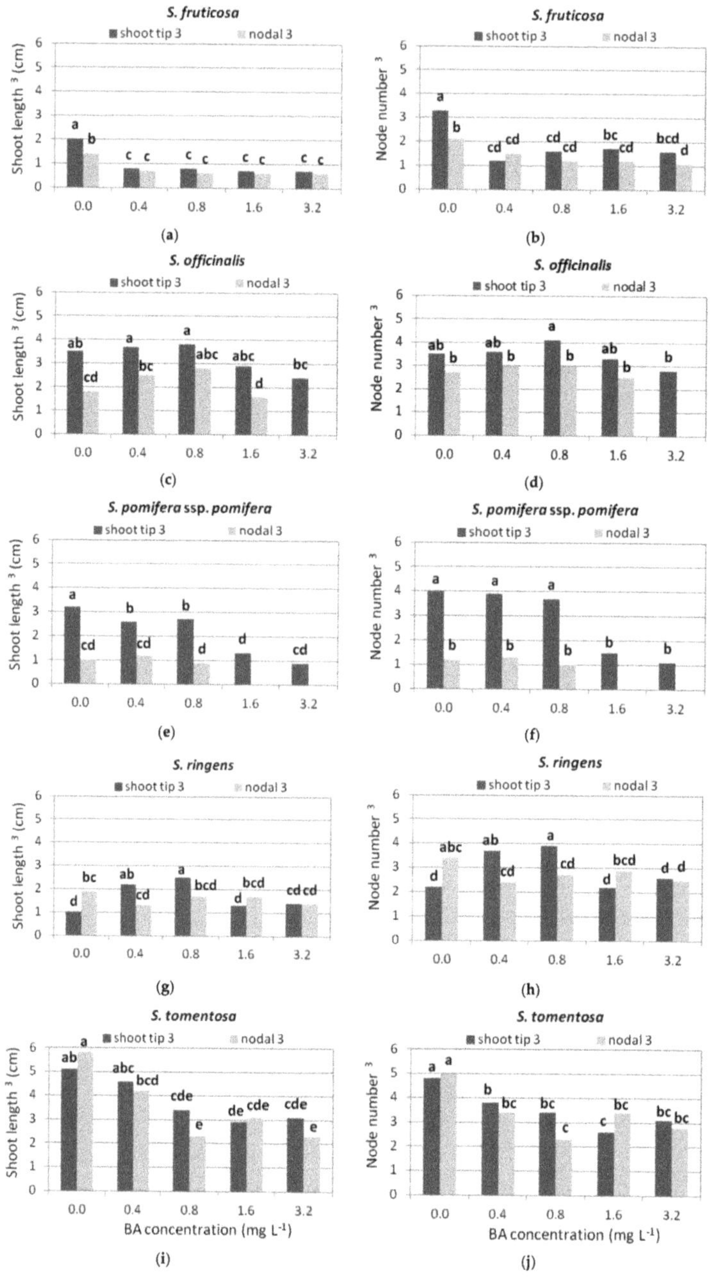

Figure 7. Effect of BA concentration on shoot length (cm) and node number per shoot, produced using shoot tip or single node (nodal) explants of each *Salvia* species shown cultured on MS medium, either hormone free or supplemented with 0.01 mg L^{-1} NAA and BA concentration marked. Only normal shoots [3] were recorded. Mean separation using Student's t, $p \leq 0.05$; means followed by the same letter are not significantly different at $p \leq 0.05$, $n = 30$. [3] Normal shoots per explant.

Both types of explants gave a similar multiplication index in Hf and low BA media, while at 1.6 and 3.2 mg L^{-1} BA, the multiplication index was significantly higher when shoot tip explants were used (Figure 8a).

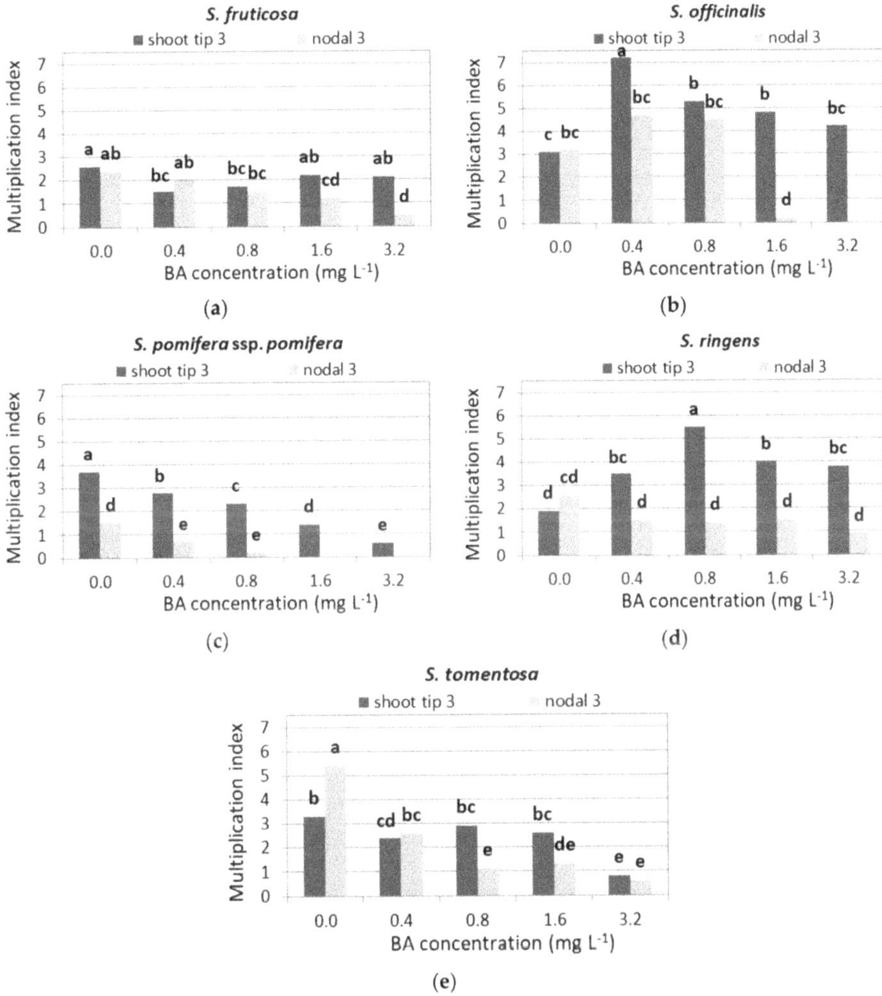

Figure 8. Effect of BA concentration on the shoot multiplication index of *Salvia fruticosa* (**a**), *S. officinalis* (**b**), *S. pomifera* spp. *pomifera* (**c**), *S. ringens* (**d**) and *S. tomentosa* (**e**), when shoot tip or single node (nodal) explants were cultured on MS medium, either hormone free or supplemented with 0.01 mg L^{-1} NAA and BA concentration marked. Mean separation using Student's t, $p \leq 0.05$; means followed by the same letter are not significantly different at $p \leq 0.05$, $n = 30$. Multiplication index = shoot-producing explants [1] (%) × mean shoot number [3] × mean node number [1]. Explants that formed both normal and hyperhydrated shoots. [3] Normal shoots per explant.

3.2.2. *S. officinalis*

In *S. officinalis*, almost all explants of both types produced shoots (Figure 4c), but some of the explants produced only hyperhydrated shoots (Figure 4d). The explants showed the highest percentage of normal shoot production on Hf medium, which gradually decreased with the gradual increase in BA concentration, while the production of hyperhydrated explants gradually increased (Figure 4d). Shoot tip explants showed higher response percentages for normal shoot production compared to single node explants, while the latter showed higher percentages of hyperhydration, except in the medium with 0.8 mg L^{-1} BA (Figure 4d). At the highest BA concentrations, hyperhydration was 50–57 and 90% for shoot tip and single node explants, respectively (Figure 4d).

Increasing the BA concentration in the medium resulted in an increase in shoot production per explant (Figures 5c and 6b), and at the same time, an increase in the hyperhydrated shoots (Figure 5d). So, at the two highest BA concentrations, the approximately five shoots formed per explant 2–4 were hyperhydrated, and in single node explants at 3.2 mg L^{-1} BA, all shoots were hyperhydrated (Figure 5d).

There was an indication that shoot length and node number per shoot were higher in shoot tip explants compared to single node explants, and this decreased at the highest BA concentration (Figure 7c,d).

The highest multiplication index was shown with 0.4 mg L^{-1} BA using shoot tip explants (Figure 8b).

3.2.3. *S. pomifera* ssp. *pomifera*

In *S. pomifera* ssp. *pomifera*, shoot tip explants showed the highest response to produce shoots (94%) on Hf medium (Figure 4e), and only 2% produced only hyperhydrated shoots (Figure 4f), while single node explants responded in a lower percentage (74%), and 12% of the responded explants produced only hyperhydrated shoots (Figure 4e,f). The gradual increase in BA concentration induced a proportional decrease in explant response, and an increase in hyperhydration (Figure 4e,f). The negative effect of BA on shoot production was stronger in single node explants, which at the two highest BA concentrations, responded at very low percentages (20%) and produced only hyperhydrated shoots (Figure 4e,f).

Single node explants produced higher shoot numbers at almost all BA concentrations compared to shoot tip explants (Figure 5e), but most of them were hyperhydrated, especially at the two highest BA concentrations (Figure 5f). Single node explants on Hf medium and shoot tip explants on the two highest BA concentrations produced the highest number of normal shoots (two shoots per explant, Figure 5f), while single node explants at the two highest BA concentrations produced double the number (3–4) shoots per explant, but they were all hyperhydrated (Figures 5f and 6c).

Shoot tip explants gave longer shoots with a higher number of nodes compared to single node explants (Figure 7e,f). Shoot length gradually decreased with increasing BA concentration (Figure 7e). The number of nodes per shoot was similar in the HF medium and in the media with the two lowest BA concentrations (Figure 7f). Both shoot length and number of nodes were significantly reduced in the media with the two highest BA concentrations (Figure 7e,f).

Shoot tip explants showed much higher multiplication indices compared to single node explants. The multiplication index was highest in the HF medium and gradually decreased with increasing BA concentration (Figure 8c).

3.2.4. *S. ringens*

In *S. ringens*, the shoot tip explants had a consistent response for shoot production in all media tested compared to single node explants, whose response decreased at BA concentrations of 0.8 mg L^{-1} and higher (Figure 4g). In addition, shoot tip explants produced normal shoots at much higher percentages (78–88%), compared to single node explants (Figure 4h). Further, hyperhydration was more pronounced in single node explants, especially at the highest BA concentration (Figure 4h).

Increasing BA up to 1.6 mg L^{-1} resulted in a slight increase in both normal and hyperhydrated shoots (Figure 5g,h). Most normal shoots were produced at 1.6 mg L^{-1} BA (Figure 5h). In the two lowest BA concentrations, shoot tip explants gave slightly longer shoots with more nodes than single node explants, and the highest elongation and number of nodes (Figures 6d and 7g,h).

Shoot tip explants resulted in higher multiplication indices than single node explants, except in the HF medium, which reached its highest value in the 0.8 mg L^{-1} BA medium (Figure 8d).

3.2.5. *S. tomentosa*

In *S. tomentosa*, shoot tip explants responded at higher percentages to produce shoots, both normal and hyperhydrated, compared to single node explants, and Hf medium induced the highest response, similar to 0.4 and 0.8 mg L^{-1} BA media, while the response percentage was almost halved in the medium with 3.2 mg L^{-1} BA (Figure 4i,j). The percentage of hyperhydrated explants was not affected by the medium plant growth regulators in both explant types (Figure 4j).

Single node explants produced a greater number of shoots in all media compared to shoot tip explants, and both explant types produced the highest number of shoots at the highest BA concentration (Figure 5i). Single node explants also produced more normal shoots compared to shoot tip explants in HF medium and in the two lowest BA concentrations, while in the two highest BA concentrations, most of the shoots produced were hyperhydrated; the number of hyperhydrated shoots in both explant types gradually increased with increasing BA concentration (Figure 5j).

The shoot length was greater in HF medium and at the lowest BA concentration (Figures 6e and 7i), and the node number was greater in HF medium and similar in all BA media (Figure 7j).

The multiplication index was highest when single node explants were cultured on HF medium, halved when 0.4 mg L^{-1} BA was added to the medium, and lowest at the highest BA concentration in both types of explant (Figure 8e).

3.3. *Effects of Cytokinin Type and Concentration on the Multiplication of S. fruticosa and S. officinalis*

Aiming to increase shoot proliferation in the commercial species *S. fruticosa* and *S. officinalis*, the effect of BA on the response of shoot tip and single node explants for shoot production was compared with that of four other cytokinins, i.e., ZEA, Kin, 2iP, and mT. Three-way ANOVA showed in most cases a significant interaction of the experimental factors (cytokinin type, cytokinin concentration, and explant type). In both species, factor means indicated that a gradual increase in cytokinin concentration induced a gradual decrease in the percentage of explants responding to produce normal shoots, and an increase in hyperhydration expressed as the percentage of explants producing only hyperhydrated shoots or the number of hyperhydrated shoots produced per explant (Tables 2 and 3). Furthermore, a gradual increase in cytokinin concentration, although causing a gradual increase in the number of normal and hyperhydrated shoots, simultaneously reduced the length and number of nodes per shoot, leading to a decrease in the multiplication index (Tables 2 and 3). Concerning the cytokinin type, in both sage species, KIN induced the highest percentage of response to produce normal shoots, and the lowest percentage of explants that produced only hyperhydrated shoots (Tables 2 and 3). In *S. fruticosa*, BA, ZEA, and mT produced the highest number of normal and hyperhydrated shoots per explant, and KIN and mT, the longest shoots; while in *S. officinalis*, BA and mT produced the highest numbers of normal and hyperhydrated shoots per explant, and ZEA produced the longest ones. mT in *S. fruticosa* and BA in *S. officinalis* achieved the highest multiplication index (Tables 2 and 3). In both sage species, shoot tip explants responded at a higher percentage than single node explants to produce normal shoots, while single node explants showed higher hyperhydration levels. Further, shoot tip explants produced slightly fewer shoots, both normal and hyperhydrated (statistically not significant), which were longer and had more nodes compared to single node explants and resulted in a slightly higher multiplication index than single node explants (Tables 2 and 3).

Table 2. The effect of the experimental factors, i.e., cytokinin type, BA, zeatin (ZEA), kinetin (KIN), 6-(γ,γ-dimethylallyamino) purine (2iP), and meta-topolin (mT), cytokinin concentration (0.4, 0.8, 1.6, and 3.2 mg L^{-1}) and explant type (shoot tip and single node) on shoot multiplication of *S. fruticosa*. All media except BA also contained 0.01 mg L^{-1} NAA.

Three-Way ANOVA	Shoot-Producing Explants [1]	Shoot-Producing Explants [2]	Mean Number of NSh [Ŧ]	Mean NSh Length [Ŧ]	Mean NSh Node Number [Ŧ]	Mean Number of HSh [ŦŦ]	Multi-Plication Index [Ŧ]
0.4	74.5	17.6	1.5	1.4 a [z]	2.1 a	0.4	2.1
0.8	66.6	23.6	1.7	1.3 ab	2.0 a	0.5	2.2
1.6	58.1	31.3	1.6	1.2 b	1.8 b	0.8	1.6
3.2	44.5	41.2	1.8	1.1 b	1.7 b	1.1	1.3
BA	57.6	31.3	2.0	0.7 c	1.4 d	0.8	1.6
ZEA	51.2	32.8	1.7	1.1 b	1.9 bc	0.9	1.5
KIN	70.9	21.0	1.4	1.7 a	2.1 ab	0.5	2.0
2iP	62.8	22.4	1.3	1.1 b	1.8 c	0.4	1.4
mT	62.1	34.6	1.8	1.6 a	2.3 a	1.0	2.6
Shoot tip	70.7	25.1	1.5	1.4 a	2.1 a	0.5	2.1
Single node	51.2	31.8	1.8	1.1 b	1.7 b	0.9	1.5
			Significance [§]				
$F_{Cytokinin\ type}$	-	-	-	***	***	-	-
$F_{Cytok\cdot\ concentration}$	-	-	-	**	***	-	-
$F_{Explant\ type}$	-	-	-	***	***	-	-
$F_{Cytok\cdot \times Concentration}$	-	-	NS	NS	NS	-	***
$F_{Cytok\cdot \times Explant\ type}$	-	-	**	NS	NS	-	*
$F_{Cytok\cdot \times Expl\cdot\ type}$	-	-	*	NS	NS	-	NS
$F_{Cytok\cdot \times Conc\cdot \times Expl\cdot\ type}$	***	***	NS	NS	NS	***	NS

[z] Mean comparison in columns within each main factor with Student's *t* test at $p \leq 0.05$; means followed by the same letter are not significantly different at $p \leq 0.05$. [§] NS or * or ** or ***, non-significant at $p \leq 0.05$ or significant at $p \leq 0.05$ or $p \leq 0.01$ or $p \leq 0.001$, respectively. [1] The explants produced normal and hyperhydrated shoots. [2] The explants produced hyperhydrated shoots only. [Ŧ] NSh = normal shoot. [ŦŦ] HSh = hyperhydrated shoot. Multiplication index = shoot-producing explants [1] (%) × mean shoot number [Ŧ] × mean node number [Ŧ].

Table 3. The effect of the experimental factors, i.e., cytokinin type (BA, ZEA, KIN, 2iP, and mT), cytokinin concentration (0.4, 0.8, 1.6, and 3.2 mg L^{-1}) and explant type (shoot tip and single node) on shoot multiplication of *S. officinalis*. All media except BA also contained 0.01 mg L^{-1} NAA.

Three-Way ANOVA	Shoot-Producing Explants [1]	Shoot-Producing Explants [2]	Mean Number of NSh [Ŧ]	Mean NSh Length [Ŧ]	Mean NSh Node Number [Ŧ]	Mean Number of HSh [ŦŦ]	Multi-Plication Index [Ŧ]
0.4	72.6	25.8	1.9	2.8 a [z]	2.9 a	0.6	3.8
0.8	58.2	41.0	2.0	2.3 b	3.1 a	0.8	3.1
1.6	43.2	55.9	2.0	2.1 b	2.8 a	1.1	2.1
3.2	33.4	65.6	1.8	1.6 c	2.1 b	1.9	1.8
BA	43.0	56.5	2.4	2.5 b	2.8 bc	2.0	3.7
ZEA	50.0	50.0	1.7	2.8 a	3.2 a	0.9	2.4
KIN	70.1	27.1	1.5	2.0 c	2.3 d	0.6	2.3
2iP	65.0	33.5	1.5	1.9 c	2.9 b	0.6	2.6
mT	30.5	68.3	2.5	1.8 c	2.4 cd	1.4	2.4
Shoot tip	60.0	42.1	1.9	2.6 a	3.2 a	0.8	3.0
Single node	46.8	52.1	1.9	1.8 b	2.3 b	1.4	2.4
			Significance [§]				
$F_{Cytokinin\ type}$	-	-	-	***	***	-	-
$F_{Cytok\cdot\ concentration}$	-	-	-	**	***	-	-
$F_{Explant\ type}$	-	-	-	***	***	-	-
$F_{Cytok\cdot \times Concentration}$	-	-	NS	NS	NS	-	***
$F_{Cytok\cdot \times Explant\ type}$	-	-	**	NS	NS	-	*
$F_{Cytok\cdot \times Expl\cdot\ type}$	-	-	*	NS	NS	-	NS
$F_{Cytok\cdot \times Conc\cdot \times Expl\cdot\ type}$	***	***	NS	NS	NS	***	NS

[z] Mean comparison in columns within each main factor with Student's *t* test at $p \leq 0.05$; means followed by the same letter are not significantly different at $p \leq 0.05$. [§] NS or * or ** or ***, non-significant at $p \leq 0.05$ or significant at $p \leq 0.05$ or $p \leq 0.01$ or $p \leq 0.001$, respectively. [1] The explants produced normal and hyperhydrated shoots. [2] The explants produced hyperhydrated shoots only. [Ŧ] NSh = normal shoot. [ŦŦ] HSh = hyperhydrated shoot. Multiplication index = shoot-producing explants [1] (%) × mean shoot number [Ŧ] × mean node number [Ŧ].

Due to interactions between experimental factors, the results were further analyzed using one-way ANOVA, and the treatment means were compared.

In *S. fruticosa*, all cytokinins, with the exception of mT, affected the percentage of explant response to produce normal and hyperhydrated shoots in a similar manner; i.e., at the lowest concentration (0.8 mg L^{-1}) they induced high response percentages for normal shoot production, such as Hf medium, while as the concentration increased, there was a proportional decrease in the response percentage for the production of normal shoots and an increase in explant hyperhydration (Table 4). mT at the lowest concentration caused a lower response percentage compared to other cytokinins and the control, which increased when 0.8 mg L^{-1} was used. As in previous experiments with BA, shoot tip explants responded at higher percentages compared to single node explants, while single node explants showed higher hyperhydration. An increase in cytokinin concentration caused a gradual increase in shoot number in both normal and hyperhydrated shoots, and a decrease in shoot length. Although ZEA at the highest concentration resulted in the highest number of shoots per explant, half of them were hyperhydrated, so that normal shoot production on ZEA media was similar to that of BA, while the other cytokinins produced slightly lower shoot numbers (Table 4 and Figure 9). Multiplication indices were higher in Hf medium, as well as those with low cytokinin concentrations, and reached the highest value in the medium, with 0.8 mg L^{-1} mT (Table 4).

Figure 9. Characteristic shoot multiplication of shoot tip and single node explants of *S. fruticosa* cultured on MS medium supplemented with 0.01 NAA (mg.L^{-1}), and cytokinin type and concentration (mg.L^{-1}) marked. Size bars= 1.0 cm.

Table 4. Effect of cytokinin type and concentration on shoot multiplication of *S. fruticosa* shoot tip or single node explants, excised from culture established from in vitro grown seedlings, in the presence of 0.01 mg L^{-1} NAA.

PGRs Concentration (mg L^{-1})	Shoot-Producing Explants $^{1/2}$ (%)	Mean Number of NSh ^{T}/HSh ^{TT}	Mean NSh Length ^{T} (cm)	Mean Node Number ^{T}	Multi-Plication Index
Shoot tip explant					
0.0 (Hf ^{TTT})	84 b ^{z}/8 n	1.0 j/0.1 n	2.2 a	3.3 a	2.8 cd
0.4 BA	80 bc/20 kl	1.6 def/0.3 lmn	0.8 fgh	1.2 ij	1.5 ijk
0.8 BA	67 fgh/27 jk	1.6 def/0.5 kl	0.8 fgh	1.6 fgh	1.7 hi
1.6 BA	67 fgh/33 hij	1.9 cde/0.8 hij	0.7 gh	1.7 efg	2.2 efg
3.2 BA	67 fgh/33 hij	2.0 bcd/0.8 hij	0.7 gh	1.6 fgh	2.1 fgh
0.4 ZEA	83 b/13 mn	1.1 ij/0.2 mn	1.6 bcd	2.5 bcd	2.3 efg
0.8 ZEA	67 fgh/21 kl	1.4 fgh/0.3 lmn	1.3 def	2.5 bcd	2.3 efg
1.6 ZEA	59 ij/29 ij	1.3 ghi/0.6 j	1.1 ef	2.3 cde	1.8 ghi
3.2 ZEA	21 op/58 b	2.2 abc/2.3 b	0.7 gh	1.2 ij	0.6 klm
0.4 KIN	92 a/8 n	1.0 j/0.2 mn	2.1 a	2.5 bcd	2.3 efg
0.8 KIN	83 b/17 lm	1.2 hij/0.3 lmn	2.0 ab	2.5 bcd	2.5 def
1.6 KIN	71 def/21 kl	1.3 ghi/0.5 kl	1.9 abc	2.0 def	1.8 ghi
3.2 KIN	58 ij/30 ij	1.5 efg/0.8 hij	1.7 bcd	2.0 def	1.7 hi
0.4 2iP	83 b/17 lm	1.0 j/0.2 mn	1.5 bcd	2.2 cde	1.8 ghi
0.8 2iP	75 cd/21 kl	1.1 ij/0.2 mn	1.2 def	1.9 efg	1.6 ij
1.6 2iP	75 cd/21 kl	1.2 hij/0.3 lmn	1.1 ef	1.8 efg	1.6 ij
3.2 2iP	71 def/29 ij	1.2 hij/0.3 lmn	0.8 fgh	2.1 def	1.8 ghi
0.4 mT	73 de/21 kl	1.7 def/0.6 j	2.0 ab	2.9 ab	3.6 b
0.8 mT	77 c/23 jkl	2.2 abc/0.6 j	1.8 bc	2.7 bc	4.6 a
1.6 mT	65 gh/35 hij	1.8 cde/0.9 gh	1.6 bcd	2.4 bcd	2.8 cd
3.2 mT	62 hi/38 fgh	2.1 abc/1.0 fgh	1.4 cde	1.7 efg	2.2 efg
Single node explant					
0.0 (Hf ^{TTT})	71 def/17 lm	1.7 def/0.3 lmn	1.4 cde	2.3 cde	2.8 cd
0.4 BA	67 fgh/20 kl	2.0 bcd/0.6 j	0.7 gh	1.5 ghi	2.0 fgh
0.8 BA	50 kl/27 jk	2.5 a/1.0 fgh	0.6 h	1.2 ij	1.5 ijk
1.6 BA	40 mn/40 ef	2.4 ab/1.0 fgh	0.6 h	1.2 ij	1.2 jk
3.2 BA	23 op/50 c	2.2 abc/1.2 ef	0.6 h	1.1 j	0.5 lm
0.4 ZEA	50 kl/29 ij	2.0 bcd/0.7 ij	1.2 def	1.7 efg	1.7 hi
0.8 ZEA	46 lm/33 hij	1.9 cde/1.0 fgh	1.1 ef	1.7 efg	1.5 ijk
1.6 ZEA	46 lm/37 ghi	2.0 bcd/1.0 fgh	1.0 efg	1.7 efg	1.6 ij
3.2 ZEA	13 p/62 a	2.0 bcd/2.4 a	0.8 fgh	1.3 hij	0.3 m
0.4 KIN	83 b/13 mn	1.4 fgh/0.3 lmn	1.6 bcd	2.0 def	2.3 efg
0.8 KIN	71 def/21 kl	1.6 def/0.4 kl	1.5 bcd	2.0 def	2.3 efg
1.6 KIN	63 ghi/25 jkl	1.6 def/0.7 ij	1.5 bcd	2.1 def	2.1 fgh
3.2 KIN	46 lm/33 hij	1.8 cde/1.0 fgh	1.4 cde	1.9 efg	1.6 ij
0.4 2iP	78 c/8 n	1.6 def/0.2 mn	1.2 def	1.4 ghi	1.7 hi
0.8 2iP	63 ghi/17 lm	1.6 def/0.2 mn	1.0 ef	1.7 efg	1.7 hi
1.6 2iP	33 n/33 hij	1.5 efg/0.8 hij	0.9 efg	1.4 ghi	0.7 klm
3.2 2iP	25 o/33 hij	1.5 efg/0.7 ij	0.9 efg	1.8 efg	0.7 klm
0.4 mT	56 jk/27 jk	1.7 def/0.6 j	1.9 abc	2.7 bcd	2.6 def
0.8 mT	67 fgh/29 ij	2.2 abc/0.9 gh	1.4 cde	2.4 bcd	3.5 b
1.6 mT	58 ij/42 de	1.7 def/1.4 d	1.2 def	1.6 fgh	1.6 ij
3.2 mT	38 mn/62 a	1.5 ghi/1.6 cd	1.0 ef	2.0 def	1.1 jk
$F_{\text{one-way ANOVA}}$	***/***	***/***	***	***	***

^{z} Mean separation in columns using Student's t, $p \leq 0.05$; means followed by the same letter are not significantly different at $p \leq 0.05$; *** significant at $p \leq 0.001$, $n = 30$. Multiplication index = shoot-producing explants [1] (%) × mean shoot number ^{T} × mean node number ^{T}. [1] The explants produced normal and hyperhydrated shoots. [2] The explants produced hyperhydrated shoots only. ^{T} NSh = normal shoot. ^{TT} HSh = hyperhydrated shoot. ^{TTT} Hf = hormone free.

In *S. officinalis*, at the two highest concentrations of cytokinins, especially in BA, the single node explants had a lower response percentage to produce normal shoots than the shoot tip explants, as they showed very high hyperhydricity percentages reaching 100%. Increasing the concentrations of all cytokinins gradually decreased the percentage of explants that gave normal shoots, and increased the percentage of hyperhydrated explants (Table 5). At the same time, there was a small increase in the number of normal and hyperhydrated shoots, and a generally small decrease in their length (Table 5 and

Figure 10). *S. officinalis* showed higher proliferation indices than *S. fruticosa*, which were highest in 0.4 mg L^{-1} BA in both shoot tip and single node explants (Table 5).

Table 5. Effect of cytokinin concentration and type on shoot multiplication of shoot tip or single node explants of *S. officinalis* excised from culture established from in vitro grown seedlings, in the presence of 0.01 mg L^{-1} NAA.

PGRs Concentration (mg L^{-1})	Shoot-Producing Explants [1/2] (%)	Mean Number of NSh ᵀ/HSh ᵀᵀ	Mean NSh Length ᵀ (cm)	Mean Node Number ᵀ	Multi-Plication Index
Shoot tip explant					
0.0 (Hf ᵀᵀᵀ)	93 a [z]/7 p	1.0 l/0.1 j	3.4 c	3.7 bc	3.4 g
0.4 BA	77 cd/23 op	2.6 de/1.0 ef	3.7 bc	3.6 bc	7.2 a
0.8 BA	50 jk/50 hi	2.6 de/1.5 de	3.8 bc	4.1 ab	5.3 c
1.6 BA	50 jk/50 hi	2.9 bc/1.6 de	2.9 cd	3.3 cd	4.8 cd
3.2 BA	43 jkl/57 gh	3.5 a/1.8 d	2.4 fg	2.8 ef	4.2 ef
0.4 ZEA	79 bcd/21 op	1.1 kl/0.2 ij	5.6 a	4.3 a	3.7 fg
0.8 ZEA	58 hi/42 jk	1.0 l/0.4 hi	2.6 de	4.2 a	2.4 jk
1.6 ZEA	55 ij/45 ij	1.2 jk/0.7 gh	2.9 cd	3.6 bc	2.4 jk
3.2 ZEA	29 mn/71 d	2.3 fgh/1.3 e	2.0 ijk	2.5 fgh	1.7 mn
0.4 KIN	75 de/17 p	1.0 l/0.2 ij	2.5 ef	2.6 fg	2.0 lm
0.8 KIN	75 de/25 n	1.1 kl/0.5 hi	2.6 de	2.8 ef	2.3 jk
1.6 KIN	71 ef/29 mn	1.1 kl/0.7 gh	2.5 ef	2.5 fgh	2.0 lm
3.2 KIN	67 fg/33 lm	1.8 hi/0.8 fg	2.5 ef	2.8 ef	3.4 g
0.4 2iP	83 b/17 p	1.0 l/0.2 ij	1.8 jk	2.5 fgh	2.1 kl
0.8 2iP	71 ef/25 n	1.0 l/0.3 ij	1.9 jk	2.9 ef	2.1 kl
1.6 2iP	62 gh/38 klm	1.0 l/0.4 hi	1.9 jk	3.3 cd	2.0 lm
3.2 2iP	58 hi/42 jk	1.0 l/0.4 hi	2.0 ijk	4.0 ab	2.3 jk
0.4 mT	67 fg/33 lm	2.5 efg/0.4 hi	2.6 de	3.5 bc	5.9 b
0.8 mT	40 kl/60 f	2.9 bc/0.6 gh	2.4 fg	3.3 cd	3.8 g
1.6 mT	20 n/80 bc	3.0 b/1.1 ef	2.3 g	2.8 ef	1.7 mn
3.2 mT	10 o/83 b	3.3 b/1.2 e	1.4 m	1.5 l	0.5 p
Single node explant					
0.0 (Hf ᵀᵀᵀ)	79 bcd/17 p	1.9 hi/0.4 hi	1.9 jk	2.4 gh	3.6 fg
0.4 BA	63 gh/33 lm	2.5 efg/1.7 d	2.5 ef	3.0 de	4.7 cd
0.8 BA	58 hi/42 jk	2.6 de/1.8 d	2.8 d	3.0 de	4.5 de
1.6 BA	3 p/97 a	2.0 gh/2.5 c	1.5 lm	2.5 fgh	0.2 p
3.2 BA	0 q/100 a	0.0 m/4.1 b	-	-	0.0 r
0.4 ZEA	67 fg/33 lm	1.8 hi/0.9 fg	3.9 b	3.3 cd	4.0 ef
0.8 ZEA	54 ij/46 ij	2.0 gh/0.8 fg	2.6 de	3.1 de	3.3 gh
1.6 ZEA	35 lmn/65 e	2.0 gh/1.3 e	1.6 lm	2.3 hij	1.6 n
3.2 ZEA	23 n/77 c	2.0 gh/1.5 de	1.5 lm	2.4 gh	1.1 o
0.4 KIN	79 bcd/17 p	1.6 i/0.3 ij	1.7 klm	1.7 kl	2.1 kl
0.8 KIN	83 b/17 p	1.7 i/0.5 hi	1.7 klm	1.9 jk	2.7 i
1.6 KIN	63 gh/33 lm	1.8 hi/0.7 gh	1.8 jk	1.7 kl	1.9 mn
3.2 KIN	54 ij/46 ij	1.9 hi/1.1 ef	1.9 jk	2.2 hij	2.3 jk
0.4 2iP	79 bcd/21 op	2.0 gh/0.4 hi	1.7 klm	2.0 ij	3.2 gh
0.8 2iP	63 gh/33 lm	2.0 gh/0.7 gh	1.9 jk	2.7 ef	3.4 g
1.6 2iP	54 ij/42 jk	1.9 hi/0.9 fg	2.0 ijk	2.8 ef	2.9 hi
3.2 2iP	50 jk/50 hi	2.0 gh/1.1 ef	2.1 hi	2.8 ef	2.8 hi
0.4 mT	57 hi/43 jk	2.7 cde/0.6 gh	2.0 ijk	2.6 fg	4.0 ef
0.8 mT	30 mn/70 d	2.7 cde/0.7 gh	1.6 lm	2.4 gh	1.9 mn
1.6 mT	20 n/80 bc	2.8 bc/1.2 e	1.7 klm	3.2 cd	1.8 mn
3.2 mT	0 q/97 a	0.0 m/5.5 a	-	-	0.0 r
$F_{\text{one-way ANOVA}}$	***/***	***/***	***	***	***

[z] Mean separation in columns using Student's t, $p \leq 0.05$; means followed by the same letter are not significantly different at $p \leq 0.05$; *** significant at $p \leq 0.001$, $n = 30$. Multiplication index = shoot-producing explants [1] (%) × mean shoot number ᵀ × mean node number ᵀ. [1] The explants produced normal and hyperhydrated shoots. [2] The explants produced hyperhydrated shoots only. ᵀ NSh = normal shoot. ᵀᵀ HSh = hyperhydrated shoot. ᵀᵀᵀ Hf = hormone free.

Figure 10. Characteristic shoot multiplication of shoot tip and single node explants of *S. officinalis* cultured on MS medium supplemented with 0.01 NAA (mg.L^{-1}) and cytokinin type and concentration (mg.L^{-1}) marked. Size bars= 1.0 cm.

3.4. *In Vitro Rooting*

3.4.1. *S. fruticosa*

One-third of *S. fruticosa* microshoots rooted on Hf 1/2 MS medium, while the highest percentage of rooting (75%) was achieved on medium supplemented with the lowest IBA concentration tested, i.e., 0.5 mg L^{-1}. An increase in IBA concentration resulted in a corresponding decrease in rooting, although the two highest IBA concentrations (2 and 4 mg L^{-1}) resulted in the same percentage of rooting (Figure 11). The number of roots formed in a microshoot increased, while the root length decreased proportionally with increasing IBA concentration in the medium (Figures 11 and 12a).

3.4.2. *S. officinalis*

S. officinalis microshoots, in contrast to *S. fruticosa*, rooted at the highest percentage (85%) on Hf medium, as well as on a medium with 0.5 mg L^{-1} IBA, and a gradual increase in IBA resulted in a corresponding reduction in rooting, but as in *S. fruticosa*, the two highest IBA concentrations led to the same percentage of rooting (Figure 11). IBA resulted in an increase in root number regardless of concentration, while the root length was similar in all media except that with the highest IBA concentration, which resulted in a halving of root length (Figures 11 and 12b).

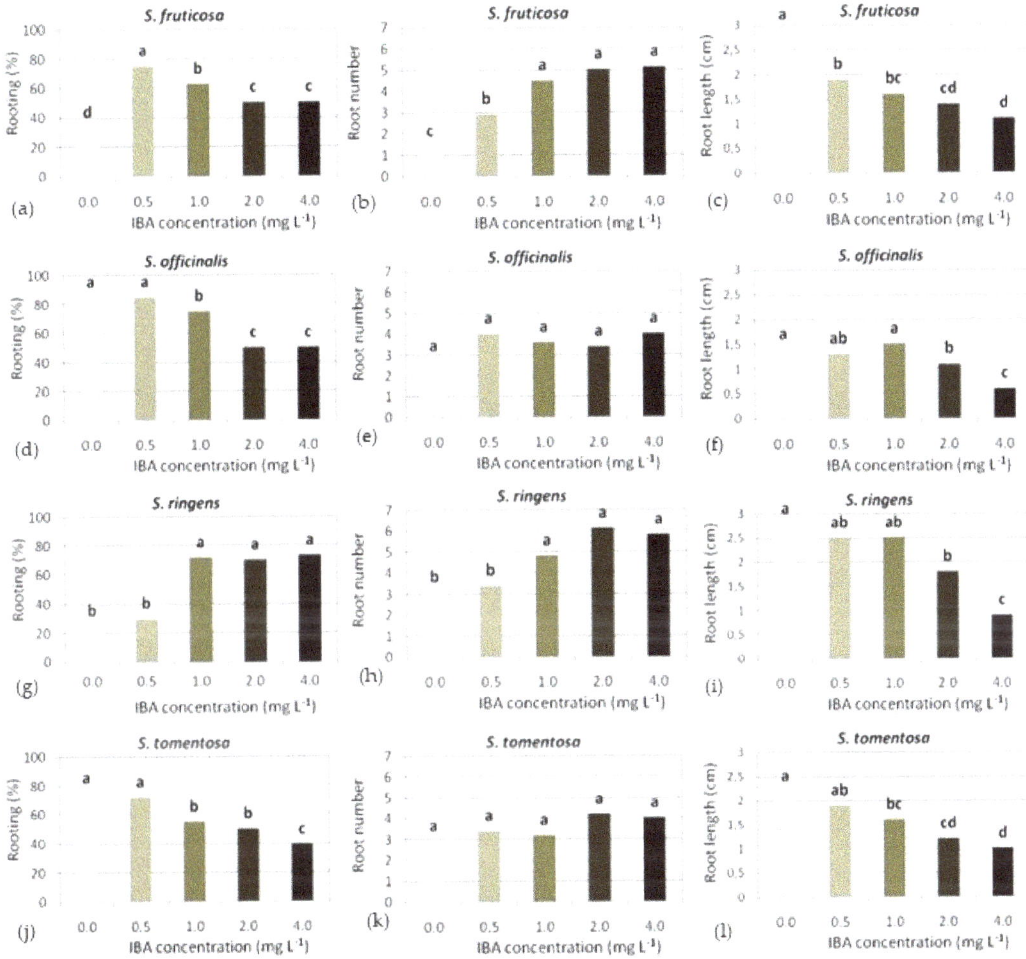

Figure 11. Effect of IBA concentration on in vitro rooting of marked *Salvia* spp. microshoots, on ½ MS medium supplemented with indole-3-butyric acid (IBA) concentration marked (*n* = 15–30).

3.4.3. *S. ringens*

In *S. ringens*, higher rooting percentages (70–73%) were induced using IBA concentrations of higher than 0.5 mg L^{-1}, which also resulted in a greater number of roots, while the root length was similar in Hf medium and those with the two lowest IBA concentrations, and decreased in those with the two highest concentrations (Figures 11 and 12c).

3.4.4. *S. tomentosa*

S. tomentosa microshoots rooted at the highest percentage (75%) on Hf medium, and a gradual increase in IBA concentration in the medium resulted in a corresponding decrease in rooting (Figure 11). Root number was not affected by IBA, while root length decreased gradually with increasing IBA concentration (Figures 11 and 12d).

Figure 12. Typical in vitro rooting of *Salvia fruticosa* (**a**), *S. officinalis* (**b**), *S. ringens* (**c**) and *S. tomentosa* (**d**) microshoots, cultured on $\frac{1}{2}$MS medium supplemented with IBA at concentration (mg.L^{-1}) marked (**left**), and ex vitro acclimatized plantlets about 6 weeks old (**right**). Size bars = 1.0 cm.

3.4.5. *S. pomifera* ssp. *pomifera*

Shoot tip explants of *S. pomifera* ssp. *pomifera* rooted at 80% when cultured on Hf MS medium, and formed numerous and long roots (Figure 6c). The small number of mother plants of this *Salvia* species available for the establishment of in vitro cultures due to the very low percentage of seed germination, as well as the low percentage of shoot multiplication, along with the high percentage of hyperhydrated shoots, did not allow for the production of sufficiently long normal shoots that could be used for rooting. As a result, in vitro rooting was not tested further for this species.

3.5. Ex Vitro Acclimatization

In all five *Salvia* spp., in vitro rooted microshoots coming from MS medium supplemented with 0.4 or 0.8 mg L^{-1} BA and 0.01 mg L^{-1} NAA, except for *S. pomifera* ssp. *pomifera*, whose microshoots were obtained from the Hf MS medium, acclimatized ex vitro at percentages of 80–95% (Figures 12 and 13).

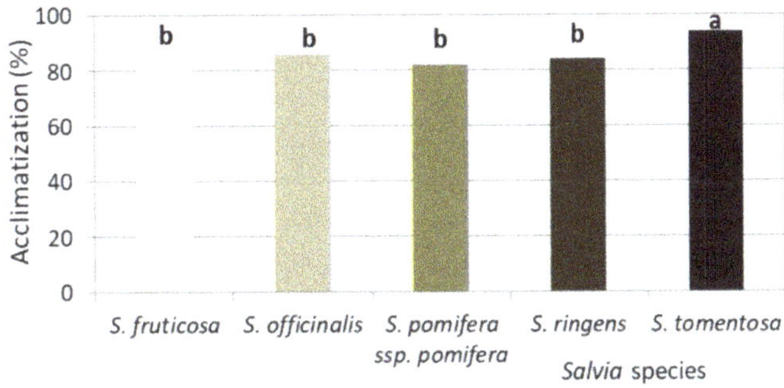

Figure 13. Ex vitro acclimatization of rooted microshoots of *Salvia* spp., on a peat–perlite substrate 1:1 (*v*/*v*) (*n* = 6–10 repetitions of 8 rooted microshoots).

4. Discussion

In vitro cultures of all five *Salvia* spp. were successfully established on Hf-MS medium from either shoot tip or single node explants excised from in vitro-grown seedlings, but explant response varied with *Salvia* species and explant type. Most shoot tip explants elongated when placed in culture, while single node explants produced two shoots, as expected due to the phyllotaxis of *Salvia* sp., confirming a previous report for *S. officinalis* [49]. Previous reports also showed reduced establishment rates due to explant necrosis, browning, and hyperhydricity [45]. In our work, almost all (96–100%) *S. officinalis* and *S. pomifera* ssp. *pomifera* explants of both types produced shoots, while in *S. fruticosa*, the response was slightly reduced (78–87%), as was the case for the shoot tip explants of *S. ringens* (87% response) too, while single node explants of *S. ringens* and both explant types of *S. tomentosa* responded at much lower percentages (36–50%) (Table 1). A problem at the establishment stage was the hyperhydricity of the produced shoots that was more pronounced in single node explants (18–50%), with the exception of *S. fruticosa*, which developed hyperhydrated shoots at a quite high percentage (17%) in shoot tip explants as well. The growth conditions mostly associated with hyperhydricity are limited aeration; high levels of total nitrogen, ammonium nitrogen, and cytokinin in the medium; and ethylene accumulation in the aerial part of the culture vessel [78–82]. As the medium of the establishment stage did not contain cytokinin, and vessels were covered with magenta caps that allowed for aeration, the most possible reason for the appearance of hyperhydricity seems to be the nitrogen richness of the MS medium. Nitrogen levels have been positively correlated with the increased occurrence of hyperhydricity in a number of species [79,82–84]. However, a low-nitrogen medium may not be suitable for shoot production, as in a previous work with *S. fruticosa*, MS medium was found superior to low-nitrogen media such as Nitsch and Nitsch and B5 [52].

Pronounced hyperhydricity has been reported in the micropropagation of several species of macchia, such as *Globularia alypum* [85], *Lithodora zahnii* [86], *Anthyllis barba-jovis* [87,88], *Calamintha cretica* [38], and *Clinopodium nepeta* [75].

At the shoot multiplication stage, in the presence of BA, shoot tip explants responded at higher percentages for shoot production compared to single node explants, with the exception of *S. officinalis*, where explants of both types responded at 100% (Figure 4). Increasing the BA concentration resulted in a corresponding decrease in explant response that was more pronounced in single node explants and in *S. pomifera* ssp. *pomifera* and *S. tomentosa*, and at the same time, it proportionally increased the number of shoots produced per explant while reducing their length and number of nodes. As in the establishment stage, the hyperhydricity was particularly pronounced in single node explants and increased proportionally with increasing BA concentration. Thus, the multiplication index in *S. pomifera* ssp. *pomifera* and *S. tomentosa* decreased with increasing BA concentration, and in *S. fruticosa* it was rather indifferent; while in *S. officinalis*, for both explant types, and in *S. ringens*, only for shoot tip explants, low BA concentrations (0,4 or 0.8 mg L^{-1}, respectively) resulted in increased multiplication indices compared to Hf-medium and a medium with higher BA content, indices that were the highest among all species.

Multiplication indices, in terms of normal shoots (excluding hyperhydrated shoots) in the most productive treatment for each species, were rather low in *S. fruticosa* (2.6) and *S. pomifera* ssp. *pomifera* (3.8), average in *S. ringens* (5.5) and *S. tomentosa* (5.4), and high in *S. officinalis* (7.2). These indices are considered satisfactory, as in all previous works on the in vitro propagation of Mediterranean sages, the multiplication indices were in almost all cases much lower than the present ones [33,44,48,50–52].

Regarding the effect of different types of cytokinin, i.e., BA, ZEA, KIN, 2iP, and mT, on the shoot multiplication of *S. fruticosa* and *S. officinalis*, it was found that mT at 0.8 mg L^{-1} induced the highest multiplication indices in *S. fruticosa*, both in shoot tip and single node explants, while in *S. officinalis*, mT at the lowest concentration of 0.4 mg L^{-1} also induced high multiplication indices, but lower than those induced by 0.4 mg L^{-1} BA. BA was superior for shoot multiplication in *S. officinalis* compared to all other cytokinins tested, in agreement with previous works, which also showed that BA or BAP in combination, as in the present study, with an auxin in low concentration, was superior to KIN [45,49,52], thidiazuron (TDZ) [50,52], 2iP, and ZEA [50] for proliferation. To the contrary, for adventitious shoots regeneration through callus, TDZ was more suitable than BA and KIN [51]. In *S. fruticosa*, a previous work had shown that BA is superior to KIN for proliferation [52], while in our work, BA, KIN, ZEA, and 2iP induced a similar response.

As discussed above for BA, increasing the concentration of cytokinin of all types in the medium resulted in an analogous increase in shoot number per explant, with a simultaneous decrease in shoot length and number of nodes, and a significant increase in hyperhydricity leading to a decrease in the multiplication indices. Single node explants showed a slightly lower response rate for shoot production, and more pronounced hyperhydricity compared to shoot tip explants in all cytokinin types, while at the highest cytokinin concentrations, the production of normal shoots was very low from single node explants due to hyperhydricity. In *S. fruticosa*, Arikat et al. [52] reported that single node explants almost failed to produce shoots when BA, KIN, or TDZ were used at concentrations of higher than 1 mg L^{-1}, but he did not report on hyperhydricity. Hyperhydricity has been reported previously in *S. officinalis* [45,48,50], as well as in *S. scarlea* [33].

Cytokinins has been shown to induce hyperhydricity in some plant species during in vitro propagation, depending on the cytokinin type and concentration [38,75,79,89,90]. Decreasing the concentration of cytokinin in the medium or using a different type of cytokinin resulted in a reduction in hyperhydricity [70,86,91], often with a simultaneous reduction in shoot production [38,75], as shown in the present work. Therefore, measures aimed at eliminating hyperhydricity should be complex and specific for each plant species.

In vitro rooting was successful (a rooting percentage of higher than 80%) in all five species. *S. officinalis*, *S. tomentosa*, and *S. pomifera* ssp. *pomifera* rooted at high percentages (75–83%) on Hf medium, or in the presence of the lowest IBA concentration tested (0.5 mg L^{-1}), while *S. ringens* and *S. fruticosa* rooted at low percentages on Hf medium (35 and 35%, respectively) and at 1.0 and 0.5 mg L^{-1} IBA, respectively, were required to achieve high rooting percentages. Our results confirmed those of Avato et al. [45], who also found that auxin supplementation was not necessary for rooting in *S. officinalis*. Arikat et al. [52] also found that *S. fruticosa* rooted only in the presence of a low concentration of auxin (0.5 mg L^{-1} IBA). An equally low concentration of IBA (0.5–1.0 mg L^{-1}) was shown to induce maximum rooting rates in the microshoots of *S. officinalis* (65–70%) [50]

and *Salvia hispanica* (65–70%) [76], while similar concentrations of IBA induced the highest percentage of rooting in *S. scarlea*, which however, did not exceed 40% [33]. IBA at low concentrations (up to 1 mg L^{-1}) was effective for the rooting of other Mediterranean medicinal Lamiaceae as well [33,67,68,71,72,75,76]. Half-strength MS medium was used in most of the above works, as in the present study, as rather low concentrations of nutrients positively affect rooting [92].

In *S. officinalis* and *S. tomentosa*, where auxin supplementation was not necessary for rooting, root number was indifferent to auxin concentration in the medium, while in *S. fruticosa* and *S. ringens*, which showed low rooting ability in auxin-free medium, increasing auxin content, in addition to higher rooting rates, also induced a higher number of roots per microshoot. In contrast, a gradual increase in IBA resulted in a gradual decrease in root length in all five species, as previously found for *S. scarlea* [66], *Teucrium capitatum* [72], and *C. cretica* [38].

Rooted microshoots of all five sage species acclimatized ex vitro at percentages of higher than 80%, retaining the characteristics of the mother plants, as has been shown for other sages [33,66–68,71,74,76] and a number of other medicinal Lamiaceae [38,69,72,75].

5. Conclusions

The five Mediterranean sages, *S. fruticosa*, *S. officinalis*, *S. ringens*, *S. tomentosa*, and *S. pomifera* spp. *pomifera*, native to Greece, showed many similarities, but also some quantitative differences in their in vitro propagation. All were successfully established on Hf-MS medium in the initial culture from either shoot tip or single node explants excised from seedlings grown in vitro, showing low proliferation indices, except for *S. officinalis*, which stood out with the highest indices. In the multiplication stage, the explant response for shoot production was high, but the number of shoots produced per explants was rather low. Supplementing the medium with various types and concentrations of cytokinin in combination with auxin did not significantly increase the proliferation indices, as the number of shoots produced per explant remained below five, and a significant proportion of them were hyperhydrated. Shoot tip explants produced normal shoots at much higher percentages compared to single node explants. All five sages rooted readily in vitro, either without or under low auxin, and they acclimatized ex vitro at high percentages. Thus, the present study led to efficient micropropagation protocols for five Mediterranean sage species native to Greece, which are expected to facilitate their sustainable exploitation in the pharmaceutical and floriculture industries.

Author Contributions: Conceptualization, M.P., G.V. and A.N.M.; methodology, M.P., G.V. and A.N.M.; validation, M.P., G.V. and A.N.M.; formal analysis, G.V. and A.N.M.; investigation, M.P., G.V. and A.N.M.; resources, M.P.; data curation, G.V. and A.N.M.; writing—original draft preparation, M.P., G.V. and A.N.M.; writing—review and editing, M.P., G.V. and A.N.M.; visualization, M.P., G.V. and A.N.M.; supervision, M.P.; project administration, M.P.; funding acquisition, M.P. All authors have read and agreed to the published version of the manuscript.

Funding: This research was co-financed by the European Regional Development Fund of the European Union and Greek national funds through the Operational Program Competitiveness, Entrepreneurship and Innovation, under the call RESEARCH—CREATE—INNOVATE (project code: T1EDK-04923, project: SALVIA-BREED-GR).

Data Availability Statement: Not applicable.

Conflicts of Interest: The authors declare no conflict of interest. The funders had no role in the design of the study; in the collection, analyses, or interpretation of data; in the writing of the manuscript; or in the decision to publish the results.

References

1. The Plant List. Version 1.1. 2013. Available online: http://www.theplantlist.org/ (accessed on 29 April 2022).
2. Walker, J.B.; Sytsma, K.J.; Treutlein, J.; Wink, M. *Salvia* (Lamiaceae) is not monophyletic: Implications for the systematics, radiation, and ecological specializations of *Salvia* and tribe Mentheae. *Am. J. Bot.* **2004**, *91*, 1115–1125. [CrossRef] [PubMed]

3. Bentham, G. Labiatae. In *Genera Plantarum*; Bentham, G., Hooker, J.D., Eds.; Reeve and Co.: London, UK, 1876; Volume 2, pp. 1160–1196.
4. Online Etymology Dictionary. Available online: https://www.etymonline.com/word/salvia (accessed on 11 December 2021).
5. Hamidpour, M.; Hamidpour, R.; Hamidpour, S.; Shahlari, M. Chemistry, Pharmacology, and Medicinal Property of Sage (*Salvia*) to Prevent and Cure Illnesses such as Obesity, Diabetes, Depression, Dementia, Lupus, Autism, Heart Disease, and Cancer. *J. Tradit. Complement. Med.* **2014**, *4*, 82–88. [CrossRef] [PubMed]
6. Bonesi, M.; Loizzo, M.R.; Acquaviva, R.; Malfa, G.A.; Aiello, F.; Tundis, R. Anti-inflammatory and antioxidant agents from *Salvia* genus (Lamiaceae): An assessment of the current state of knowledge. *Anti-Inflamm. Anti-Allergy Agents Med. Chem.* **2017**, *16*, 70–86. [CrossRef] [PubMed]
7. Lopresti, A.L. *Salvia* (Sage): A review of its potential cognitive-enhancing and protective effects. *Drugs R&D* **2017**, *17*, 53–64. [CrossRef]
8. Hao, D.C.; Ge, G.B.; Xiao, P.G. Anticancer drug targets of *Salvia* phytometabolites: Chemistry, biology and omics. *Curr. Drug Targets* **2018**, *19*, 1–20. [CrossRef]
9. Ververis, A.; Savvidou, G.; Ioannou, K.; Nicolaou, P.; Christodoulou, K.; Plioukas, M. Greek sage exhibits neuroprotective activity against amyloid beta-induced toxicity. *Evid. Based Complement. Altern. Med.* **2020**, *2020*, 2975284. [CrossRef]
10. Afonso, A.F.; Pereira, O.R.; Cardoso, S.M. *Salvia* species as nutraceuticals: Focus on antioxidant, antidiabetic and anti-obesity properties. *Appl. Sci.* **2021**, *11*, 9365. [CrossRef]
11. Mervić, M.; Štefan, M.B.; Kindl, M.; Blažeković, B.; Marijan, M.; Vladimir-Knežević, S. Comparative Antioxidant, Anti-Acetylcholinesterase and Anti-α-Glucosidase Activities of Mediterranean *Salvia* Species. *Plants* **2022**, *11*, 625. [CrossRef]
12. Karalija, E.; Dahija, S.; Tarkowski, P.; Ćavar Zeljkovic, S. Influence of Climate-Related Environmental Stresses on Economically Important Essential Oils of Mediterranean *Salvia* sp. *Front. Plant Sci.* **2022**, *13*, 864807. [CrossRef]
13. Jiang, S. Therapeutic landscapes and healing gardens: A review of Chinese literature in relation to the studies in western countries. *Front. Archit. Res.* **2014**, *3*, 141–153. [CrossRef]
14. Savi, T.; Andri, S.; Nardini, A. Impact of different green roof layering on plant water status and drought survival. *Ecol. Eng.* **2013**, *57*, 188–196. [CrossRef]
15. Raimondo, F.; Trifilò, P.; Lo Gullo, M.A.; Andri, S.; Savi, T.; Nardini, A. Plant performance on Mediterranean green roofs: Interaction of species-specific hydraulic strategies and substrate water relations. *AoB Plants* **2015**, *7*, plv007. [CrossRef]
16. Martini, A.N.; Tassoula, L.; Papafotiou, M. Adaptation of *Salvia fruticosa, S. officinalis, S. ringens* and interspecific hybrids in an extensive green roof under two irrigation frequencies. *Notulae* **2022**, *50*, 12767. [CrossRef]
17. Papafotiou, M.; Martini, A.N.; Tassoula, L.; Stylias, E.G.; Kalantzis, A.; Dariotis, E. Acclimatization of Mediterranean native sages (*Salvia* spp.) and interspecific hybrids in an urban green roof under regular and reduced irrigation. *Sustainability* **2022**, *14*, 4978. [CrossRef]
18. Rahimi, E.; Barghjelveh, S.; Dong, P. A review of diversity of bees, the attractiveness of host plants and the effects of landscape variables on bees in urban gardens. *Agric. Food Secur.* **2022**, *11*, 6. [CrossRef]
19. Dimopoulos, P.; Raus, T.; Bergmeier, E.; Constantinidis, T.; Iatrou, G.; Kokkini, S.; Strid, S.; Tzanoudakis, D. *Vascular Plants of Greece: An Annotated Checklist*; Botanischer Garten und Botanisches Museum Berlin-Dahlem; Hellenic Botanical Society: Berlin, Germany, 2013.
20. Chatzopoulou, F.M.; Makris, A.M.; Argiriou, A.; Degenhardt, J.; Kanellis, A.K. EST analysis and annotation of transcripts derived from a trichome-specific cDNA library from *Salvia fruticosa*. *Plant Cell Rep.* **2010**, *29*, 523–534. [CrossRef]
21. Ghorbani, A.; Esmaeilizadeh, M. Pharmacological properties of *Salvia officinalis* and its components. *J. Tradit. Complement Med.* **2017**, *7*, 433–440. [CrossRef]
22. Kemp, S.; Hadley, P.; Blanuša, T. The influence of plant type on green roof rainfall retention. *Urban Ecosyst.* **2019**, *22*, 355–366. [CrossRef]
23. Göçer, H.; Yetişir, H.; Ulaş, A.; Arslan, M.; Aydin, A. Plant growth, ion accumulation and essential oil content of *Salvia officinalis* Mill. and *S. tomentosa* L. grown under different salt stress. *KSU J. Agric. Nat.* **2021**, *24*, 505–514.
24. Papafotiou, M.; Martini, A.N.; Papanikolaou, E.; Stylias, E.G.; Kalantzis, A. Hybrids Development between Greek *Salvia* Species and Their Drought Resistance Evaluation along with *Salvia fruticosa*, under Attapulgite-Amended Substrate. *Agronomy* **2021**, *11*, 2401. [CrossRef]
25. Putievsky, E.; Ravid, U.; Diwan-Rinzler, N.; Zohary, D. Genetic affinities and essential oil composition of *Salvia officinalis* L., *S. fruticosa* Mill., *S. tomentosa* Mill. and their hybrids. *Flavour Fragr. J.* **1990**, *5*, 121–123. [CrossRef]
26. Karousou, R.; Vokou, D.; Kokkini, S. Distribution and essential oils of *Salvia pomifera* subsp. *pomifera* (Labiatae) on the island of Crete (S Greece). *Biochem. Syst. Ecol.* **1998**, *26*, 889–897. [CrossRef]
27. Haznedaroglu, M.; Karabay, N.; Zeybek, U. Antibacterial activity of *Salvia tomentosa* essential oil. *Fitoterapia* **2001**, *72*, 829–831. [CrossRef] [PubMed]
28. Tepe, B.; Daferera, D.; Sokmen, A.; Sokmen, M.; Polissiou, M. Antimicrobial and antioxidant activities of the essential oil and various extracts of *Salvia tomentosa* Miller (Lamiaceae). *Food Chem.* **2005**, *90*, 333–340. [CrossRef]
29. Arslan, M. Cultivation potential of *Salvia tomentosa* and *S. aramiensis* under the Eastern Mediterranean conditions. *Sci. Pap. Ser. A Agron.* **2016**, *59*, 174–177.

30. Hanlidou, E.; Karousou, R.; Lazari, D. Essential-oil diversity of *Salvia tomentosa* Mill. in Greece. *Chem. Biodivers.* **2014**, *11*, 1205–1215. [CrossRef]
31. Bardakci, H.; Servi, H.; Polatoglu, K. Essential oil composition of *Salvia candidissima* Vahl. occidentalis Hedge, *S. tomentosa* Miller and *S. heldreichiana* Boiss. Ex Bentham from Turkey. *J. Essent. Oil-Bear. Plants* **2019**, *22*, 1467–1480. [CrossRef]
32. Özcan, M.M.; Figueredo, G.; Chalchat, J.C.; Chalard, P.; Tugay, O.; Ceylan, D.A. Chemical constituents of essential oils of *Salvia heldreichiana* Boiss. Ex Bentham and *Salvia tomentosa* Mill. *J. Agroaliment. Process. Technol.* **2019**, *25*, 106–110.
33. Grigoriadou, K.; Trikka, F.A.; Tsoktouridis, G.; Krigas, N.; Sarropoulou, V.; Papanastasi, K.; Maloupa, E.; Makris, A.M. Micropropagation and cultivation of *Salvia sclarea* for essential oil and sclareol production in northern Greece. *In Vitro Cell. Dev. Biol. Plant* **2020**, *56*, 51–59. [CrossRef]
34. Thanos, C.A.; Doussi, M.A. Ecophysiology of seed germination in endemic labiates of Crete. *Isr. J. Plant Sci.* **1995**, *43*, 227–237. [CrossRef]
35. Tutin, T.G.; Heywood, V.H.; Burges, N.A.; Moore, D.M.; Valentine, D.H.; Walters, S.M.; Webb, D.A. *Flora Europaea, Volume 3 Diapenstaceae to Myoporaceae*; Cambridge University Press: Cambridge, UK, 1972; pp. 188–190.
36. Guner, A.; Ozhatay, N.; Ekim, T.; Baser, K.H.C. *Flora of Turkey and the East Aegean Islands, Supplement II*; Edinburg University Press: Edinburg, UK, 2000; Volume 11.
37. Martini, A.N.; Bertsouklis, K.F.; Vlachou, G.; Dariotis, E.; Papafotiou, M. Comparative evaluation of rooting cuttings of five Mediterranean sage species (*Salvia* sp.) native to Greece. *Acta Hortic.* **2020**, *1298*, 587–592. [CrossRef]
38. Vlachou, G.; Papafotiou, M.; Bertsouklis, K.F. Seed germination, micropropagation from adult and juvenile origin explants and address of hyperhydricity of the Cretan endemic herb *Calamintha cretica*. *Not. Bot. Horti. Agrobot.* **2020**, *48*, 1504–1518. [CrossRef]
39. Kanellou, E.; Vlachou, G.; Martini, A.N.; Bertsouklis, K.F.; Papafotiou, M. Seed germination of five sage species (*Salvia* sp.) of populations native to Greece. *Acta Hortic.* **2022**, *1345*, 439–444. [CrossRef]
40. Jaafar, A.; Mohsen, E.; Hossein, A.R.; Ali, A.J.; Mostafa, E.; Yaser, S.M.; Mohammad, A.S.B.G. Seed germination as the major conservation issue of endemic Iranian salvia species. *J. Med. Plants Res.* **2012**, *6*, 37–46. [CrossRef]
41. Jaafar, A.; Seyed, H.T.; Mohsen, E.; Hossein, A.R.; Ahamad, R.D.T.; Mehdi, M.; Mostafa, E.; Amir, H.D.; Asma, B. Effect of drought stress on germination and seedling growth of *Salvia* species. *Afr. J. Agric. Res.* **2012**, *7*, 5719–5725. [CrossRef]
42. Kintzios, S.; Nikolau, A.; Skoula, M. Somatic embryogenesis and in vitro rosmarinic acid accumulation in *Salvia officinalis* and *S. fruticosa* leaf callus cultures. *Plant Cell Rep.* **1999**, *18*, 462–466. [CrossRef]
43. Santos-Gomes, P.C.; Seabra, R.M.; Andrade, P.B.; Ferreira, M.F. Phenolic antioxidant compounds produced by in vitro shoots of sage (*Salvia officinalis* L.). *Plant Sci.* **2002**, *162*, 981–987. [CrossRef]
44. Grzegorczyk, I.; Bilichowski, I.; Mikiciuk-Olasik, E.; Wysokinska, H. The effect of triacontanol on shoot multiplication and production of antioxidant compounds in shoot cultures of *Salvia officinalis* L. *Acta Soc. Bot. Pol.* **2006**, *75*, 11–15. [CrossRef]
45. Avato, P.; Fortunato, I.M.; Ruta, C.; D' Elia, R. Glandular hairs and essential oils in micropropagated plants of *Salvia officinalis* L. *Plant Sci.* **2005**, *169*, 29–36. [CrossRef]
46. Tawfik, A.A.; Mohamed, F.M. Regeneration of salvia (*Salvia officinalis* L.) via induction of meristematic callus. *In Vitro Cell. Dev. Biol. Plant* **2007**, *43*, 21–27. [CrossRef]
47. Gostin, I. Effects of different plant hormones on *Salvia officinalis* cultivated in vitro. *Int. J. Bot.* **2008**, *4*, 430–436.
48. Grzegorczyk, I.; Wysokinska, H. Liquid shoot culture of *Salvia officinalis* L. for micropropagation and production of antioxidant compounds; effect of triacontanol. *Acta Soc. Bot. Pol.* **2008**, *77*, 99–104. [CrossRef]
49. Cristea, O.T.; Prisecaru, M.; Ambarus, S.; Calin, M.; Brezeanu, C.; Brezeanu, M.; Florin, G. Vegetative multiplication of *Salvia officinalis* L. for the obtaining of true-to-type plants. *Biologie* **2014**, *23*, 104–107.
50. Petrova, M.; Nikolova, M.; Dimitrova, L.; Zayova, E. Micropropagation and evaluation of flavonoid content and antioxidant activity of *Salvia officinalis* L. *Genet. Plant Physiol.* **2015**, *5*, 48–60.
51. Mohamed, M.A.-H.; Aly, M.K.; Ahmed, E.T.; Abd El-latif, S.A.H. Effect of plant growth regulators on organogenesis of *Salvia officinalis* L. plants. *Minia J. Agric. Res. Develop.* **2019**, *3*, 401–414. [CrossRef]
52. Arikat, N.A.; Jawad, F.M.; Karam, N.S.; Shibli, R.A. Micropropagation and accumulation of essential oils in wild sage (*Salvia fruticosa* Mill.). *Sci. Hortic.* **2004**, *100*, 193–202. [CrossRef]
53. Martini, A.N.; Vlachou, G.; Papafotiou, M. Effect of Explant Origin and Medium Plant Growth Regulators on In Vitro Shoot Proliferation and Rooting of *Salvia tomentosa*, a Native Sage of the Northeastern Mediterranean Basin. *Agronomy* **2022**, *12*, 1889. [CrossRef]
54. Kitto, S.L. Commercial micropropagation. *HortScience* **1997**, *32*, 1012–1014. [CrossRef]
55. Sidhu, Y. In vitro micropropagation of medicinal plants by tissue culture. *Plymouth Stud. Sci.* **2011**, *4*, 432–449.
56. Grigoriadou, K.; Krigas, N.; Sarropoulou, V.; Papanastasi, K.; Tsoktouridis, G.; Maloupa, E. In vitro propagation of medicinal and aromatic plants: The case of selected Greek species with conservation priority. *In Vitro Cell. Dev. Biol. Plant* **2019**, *55*, 635–646. [CrossRef]
57. Murashige, T.; Skoog, F. A revised medium for rapid growth and bioassays with tobacco tissue cultures. *Physiol. Plant.* **1962**, *15*, 473–497. [CrossRef]
58. Grzegorczyk, I.; Wysokinska, H. Micropropagation of *Salvia officinalis* L. by shoot tips. *Biotechnologia* **2004**, *2*, 212–218.
59. CachiŢĂ-Cosma, D. The effect of the nature and origin of explants on micropropagation. In *High-Tech and Micropropagation I. Biotechnology in Agriculture and Forestry*; Bajaj, Y.P.S., Ed.; Springer: Berlin/Heidelberg, Germany, 1991; Volume 17.

60. Papafotiou, M.; Martini, A. Effect of position and orientation of leaflet explants with respect to plant growth regulators on micropropagation of *Zamioculcas zamiifolia* Engl. (ZZ). *Sci. Hortic.* **2009**, *120*, 115–120. [CrossRef]
61. Kartsonas, E.; Papafotiou, M. Mother plant age and seasonal influence on in vitro propagation of *Quercus euboica* Pap., an endemic, rare and endangered oak species of Greece. *Plant Cell Tissue Organ Cult. (PCTOC)* **2007**, *90*, 111–116. [CrossRef]
62. Martini, A.N.; Papafotiou, M.; Vemmos, S.N. Season and explant origin affect phenolic content, browning of explants, and micropropagation of ×*Malosorbus florentina* (Zucc.) Browicz. *HortScience* **2013**, *48*, 102–107. [CrossRef]
63. Vlachou, G.; Papafotiou, M.; Bertsouklis, K.F. In Vitro Propagation of *Calamintha nepeta*. *Acta Hort.* **2016**, *1113*, 189–194. [CrossRef]
64. Vlachou, G.; Papafotiou, M.; Bertsouklis, K.F. In Vitro Propagation of *Ballota acetabulosa*. *Acta Hort.* **2016**, *1113*, 171–174. [CrossRef]
65. Cuenca, S.; Amo-Marco, J.B. In vitro propagation of two Spanish endemic species of *Salvia* through bud proliferation. *In Vitro Cell. Dev. Biol. Plant* **2000**, *36*, 225–229. [CrossRef]
66. Ghanbar, T.; Hosseini, B.; Jabbarzadeh, Z.; Farokhzad, A.; Sharafi, A. High-frequency in vitro direct shoots regeneration from axillary nodal and shoot tip explants of clary sage (*Salvia sclarea* L.). *Bulg. J. Agric. Sci.* **2016**, *22*, 73–78.
67. Raja, H.D.; Arockiasamy, D.I. In vitro propagation of *Mentha viridis* L. from nodal and shoot tip explants. *Plant Tissue Cult. Biotechnol.* **2008**, *18*, 1–6. [CrossRef]
68. Sujana, P.; Naidu, C.V. High frequency rapid plant regeneration from shoot tip and nodal explants of *Mentha piperita* (L.)—An important multipurpose medicinal plant. *J. Phytol.* **2011**, *3*, 9–13.
69. Moharami, L.; Hosseini, B.; Ghotbi Ravandi, E.; Jafari, M. Effects of plant growth regulators and explant types on in vitro direct plant regeneration of *Agastache foeniculum*, an important medicinal plant. *In Vitro Cell. Dev. Biol. Plant* **2014**, *50*, 707–711. [CrossRef]
70. Sharma, U.; Mohan, J.S.S. Reduction of vitrification in in vitro raised shoots of *Chlorophytum borilivinum* (Sant. and Fernand.), a rare potent medicinal herb. *Indian J. Exp. Biol.* **2006**, *44*, 499–505. [PubMed]
71. Islam, A.T.M.R.; Islam, M.M.; Alam, M.F. Rapid in vitro clonal propagation of herbal spice, *Mentha piperita* L. using shoot tip and nodal explants. *Res. Plant Biol.* **2017**, *5*, 43–50. [CrossRef]
72. Papafotiou, M.; Martini, A.N. In vitro seed and clonal propagation of the Mediterranean aromatic and medicinal plant *Teucrium capitatum*. *HortScience* **2016**, *51*, 403–411. [CrossRef]
73. Papafotiou, M.; Bertsouklis, K.F.; Trigka, M. Micropropagation of *Arbutus unedo*, *A. andrachne*, and their natural hybrid, *A.* x *andrachnoides* from seedling explants. *J. Hort. Sci. Biot.* **2013**, *88*, 768–775. [CrossRef]
74. Sharma, S.; Shahzad, A.; Kumar, J.; Anis, M. In vitro propagation and synseed production of scarlet salvia (*Salvia splendens*). *Rend. Fis. Acc. Lincei* **2014**, *25*, 359–368. [CrossRef]
75. Vlachou, G.; Papafotiou, M.; Bertsouklis, K. Studies on seed germination and micropropagation of *Clinopodium nepeta*: A medicinal and aromatic plant. *HortScience* **2019**, *54*, 1558–1564. [CrossRef]
76. Zayova, E.; Nikolova, M.; Dimitrova, L.; Petrova, M. Comparative study of in vitro, ex vitro and in vivo propagated *Salvia hispanica* (Chia) plants: Morphometric analysis and antioxidant activity. *AgroLife Sci. J.* **2016**, *5*, 166–174.
77. Vlachou, G.; Martini, A.N.; Dariotis, E.; Papafotiou, M. Comparative evaluation of seed germination of five Mediterranean sage species (*Salvia* sp.) native to Greece. *Acta Hortic.* **2020**, *1298*, 593–598. [CrossRef]
78. Lai, C.C.; Lin, H.M.; Nalawade, S.M.; Fang, W.; Tsay, H.S. Hyperhydricity in shoot cultures of *Scrophularia yoshimurae* can be effectively reduced by ventilation of culture vessels. *J. Plant Physiol.* **2005**, *162*, 355–361. [CrossRef] [PubMed]
79. Ivanova, M.; van Staden, J. Effect of ammonium ions and cytokinins on hyperhydricity and multiplication rate of in vitro regenerated shoots of *Aloe polyphylla*. *Plant Cell Tissue Organ Cult. (PCTOC)* **2008**, *92*, 227–231. [CrossRef]
80. Ivanova, M.; van Staden, J. Influence of gelling agent and cytokinins on the control of hyperhydricity in *Aloe polyphylla*. *Plant Cell Tissue Organ Cult. (PCTOC)* **2011**, *104*, 13–21. [CrossRef]
81. Badr-Elden, A.M.; Nower, A.A.; Ibrahim, I.A.; Ebrahim, M.K.; Abd Elaziem, T.M. Minimizing the hyperhydricity associated with in vitro growth and development of watermelon by modifying the culture conditions. *Afr. J. Biotechnol.* **2012**, *11*, 8705–8717. [CrossRef]
82. Pence, V.C.; Finke, L.R.; Niedz, R.P. Evaluating a DOE screen to reduce hyperhydricity in the threatened plant, *Cycladenia humilis* var. *jonesii*. *In Vitro Cell. Dev. Biol. Plant* **2020**, *56*, 215–229. [CrossRef]
83. Reyes-Vera, I.; Potenza, C.; Barrow, J. Hyperhydricity reversal and clonal propagation of four-wing saltbush (*Atriplex canescens*, Chenopodiaceae) cultivated in vitro. *Aust. J. Bot.* **2008**, *56*, 358–362. [CrossRef]
84. Wu, H.; Yu, X.; Teixeira da Silva, J.A.; Lu, G. Direct shoot induction of *Paeonia lactiflora* 'Zhong Sheng Fen'and rejuvenation of hyperhydric shoots. *N. Zeal. J. Crop Hortic. Sci.* **2011**, *39*, 271–278. [CrossRef]
85. Bertsouklis, K.; Papafotiou, M.; Balotis, G. Effect of medium on in vitro growth and ex vitro establishment of *Globularia alypum* L. *Acta Hortic.* **2003**, *616*, 177–180. [CrossRef]
86. Papafotiou, M.; Kalantzis, A. Studies on in vitro propagation of *Lithodora zahnii*. *Acta Hortic.* **2009**, *813*, 465–470. [CrossRef]
87. Vlachou, G.; Papafotiou, M.; Bertsouklis, K.F. Studies on in vitro propagation of *Anthyllis barba-jovis* L. *Acta Hortic.* **2017**, *1155*, 317–320. [CrossRef]
88. Trigka, M.; Papafotiou, M. In vitro propagation of *Anthyllis barba-jovis* from seedling tissues. *Acta Hortic.* **2017**, *1189*, 473–748. [CrossRef]
89. Kataeva, N.V.; Alexandrova, I.G.; Butenko, R.G.; Dragavtceva, E.V. Effect of applied and internal hormones on vitrification and apical necrosis of different plants cultured in vitro. *Plant Cell Tissue Organ Cult. (PCTOC)* **1991**, *27*, 149–154. [CrossRef]
90. Ravanfar, S.A.; Salim, S.; Aziz, M.A.; Abdullah, S.N.A.; Rashid, A.A. Influence of phenyl-urea and adenine-type cytokinins on direct adventitious shoot regeneration of cabbage (*Brassica oleracea* subsp. *capitata*) "KCross". *Biotechnology* **2014**, *31*, 1–6. [CrossRef]

91. Liu, M.; Jiang, F.; Kong, X.; Tian, J.; Wu, Z.; Wu, Z. Effects of multiple factors on hyperhydricity of *Allium sativum* L. *Sci. Hortic.* **2017**, *217*, 285–296. [CrossRef]
92. Murashige, T. Principles of rapid propagation. In *Propagation of Higher Plants through Tissue Culture: A Bridge between Research and Application*; Hughes, K.W., Henke, R., Constantin, M., Eds.; Tech. Information Center, U.S. Dept. of Energy: Oak Ridge, TN, USA, 1979; pp. 14–24.

horticulturae

MDPI

Article

In Vitro Evaluation of the Effects of BAP Concentration and Pre-Cooling Treatments on Morphological, Physiological, and Biochemical Traits of Different Olive (*Olea euorpea* L.) Cultivars

Sadia Khatoon [1], Wencong Liu [1,2,*], Chuan-bo Ding [1], Xinglong Liu [1], Yinan Zheng [1], Yue Zhang [1], Xueyan Chen [1], Muhammad Rauf [3], Fahad Alghabari [4] and Zahid Hussain Shah [5,*]

1 College of Chinese Medicinal Materials, Jilin Agriculture University, Changchun 453003, China
2 College of Resources and Environment Sciences, Jilin Agriculture University, Changchun 453003, China
3 Vegetable Research Station Lal Esan, Karor 31100, Pakistan
4 Department of Arid Land Agriculture, King Abdulaziz University, Jeddah 21589, Saudi Arabia
5 Department of Plant Breeding and Genetics, Pir Mehr Ali Shah, Arid Agriculture University, Rawalpindi 46000, Pakistan
* Correspondence: liuwencong@jlau.edu.cn (W.L.); shahzahid578@hotmail.com (Z.H.S.)

Abstract: In vitro propagation of olive (*Olea euorpea* L.) always remained a challenging task due to its woody nature and oxidation of culture. The current study intended to optimize shoot induction and proliferation protocol for different cultivars ("Leccino", "Gemlik", "Moraiolo" and "Arbosana") of olive-on-olive media (OM) provided with different concentrations (0, 0.5, 1.5, and 2.5 mgL^{-1}) of 6-benzylaminopurine (BAP) by pre-exposing their explants (nodal segments) with different regimes (0, 24, and 48 h) of cooling. The impacts of treatments were evaluated on morphological (shoot induction percentage, primary shoot length, number of leaves $shoot^{-1}$, and number of shoots per $explant^{-1}$), physiological (total chlorophyll, carotenoids, CO_2 assimilation, and proline), biochemical (primary and secondary metabolites) attributes of cultivars after 50 to 60 days of culture. Data recorded were subjected to statistical analysis. All traits depicted significant increases in all genotypes with increasing pre-cooling treatments and increasing supplementations of 6-benzylaminopurine (BAP). This increase was the highest for the interaction of 48 h pre-cooling and 2.5 mgL^{-1} BAP concentration. Moreover, correlation analysis of all traits revealed significant paired association among them in a positive direction, while principal component analysis (PCA) revealed the extent of association varied with types of treatments and the nature of genotypes. Among cultivars, Arbosana depicted more dramatic changes in morphological traits, physiological attributes, and biochemical contents due to varying interactions of pre-cooling and BAP treatments as compared to Moraiolo, Gemlik, and Leccino with in vitro systems.

Keywords: explants; shoot induction; metabolites; micropropagation; correlation

Citation: Khatoon, S.; Liu, W.; Ding, C.-b.; Liu, X.; Zheng, Y.; Zhang, Y.; Chen, X.; Rauf, M.; Alghabari, F.; Shah, Z.H. In Vitro Evaluation of the Effects of BAP Concentration and Pre-Cooling Treatments on Morphological, Physiological, and Biochemical Traits of Different Olive (*Olea euorpea* L.) Cultivars. *Horticulturae* **2022**, *8*, 1108. https://doi.org/10.3390/horticulturae8121108

Academic Editor: Giuseppe Ferrara

Received: 18 October 2022
Accepted: 22 November 2022
Published: 25 November 2022

Publisher's Note: MDPI stays neutral with regard to jurisdictional claims in published maps and institutional affiliations.

1. Introduction

Olive (*Olea euorpea* L.) is an important plant of the Mediterranean region and an important source of traditional landscape, food, and oil. Various traditional methods of propagation such as grafting, cuttings, and leafy stem rooting are commonly used for olive multiplication [1]. These methods ensure the preservation of genetic traits but have numerous limitations such as season dependence, low success rate, nutrient requirement, the hygienic status of the mother plant, and variable response from cultivar to cultivar [2]. Moreover, these methods require large spaces and extensive areas to establish plant nurseries and do not guarantee the production of disease-free plants. From this perspective, the in vitro propagation method is one of the best choices for the commercial propagation of olives with a high survival rate throughout the season [3]. In addition, the multiplication of olives using tissue culture is not an easy task, owing to some challenges such as explants oxidation, high phenolic contents, the laborious establishment of shoot culture, cultivars dependency, and difficulties in disinfection [4]. Although in the past various shoot induction protocols of olives were optimized within micropropagation systems, all were cultivar dependent [1,5,6]. Successful in vitro propagation of olive is highly associated with type and culture medium composition. For example, Revilla et al. [7] successfully propagated olive using Kuniyuki walnut medium supplemented with 6-benzylaminopurine (BAP) as a growth regulator while sucrose was a carbohydrate source. Likewise, Piexe et al. [1] used BAP and coconut water for successful shoot induction in olive. Correspondingly, Chaari-Rkhis et al. [6] found olive medium (OM) [8] with supplementation of growth regulators is more effective in boosting the shooting and proliferation of different olive cultivars. In the same way, Hamooh and Shah [9] found zeatin as an active hormonal supplement in enhancing the shooting traits of different olive cultivars. Furthermore, Ali et al. [5] explicated that individual and combined application of BAP within in vitro olive medium significantly improves the growth-related traits of olive explants. Plant growth regulators such as BAP have a tendency to initiate intracellular processes that substantially improve related physiological and biochemical activities within the plant system [10]. Moreover, organic substances such as humic acids play an important role in plant tissue culture and micropropagation since they can act as growth by reducing the mutagenic effects of various chemical compounds [11–14]. Hence, a culture medium supplemented with proper hormonal concentrations is more effective in inducing growth and differentiation within in vitro culture system. The growth regulators such as BAP not only accelerate the differentiation processes of culture but also regulate the post-differentiation events [15]. Plantlets established within a nutrient medium containing a proper concentration of growth regulators show robust physiological and metabolic activities that consequently result in their early establishment within in vitro culture [16].

Hitherto, very limited studies have been conducted to elucidate the impacts of in vitro hormonal treatments on the dynamic association of physiological and metabolic traits with morphological traits of plant cultures [17]. Therefore, the effective role of hormones always remained a potential consideration while optimizing a protocol for any micropropagation system [10]. Apart from this, the choice of explants for olive propagation is also important for the successful establishment of culture [6]. Although in previous studies successful cultures of olive were established on OM using axillary buds as explants source, observed shoot induction percentage was comparatively low in explants that were not provided by proper pre-culturing treatments [6,18,19]. In addition, perennial plants including olives protect their delicate reproductive meristematic tissues in specialized structures known as buds, whose growth is strictly regulated by dormancy mechanisms imposing physiological constraints on their growth until optimum conditions return [20]. Dormancy is released in buds or primrodia by prolonged intervals of chilling. Moreover, axillary buds become dormant due to various endogenous factors such as the accumulation of ABA [21]. Infact, ABA masks intercellular communication during dormancy due to enhanced activity of callose synthase causing a high accumulation of callose that results in blockage of plasmodesmata [22]. Vernalin has the tendency to dissolve callose content that restores cellular communication and breaks bud dormancy by reciprocating the effect of ABA [23]. Moreover, chilling treatments break bud dormancy due to the increase in the production of vernalin, as reported by Leida et al. [19] in peach plants. In this context, the pre-cooling treatment is effective in breaking the dormancy owing to its tendency to counter the effect of ABA by stimulating the activity of vernalin [24]. Moreover, nodal segments are considered better explants than axillary buds due to their comparatively high resistance against oxidation due to the deposition of a high content of phenolic compounds [3]. The oxidation of phenolics is harmful in plant tissue culture as it leads to the browning of explants and medium that stops cell division and finally results in the failure of tissue culture [25]. Oxidative browning of explants can be prevented by controlling the leaching of phenolic compounds from plant tissue by providing them with some pre-treatments such as cooling [26]. Although high phenolic contents are a great obstacle in devising successful shooting protocols for woody plants, their effect can be nullified by pre-culturing treatments [27]. In this perspective, nodal explants can get rid of seasonal dormancy by properly exposing them to the proper durations of pre-cooling. To date, no comprehensive study has been conducted for the optimization of olive in vitro propagation protocol at a physiological, molecular, and biochemical level using different pre-cooling treatments for explants and varying concentrations of BAP. In this regard, the current study was conducted to optimize shoot induction protocol for olives using four different cultivars ("Leccino", "Gemlik", "Moraiolo" and "Arbosana"), by culturing their pre-cooled explants on OM using different concentrations of BAP. Furthermore, the effects of these treatments were assessed on the physiological, biochemical, and growth traits of plantlets grown within the micropropagation system.

2. Materials and Methods

In the current study, four different cultivars of olive such as "Leccino" (Italy), "Gemlik" (Turkey), "Moraiolo" (Italy), and "Arbosana" (Spain) were evaluated for the optimization of shoot induction protocol in a three replicate experiment using three factorial arrangements in randomized complete block design (RCBD) at plant tissue culture lab of Jilin Agricultural University, China.

2.1. Explant Disinfection, Media Preparation, and Treatments

The nodal segments obtained from soft, healthy, and lateral one-year-old branches of olive cultivars were used as explants. After detaching the leaves, branches were divided into 10 cm long pieces that were thoroughly washed for 30 min under running tap water. Subsequently, branch segments were cut into nodal segments of size 1–1.5 cm as explicated Chaari-Rkhis et al. [6]. The nodal segments were surface sterilized by treating them with 70% ethanol for 1 to 2 min. Afterward, they were dipped in 15% sodium hypochlorite solution for 10 min. Finally, the nodal segments were rinsed four times using sterilized distilled water and each step was conducted for 5 min. All sterilization steps were carried out under laminar air flow hood by sustaining sterilized atmosphere. In addition, Rugini [8] olive medium (OM) (PhytoTech labs, Lenexa, KS, USA) with optimized pH (5.75 $\pm$ 0.5) was prepared and supplemented with different concentrations (0, 0.5, 1.5, and 2.5 mgL^{-1}) of 6-benzylaminopurine (BAP) (PhytoTech labs, USA) by following manufacturer's instruction within 300 mL glass bottle. Before adding BAP, media was autoclaved as per method of Chaar-Rkhis et al. [6] at 121 °C under 15 psi pressure for 15 min. Afterward, media was cooled to 50 °C and filter sterilized BAP solution was added. Before culturing, explants were wrapped in aluminum foil were subjected to pre-cooling treatments of 0, 24, and 48 h at 4 to 8 °C in a chiller. Afterward, 3 to 4 explants (nodal segments) from each cultivar were cultured on Rugini [8] olive medium (OM) (PhytoTech labs, USA) in glass bottles under laminar flow chamber. For each treatment, three bottles were used.

2.2. In Vitro Propagation and Data Collection

The cultured bottles were kept in growth chamber under controlled conditions. The cultures were provided with 2500 lux light intensity, 25 °C temperature, and 16 h photoperiod. The data for growth, physiological, and biochemical parameters were collected from 50 to 60 days old plantlets of olives. For this purpose, we followed destructive sampling. Furthermore, cultures facing browning were immediately shifted within fresh medium.

2.3. Measurement of Growth Traits

The morphological parameters were evaluated on the basis of percentage of the induced shoots (PIS), length of primary shoot (cm) (LPS), number of leaves per shoot (NLPS), and number of shoots per explant (NSPE). The data for PIS were recorded for each cultivar by applying percentage formula while the data obtained for LPS and NLPS were estimated on average basis from randomly selected five primary shoots from each treatment before subjecting to statistical analysis. In the same way, data were recorded for NSPE.

2.4. Assessment of Physiological Parameters

Among physiological parameters such as total chlorophyll, carotenoids, and proline were estimated following the methods used by Mahmood et al. [28]. In addition, proline content was estimated based on reactivity with ninhydrin using UV–Vis spectrophotometer (DeNovix, Wilmington, DE, USA). On the other hand, assimilation rate of CO_2 (A_{CO2}) was measured using specific apparatus IRGA (ADC BioScientific, Hoddesdon, UK).

2.5. Assessment ofBiochemical Contents

2.5.1. Quantification of Primary Metabolites

Leaf carbohydrates were quantified using the method of Boussadia et al. [29]. The carbohydrate contents such as fructose, glucose, and sucrose were extracted using ethanol reagent and extract was subjected to centrifugation at 5000 rpm for 10 min. Afterward, metabolites were quantified using high pH anion-exchange chromatography. On the other hand, starch content was estimated using acid hydrolysis method used by Chow and Landhausser [30]. For this purpose, dried 50 mg leaf samples were extracted using 80% hot ethanol that subsequently followed enzymatic digestion. Afterward, quantification was performed at 525 after adding H_2SO_4.

2.5.2. Quantification of Secondary Metabolites

Among secondary metabolites, total phenols were estimated following the method used by Siddiqui et al. [31]. For this purpose, Folin–Ciocalteu reagent was used, with gallic acid as the standard. In addition, total flavonoids were calculated following the procedure by Zhao et al. [32] and quantified using colorimetric assay method, with rutin as standard. Total tannins content was determined by following the protocol of Fadda and Mulas [33]. For this purpose, 2 mL of extracted sample was treated with 10 mL vanillin hydrochloride and for quantification; absorbance was recorded at 500 nm. Total alkaloid contents were determined by following the procedure optimized by Li et al. [34]. In this regard, the prepared standard solution of leaf extract was quantified using UV–Vis spectrophotometer (DeNovix, USA), and absorbance was recorded as 418 nm.

2.6. Statistical Analysis

A tri-replicate experiment was conducted in randomized complete block design (RCBD) using three factorial arrangements with pre-cooling as one factor, varieties as second factor, and BAP concentrations as third factor. The data collected were evaluated statistically by applying analysis of variance (ANOVA) at a 5% probability level, with the help of computer-based software Statistix ver. 8.1 (McGraw-Hill 2008). Furthermore, correlation and principal component analysis (PCA) were conducted with the help of computer-based statistical tool RStudio version 1.3.959 (RStudio Team 2020) using the PerformaceAnalytics, FactoMineR, factoextra, devtools, ggplot2, ggpubr, gplots, and pheatmap packages of R version 4.1.0 (R Core Team 2021).

3. Results

3.1. Growth Traits

All individual factors of treatments significantly ($p \leq 0.01$) affected the mean values of growth-related traits such as LPS, PIS, NLPS, and NSPE in all olive cultivars (Table 1). All growth traits showed statistically significant ($p \leq 0.01$) increase with changing regimes of pre-cooling and increasing concentrations of BAP (Table 1). Moreover, among cultivars "Arbosana" (V4) plantlets recorded a maximum improvement in the aforementioned growth traits followed by "Moraiolo" (V3), "Leccino" (V2), and "Gemlik" (V1) as indicated in Table 1. In addition, V4 revealed a maximum mean value for LPS (3 cm) that was significantly ($p \leq 0.01$) different from the means of other varieties. Correspondingly, V4 exhibited significantly ($p \leq 0.01$) high PIS (85%) as compared to V3 (75%), V2 (70%), and V1 (60%) as shown in Table 1. Similarly, under the same conditions, V4 depicted statistically ($p \leq 0.01$) distinct improvement in NLPS (5) and NSPE (4). In addition, among two-way interactions, the interaction of BAP with pre-cooling

($T \times L_B$) and cultivars ($V \times L_B$) illustrated statistically distinct ($p \leq 0.05$) change in all growth traits (Tables 5 and 6). At all regimes of pre-cooling, BAP manifested maximum improvements in growth-related traits of olive cultivars at concentration 2.5 mgL^{-1} (L4) (Table 6); however, this improvement was more dramatic at 48 h (T3) interval of pre-cooling (Table 5). Overall, the performance of explants from all cultivars was notably different under varying concentrations of BAP and changing regimes of pre-cooling. Likewise, the two-way interaction $T \times V$, and the three-way interaction $T \times V \times L_B$ showed no significant effect on the growth traits of cultivars.

Table 1. Effect of different pre-cooling treatments and varying concentrations of BAP on growth traits of different olive (*Olea euorpea* L.) genotypes cultured within micropropagation system using OM.

Treatments	LPS (cm)	PIS	NLPS	NSPE
		Pre-cooling (T) (Hours)		
T1 (0 h)	1.97 ± 0.11 ^{b}	70.02 ± 2.00 ^{c}	3.71 ± 0.15 ^{c}	2.045 ± 0.10 ^{c}
T2 (24 h)	2.27 ± 0.13 ^{ab}	78.05 ± 1.87 ^{b}	3.93 ± 0.17 ^{b}	2.56 ± 0.13 ^{b}
T3 (48 h)	2.55 ± 0.15 ^{a}	83.06 ± 1.62 ^{a}	4.28 ± 0.20 ^{a}	4.00 ± 0.19 ^{a}
LSD	0.35	1.96	0.13	0.14
		Varieties (V)		
V_1 (Leccino)	1.98 ± 0.15 ^{c}	65 ± 2.23 ^{d}	3.58 ± 0.18 ^{d}	2.05 ± 0.13 ^{d}
V_2 (Gemlik)	2.26 ± 0.14 ^{b}	70 ± 2.49 ^{c}	3.88 ± 0.19 ^{c}	2.43 ± 0.18 ^{c}
V_3 (Morailo)	2.34 ± 0.15 ^{b}	75 ± 2.74 ^{b}	4.09 ± 0.20 ^{b}	3.60 ± 0.20 ^{b}
V_4 (Arbosana)	3.00 ± 0.17 ^{a}	85 ± 3.13 ^{a}	5.00 ± 0.23 ^{a}	4.00 ± 0.21 ^{a}
LSD	0.10	2.70	0.08	0.08
		BAP (L_B) (mgL^{-1})		
L1 (0)	1.05 ± 0.034 ^{d}	68 ± 0.073 ^{d}	2.48 ± 0.045 ^{d}	2.5 ± 0.021 ^{d}
L2(0.5)	2.03 ± 0.083 ^{c}	75 ± 0.80 ^{c}	3.40 ± 0.09 ^{c}	3 ± 0.093 ^{c}
L3 (1.5)	2.32 ± 0.088 ^{b}	79 ± 1.93 ^{b}	4.18 ± 0.10 ^{b}	3.25 ± 0.10 ^{b}
L4(2.5)	2.55 ± 0.083 ^{a}	84 ± 1.87 ^{a}	4.58 ± 0.98 ^{a}	4.01 ± 0.14 ^{a}
LSD	0.08	1.30	0.11	0.08
Significance				
T	*	*	**	*
V	**	**	**	**
L_B	**	**	**	**
$T \times V$	ns	ns	ns	ns
$T \times L_B$	*	*	*	*
$V \times L_B$	*	*	*	*
$T \times V \times L_B$	ns	ns	ns	ns

Means with same letter (s) in each column indicate no significant difference in traits due to treatments. *, ** represent significant differences at $p \leq 0.05$ and $p \leq 0.01$, respectively, while "ns" represents non-significant difference (LPS, length of primary shoot; PIS, percentage of induced shoots; NLPS, number of leaves per shoot; NSPE, number of shoots per explant).

3.2. Physiological Parameters

All traits except proline depicted a significant increase with increasing regimes of pre-cooling and BAP concentrations (Table 2). In addition, among cultivars "Arbosana" (V4) illustrated a maximum increase in all physiological traits followed by "Moraiolo" (V3), "Gemlik" (V2) and "Leccino" (V1). Furthermore, V4 exhibited maximum chlorophyll content (50 µg cm^{-2}) that was slightly higher than the mean chlorophyll contents of other varieties (Table 2). Correspondingly, V4 recorded the highest (8 µmol m^{-2}s^{-1}) while V1 recorded the lowest mean value (7 µmol m^{-2}s^{-1}) for A_{CO2} at 2.5 mgL^{-1} (L4) concentration of BAP in plantlets whose explants were treated with 48 h interval of pre-cooling. In addition, under analogous conditions, V4 depicted a statistically significant ($p \leq 0.01$) increase in carotenoids (5.29 µg cm^{-2}) and a decline in proline (38.7 µg g^{-1} FW) contents. Among physiological traits, chlorophyll and carotenoids showed significant ($p \leq 0.05$) variation due to the two-way interaction of BAP with pre-cooling treatments ($T \times L_B$)

and cultivars (V × L_B) (Tables 5 and 6). All cultivars pre-treated with a 48 h regime of cooling illustrated maximum chlorophyll and carotenoid contents at a BAP concentration of 2.5 mgL^{-1} (Tables 5 and 6). In general, explants from all cultivars pre-treated with different cooling intervals responded differently under provided in vitro conditions due to varying concentrations of BAP. All traits showed no significant variation due to two-way interaction T × V and three-way interaction T × V × L_B.

Table 2. Effect of different pre-cooling treatments and varying concentrations of BAP on physiological parameters of different olive (*Olea euorpea* L.) genotypes cultured within micropropagation system using OM.

Treatments	Total Chlorophyll ($\mu g\ cm^{-2}$)	Carotenoids ($\mu g\ cm^{-2}$)	A_{CO2} ($\mu mol\ m^{-2} s^{-1}$)	Proline ($\mu g\ g^{-1}$ FW)
		Pre-cooling (T) (Hours)		
T1 (0 h)	41.83 ± 2.52 [b]	4.53 ± 0.20 [c]	6.93 ± 0.37 [c]	42.00 ± 2.10 [a]
T2 (24 h)	47.96 ± 3.07 [a]	5.05 ± 0.25 [b]	7.47 ± 0.39 [b]	39.70 ± 1.76 [b]
T3 (48 h)	51.55 ± 3.45 [a]	5.52 ± 0.32 [a]	8.09 ± 0.42 [a]	38.87 ± 1.76 [b]
LSD	5.12	0.38	0.32	1.91
		Varieties (V)		
V_1 (Leccino)	42 ± 2.00 [c]	4.81 ± 0.28 [c]	6.59 ± 0.40 [c]	41.32 ± 2.35 [a]
V_2 (Gemlik)	46.57 ± 3.37 [b]	4.93 ± 0.35 [a]	7.41 ± 0.44 [b]	39.74 ± 2.29 [b]
V_3 (Morailo)	48.53 ± 3.76 [ab]	5.09 ± 0.30 [b]	7.76 ± 0.48 [ab]	41.00 ± 2.14 [a]
V_4 (Arbosana)	50.37 ± 4.11 [a]	5.29 ± 0.28 [c]	7.93 ± 0.50 [a]	38.70 ± 2.08 [c]
LSD	2.74	0.11	0.36	0.59
BAP (L_B) (mgL^{-1})				
L1 (0)	19.70 ± 0.26 [d]	2.71 ± 0.050 [d]	3.82 ± 0.10 [d]	48.66 ± 0.63 [a]
L2(0.5)	27.62 ± 0.70 [c]	4.17 ± 0.12 [c]	5.82 ± 0.12 [c]	46.62 ± 0.62 [b]
L3 (1.5)	32.53 ± 0.83 [b]	4.84 ± 0.11 [b]	6.40 ± 0.12 [b]	43.37 ± 0.61 [c]
L4(2.5)	37.68 ± 0.77 [a]	5.44 ± 0.13 [a]	6.95 ± 0.12 [a]	23.41 ± 0.20 [d]
LSD	0.60	0.14	0.17	1.24
Significance				
T	*	*	**	*
V	**	**	*	**
L	**	**	**	**
T × V	ns	ns	ns	ns
T × L_B	*	*	ns	*
V × L_B	*	*	ns	ns
T × V × L_B	ns	ns	ns	ns

Means with same letter (s) in each column indicate no significant difference in traits due to treatments. *, ** represent significant differences at $p \leq 0.05$ and $p \leq 0.01$, respectively, while "ns" represents non-significant difference.

3.3. *Biochemical Contents*

3.3.1. Primary Metabolites

All primary metabolic contents including fructose, glucose, sucrose, and starch varied significantly ($p \leq 0.01$, $p \leq 0.05$) due to the individual effect of olive cultivars' pre-cooling regimes and varying concentrations of BAP within OM (Table 3). All metabolites depicted statistically significant ($p \leq 0.01$) increase at the pre-cooling treatment of 48 h (T3) and BAP concentration of 2.5 mgL^{-1} (L4) as shown in Table 3. On the other hand, among cultivars, "Arbosana" (V4) revealed the highest rise in all primary metabolites followed by "Moraiolo" (V3), "Gemlik" (V2), and "Leccino" (V1). Moreover, V4 recorded maximum fructose (0.04 mg g^{-1} DW), sucrose (0.5 mg g^{-1} DW), and starch (0.7 mg g^{-1} DW) contents as compared to other cultivars. In the same way, V3 demonstrated the maximum (1.42 mg g^{-1} DW) and V1 recorded the minimum (1 mg g^{-1} DW) increase in the amount of glucose. In addition, among two-way interactions T × L_B significantly ($p \leq 0.05$) altered the mean values of glucose, sucrose, and fructose contents while V × L_B significantly ($p \leq 0.05$) altered the mean values of fructose and sucrose content (Tables 5 and 6). All pre-cooled olive cultivars showed a maximum rise in the quantities of primary metabolites when OM augmented with BAP at 2.5 mgL^{-1} (L4), however, this incline was more dramatic in all cultivars that followed the pre-cooling treatment of 48 h (T3) as concluded by Tables 5 and 6. As a whole, explants from all cultivars pre-treated with different cooling regimes behaved differently under provided in vitro conditions due to varying levels of BAP. On the other hand, all primary metabolites showed no significant variation due to two-way interaction T × V and three-way interaction p× V × L_B.

Table 3. Effect of different pre-cooling treatments and varying concentrations of BAP on primary metabolites of different olive genotypes cultured within micropropagation system using OM.

Treatments	Starch mg g^{-1} DW	Glucose mg g^{-1} DW	Fructose (mg g^{-1} DW)	Sucrose (mg g^{-1} DW)
Pre-cooling (T) (Hours)				
T1 (0 h)	0.33 ± 0.05 ^{b}	1.04 ± 0.10 ^{c}	0.029 ± 0.002 ^{c}	0.29 ± 0.03 ^{c}
T2 (24 h)	0.53 ± 0.06 ^{ab}	1.19 ± 0.10 ^{b}	0.032 ± 0.003 ^{b}	0.41 ± 0.04 ^{b}
T3 (48 h)	0.62 ± 0.06 ^{a}	1.39 ± 0.12 ^{a}	0.035 ± 0.004 ^{a}	0.51 ± 0.05 ^{a}
LSD	0.21	0.08	0.002	0.07
Varieties (V)				
V_1 (Leccino)	0.42 ± 0.06 ^{b}	1.07 ± 0.11 ^{c}	0.024 ± 0.003 ^{b}	0.32 ± 0.04 ^{d}
V_2 (Gemlik)	0.41 ± 0.06 ^{b}	1.13 ± 0.11 ^{bc}	0.027 ± 0.004 ^{b}	0.38 ± 0.05 ^{c}
V_3 (Morailo)	0.47 ± 0.07 ^{b}	1.42 ± 0.11 ^{a}	0.028 ± 0.004 ^{b}	0.42 ± 0.05 ^{b}
V_4 (Arbosana)	0.68 ± 0.08 ^{a}	1.20 ± 0.15 ^{b}	0.039 ± 0.005 ^{a}	0.48 ± 0.06 ^{a}
LSD	0.11	0.08	0.08	0.03
BAP (L_B) (mgL^{-1})				
L1 (0)	0.09 ± 0.01 ^{d}	0.21 ± 0.02 ^{d}	0.007 ± 0.0003 ^{d}	0.05 ± 0.004 ^{d}
L2 (0.5)	0.19 ± 0.01 ^{c}	0.55 ± 0.01 ^{c}	0.020 ± 0.0014 ^{c}	0.19 ± 0.014 ^{c}
L3 (1.5)	0.32 ± 0.02 ^{a}	0.67 ± 0.02 ^{b}	0.031 ± 0.0020 ^{b}	0.26 ± 0.009 ^{b}
L4 (2.5)	0.26 ± 0.01 ^{b}	0.83 ± 0.02 ^{a}	0.040 ± 0.0024 ^{a}	0.31 ± 0.010 ^{a}
LSD	0.02	0.02	0.002	0.006
Significance				
T	**	*	**	*
V	**	**	**	*
L_B	**	**	**	**
T × V	ns	ns	ns	ns
T × L_B	ns	*	*	*
V × L_B	ns	ns	*	*
T × V × L_B	ns	ns	ns	ns

Means with same letter (s) in each column indicate no significant difference in traits due to treatments. *, ** represent significant differences at $p \leq 0.05$ and $p \leq 0.01$, respectively, while "ns" represents non-significant difference.

3.3.2. Secondary Metabolites

Among individual treatment factors, pre-cooling regimes significantly ($p \leq 0.05$, $p \leq 0.01$) increased alkaloids and phenolic contents (Table 4). On the other hand, varying BAP concentrations significantly ($p \leq 0.05$) impacted the amount of secondary metabolic contents including alkaloids, phenols, tannins, and flavonoids (Table 4). Moreover, among cultivars "Arbosana" (V4) manifested a more dramatic increase in the concentrations of alkaloids, phenols, and flavonoids followed by "Moraiolo" (V3), "Gemlik" (V2) and "Leccino" (V1). In addition, V4 illustrated the maximum alkaloids (2.5 mg g^{-1} DW) that were a bit higher than the alkaloid content of other cultivars (Table 4). Following the analogous trend, V4 revealed a more dramatic rise in flavonoid (1.54 mg g^{-1} DW) and phenol content (1.5 mg g^{-1} DW) unlike other cultivars (Table 4). Among two-way interactions T × L_B and V × L_B significantly altered the mean values of alkaloids and flavonoids (Tables 5 and 6). The BAP made a statistically distinct ($p \leq 0.05$) rise in alkaloids and flavonoid contents of all olive cultivars at a concentration of 2.5 mgL^{-1} (L4) under all pre-cooling treatments; however, this rise was at the maximum at 48 h (Tables 5 and 6). Generally, plantlets from all cultivars whose explants followed different pre-cooling treatments before culturing showed variable responses within in vitro conditions on OM augmented with different concentrations of BAP. In addition, two-way interaction T × V and three-way interaction T × V × L_B showed no significant effect on the concentration of secondary metabolites.

Table 4. Effect of different pre-cooling treatments and varying concentrations of BAP on secondary metabolites of different olive genotypes cultured within micropropagation system using OM.

Treatments	Alkaloids mg (g DW)$^{-1}$	Flavonoids mg (g DW)$^{-1}$	Tannins mg (g DW)$^{-1}$	Phenols mg (g DW)$^{-1}$
		Pre-cooling (T) (Hours)		
T1 (0 h)	1.52 ± 0.074 ^{c}	1.35 ± 0.067 ^{b}	0.63 ± 0.015 ^{a}	0.70 ± 0.025 ^{a}
T2(24 h)	1.75 ± 0.077 ^{b}	1.44 ± 0.077 ^{ab}	0.65 ± 0.014 ^{a}	0.85 ± 0.026 ^{a}
T3 (48 h)	2.00 ± 0.09 ^{a}	1.60 ± 0.082 ^{a}	0.72 ± 0.015 ^{a}	0.93 ± 0.026 ^{a}
LSD	0.18	0.15	0.17	0.14
		Varieties (V)		
V_1 (Leccino)	1.5 ± 0.093 ^{d}	1.39 ± 0.084 ^{c}	1.45 ± 0.084 ^{a}	1.00 ± 0.049 ^{d}
V_2 (Gemlik)	1.70 ± 0.10 ^{c}	1.42 ± 0.080 ^{bc}	1.43 ± 0.080 ^{a}	1.03 ± 0.055 ^{bc}
V_3 (Morailo)	1.84 ± 0.10 ^{b}	1.50 ± 0.095 ^{ab}	1.49 ± 0.095 ^{a}	1.08 ± 0.059 ^{ab}
V_4 (Arbosana)	2.5 ± 0.01 ^{a}	1.54 ± 0.098 ^{a}	1.50 ± 0.098 ^{a}	1.5 ± 0.064 ^{a}
LSD	0.045	0.09	0.1	0.06
BAP (L_B) (mgL^{-1})				
L1 (0)	1.02 ± 0.031 ^{d}	0.84 ± 0.035 ^{d}	0.68 ± 0.017 ^{d}	0.73 ± 0.026 ^{d}
L2 (0.5)	1.64 ± 0.054 ^{b}	1.20 ± 0.049 ^{c}	0.80 ± 0.027 ^{c}	1.04 ± 0.035 ^{c}
L3 (1.5)	1.48 ± 0.054 ^{c}	1.42 ± 0.054 ^{b}	1.01 ± 0.032 ^{b}	1.27 ± 0.031 ^{a}
L4(2.5)	1.85 ± 0.062 ^{a}	1.62 ± 0.068 ^{a}	1.11 ± 0.027 ^{a}	1.16 ± 0.033 ^{b}
LSD	0.023	0.040	0.055	0.028
Significance				
T	*	**	ns	ns
V	**	**	ns	*
L_B	**	**	**	**
T × V	ns	ns	ns	ns
T × L_B	*	*	ns	*
V × L_B	*	*	ns	ns
T × V × L_B	ns	ns	ns	ns

Means with same letter (s) in each column indicate no significant difference in traits due to treatments. *, ** represent significant differences at $p \leq 0.05$ and $p \leq 0.01$, respectively, while "ns" represents non-significant difference.

Table 5. Traits showing statistically significant ($p \leq 0.05$) difference due to interaction effect of pre-cooling regimes and BAP concentrations ($T \times L_B$).

PCT (T) (hours)	BAP (L_B) (mgL^{-1})	LPS (cm)	PIS	NLPS	NSPE	Chlorophyll ($\mu g\ cm^{-2}$)	Carotenoids ($\mu g\ cm^{-2}$)	Glucose $mg\ g^{-1}$ DW	Fructose $mg\ g^{-1}$ DW	Sucrose $mg\ g^{-1}$ DW	Proline $\mu g\ g^{-1}$ FW	Alkaloids $mg\ g^{-1}$ DW	Flavanoids $mg\ g^{-1}$ DW	Phenols $mg\ g^{-1}$ DW
T1 (0 h)	L1 (0)	1.00 ± 0.08	36.7 ± 1.41	2.45 ± 0.08	1.06 ± 0.03	18.5 ± 0.37	2.62 ± 0.07	0.16 ± 0.01	0.005 ± 0.0007	0.02 ± 0.001	46.7 ± 0.3	0.85 ± 0.03	0.75 ± 0.03	0.69 ± 0.02
	L2 (0.5)	1.69 ± 0.12	46.0 ± 2.32	3.15 ± 0.17	1.74 ± 0.10	23.8 ± 0.39	3.60 ± 0.11	0.47 ± 0.009	0.012 ± 0.001	0.10 ± 0.007	44.7 ± 0.44	1.22 ± 0.052	0.94 ± 0.05	0.85 ± 0.034
	L3 (1.5)	1.94 ± 0.01	52.4 ± 2.94	3.79 ± 0.14	2.12 ± 0.11	27.5 ± 0.41	4.22 ± 0.9	0.58 ± 0.010	0.020 ± 0.002	0.21 ± 0.003	42.0 ± 0.90	1.420 ± 0.058	1.13 ± 0.05	1.001 ± 0.050
	L4 (2.5)	2.19 ± 0.07	57.0 ± 2.77	4.22 ± 0.15	2.40 ± 0.12	33.5 ± 0.52	4.81 ± 0.14	0.74 ± 0.011	0.028 ± 0.002	0.25 ± 0.004	22.8 ± 0.37	1.58 ± 0.06	1.28 ± 0.06	1.11 ± 0.04
T2 (24 h)	L1 (0)	1.04 ± 0.02	42.0 ± 0.73	2.49 ± 0.05	1.21 ± 0.02	19.5 ± 0.31	2.78 ± 0.06	0.22 ± 0.005	0.008 ± 0.0003	0.05 ± 0.008	48.5 ± 0.52	1.08 ± 0.03	1.001 ± 0.04	0.87 ± 0.034
	L2 (0.5)	2.04 ± 0.06	55.0 ± 1.88	3.25 ± 0.08	2.23 ± 0.12	27.4 ± 0.30	4.03 ± 0.10	0.56 ± 0.005	0.020 ± 0.001	0.21 ± 0.011	45.4 ± 0.72	1.44 ± 0.04	1.20 ± 0.03	1.07 ± 0.04
	L3 (1.5)	2.29 ± 0.01	61.2 ± 1.88	4.25 ± 0.16	2.57 ± 0.12	33.3 ± 0.55	4.94 ± 0.12	0.68 ± 0.006	0.034 ± 0.001	0.26 ± 0.01	42.8 ± 0.48	1.58 ± 0.05	1.40 ± 0.05	1.22 ± 0.03
	L4 (2.5)	2.57 ± 0.07	65.3 ± 1.54	4.70 ± 0.16	2.98 ± 0.12	37.4 ± 0.67	5.40 ± 0.9	0.85 ± 0.001	0.043 ± 0.0021	0.31 ± 0.010	24.7 ± 0.31	1.77 ± 0.05	1.61 ± 0.07	1.31 ± 0.03
T3 (48 h)	L1 (0)	1.13 ± 0.03	41.3 ± 0.59	2.50 ± 0.01	1.20 ± 0.01	20.8 ± 0.29	2.72 ± 0.10	0.23 ± 0.008	0.007 ± 0.0004	0.07 ± 0.005	50.5 ± 0.51	1.12 ± 0.03	1.04 ± 0.03	0.93 ± 0.025
	L2 (0.5)	2.37 ± 0.11	62.0 ± 2.15	3.79 ± 0.16	2.33 ± 0.17	31.4 ± 0.72	4.87 ± 0.13	0.63 ± 0.009	0.025 ± 0.001	0.27 ± 0.007	49.5 ± 0.74	1.77 ± 0.04	1.46 ± 0.03	1.20 ± 0.02
	L3 (1.5)	2.67 ± 0.13	68.4 ± 2.50	4.53 ± 0.12	2.93 ± 0.20	36.5 ± 0.67	5.37 ± 0.13	0.75 ± 0.009	0.037 ± 0.002	0.32 ± 0.008	45.0 ± 0.84	1.94 ± 0.04	1.69 ± 0.02	1.28 ± 0.01
	L4 (2.5)	2.89 ± 0.14	72.8 ± 2.58	4.83 ± 0.11	3.57 ± 0.28	41.8 ± 0.63	6.12 ± 0.17	0.91 ± 0.011	0.050 ± 0.002	0.36 ± 0.011	22.5 ± 0.39	2.20 ± 0.03	1.97 ± 0.04	1.39 ± 0.01
	LSD	0.1388	2.2443	0.1944	0.1422	1.0253	0.2206	0.0289	0.0028	0.0102	2.1257	0.0421	0.0700	0.0500

* Only traits showing significant difference are indicated in table, while means with difference greater than LSD are significantly different at $p \leq 0.05$. (LPS, length of primary shoot; PIS, percentage of induced shoots; NLPS, number of leaves per shoot; NSPE, number of shoots per explant).

Table 6. Traits showing statistically significant ($p \leq 0.05$) difference due to interaction effect of varieties and BAP concentrations ($V \times L_B$).

Varieties (V)	BAP (L_B) (mgL^{-1})	LPS	PIS	NLPS	NSPE	Chlorophyll ($\mu g\ cm^{-2}$)	Carotenoids ($\mu g\ cm^{-2}$)	Fructose $mg\ g^{-1}$ DW	Sucrose $mg\ g^{-1}$ DW	Alkaloids $mg\ g^{-1}$ DW	Flavanoids $mg\ g^{-1}$ DW
Leccino (V1)	L1 (0)	0.87 ± 0.07	39.1 ± 2.16	2.27 ± 0.07	1.11 ± 0.04	11.1 ± 0.30	2.64 ± 0.10	0.006 ± 0.0008	0.03 ± 0.008	0.93 ± 0.06	0.81 ± 0.04
	L2 (0.5)	1.67 ± 0.15	47.1 ± 2.95	2.91 ± 0.15	1.64 ± 0.12	18.1 ± 1.18	3.87 ± 0.23	0.016 ± 0.002	0.17 ± 0.02	1.29 ± 0.10	1.06 ± 0.07
	L3 (1.5)	1.92 ± 0.12	53.1 ± 3.08	3.66 ± 0.14	1.94 ± 0.01	22.4 ± 1.37	4.62 ± 0.1	0.026 ± 0.003	0.24 ± 0.01	1.48 ± 0.10	1.24 ± 0.07
	L4 (2.5)	2.21 ± 0.10	57.2 ± 3.38	4.07 ± 0.15	2.26 ± 0.11	25.6 ± 1.44	5.10 ± 0.1	0.035 ± 0.005	0.28 ± 0.01	1.65 ± 0.13	1.41 ± 0.07
Gemlik (V2)	L1 (0)	1.9 ± 0.03	39.4 ± 1.27	2.49 ± 0.04	1.17 ± 0.04	11.6 ± 0.20	2.87 ± 0.04	0.006 ± 0.0007	0.05 ± 0.001	0.97 ± 0.05	0.94 ± 0.05
	L2 (0.5)	1.97 ± 0.13	54.6 ± 2.41	3.22 ± 0.08	2.01 ± 0.14	18.7 ± 1.39	3.97 ± 0.25	0.015 ± 0.002	0.19 ± 0.02	1.43 ± 0.08	1.17 ± 0.06
	L3 (1.5)	2.22 ± 0.11	60.2 ± 2.49	4.11 ± 0.18	2.46 ± 0.1	23.1 ± 1.57	4.62 ± 0.24	0.029 ± 0.003	0.26 ± 0.01	1.54 ± 0.09	1.34 ± 0.05
	L4 (2.5)	2.51 ± 0.10	65.0 ± 2.40	4.49 ± 0.14	3.01 ± 0.2	27.0 ± 1.52	5.27 ± 0.23	0.038 ± 0.004	0.29 ± 0.01	1.79 ± 0.13	1.53 ± 0.05
Morailo (V3)	L1 (0)	1.12 ± 0.04	42.4 ± 1.42	2.59 ± 0.03	1.16 ± 0.02	11.2 ± 0.32	2.69 ± 0.07	0.007 ± 0.002	0.05 ± 0.001	1.02 ± 0.03	0.94 ± 0.05
	L2 (0.5)	2.19 ± 0.12	56.1 ± 3.40	3.61 ± 0.16	2.24 ± 0.11	19.6 ± 1.53	4.34 ± 0.20	0.020 ± 0.003	0.20 ± 0.02	1.58 ± 0.09	1.23 ± 0.04
	L3 (1.5)	2.42 ± 0.16	62.4 ± 3.61	4.32 ± 0.11	2.76 ± 0.16	24.1 ± 1.91	4.80 ± 0.17	0.017 ± 0.004	0.27 ± 0.01	1.73 ± 0.07	1.47 ± 0.02
	L4 (2.5)	2.62 ± 0.16	67.0 ± 2.71	4.81 ± 0.14	3.23 ± 0.27	28.2 ± 1.73	5.70 ± 0.25	0.040 ± 0.003	0.32 ± 0.02	1.92 ± 009	1.71 ± 0.03
Arbosana (V4)	L1 (0)	1.12 ± 0.04	41.2 ± 0.94	2.57 ± 0.11	1.19 ± 0.03	11.6 ± 0.20	2.62 ± 0.13	0.007 ± 0.0006	0.06 ± 0.001	1.13 ± 0.05	1.03 ± 0.03
	L2 (0.5)	2.29 ± 0.14	59.7 ± 4.06	3.86 ± 0.15	2.51 ± 0.15	20.6 ± 1.52	4.50 ± 0.2	0.023 ± 0.002	0.21 ± 0.02	1.60 ± 0.09	1.35 ± 0.07
	L3 (1.5)	2.62 ± 0.16	67.0 ± 4.48	4.67 ± 0.13	3.01 ± 0.17	25.2 ± 1.95	5.24 ± 0.2	0.037 ± 0.002	0.28 ± 0.01	1.82 ± 0.09	1.57 ± 0.06
	L4 (2.5)	2.86 ± 0.17	71.1 ± 4.45	4.96 ± 0.14	3.44 ± 0.26	28.7 ± 2.11	5.61 ± 0.34	0.048 ± 0.004	0.33 ± 0.02	2.04 ± 0.08	1.84 ± 0.05
	LSD	0.1610	2.588	0.2234	0.1644	1.2170	0.2659	0.0045	0.0118	0.0482	0.0809

* Only traits showing significant difference are indicated in table, while means with difference greater than LSD are significantly different at $p \leq 0.05$. (LPS, length of primary shoot; PIS, percentage of induced shoots; NLPS, number of leaves per shoot; NSPE, number of shoots per explant.

3.4. Correlation and Principal Component Analysis (PCA)

Correlation analysis illustrated a significantly high paired association among growth traits, physiological parameters, and biochemical contents of all olive cultivars subjected to different pre-cooling regimes and varying in vitro concentrations of BAP (Table 7). All primary and secondary metabolites revealed a significantly high paired association with physiological traits such as chlorophyll and A_{CO2}. Furthermore, all growth traits of plantlets depicted a strong paired association between themselves, and physiological traits as illustrated in Table 1. On the other hand, principal component analysis (PCA) deciphered varying responses of traits with respect to varying concentrations of BAP, different regimes of pre-cooling, and types of cultivars. The response of all traits under study revealed analogous behavior at varying doses of BAP such as L2 (0.5 mgL^{-1}), L3 (1.5 mgL^{-1}), and L4 (2.5 mgL^{-1}) as compared to control (Figure 1). However, the divergence of trait clusters from graph origin was different at all applied concentrations of BAP which revealed a differential extent of association of traits at each concentration. Overall, this divergence of traits cluster from origin was the highest for 2.5 mgL^{-1} (L4), which was an indicator of a promising increase in the strength of traits association. In addition, the PCA graph for pre-cooling treatments revealed the analogous response of all traits at treatments T2 and T3 as compared to control (T1) as illustrated in Figure 2. Correspondingly, the differential divergence of trait clusters from the origin at each pre-cooling treatment showed the varying extent of association among traits that was maximum at T3. This indicated variable associated response of traits with changing regimes of pre-cooling. Furthermore, the PCA graph for all olive cultivars revealed the complementary response of all traits as explicated by the merged circles of clusters (Figure 3). However, the divergence cultivars with respect to the proximity of traits from origin were different, which indicated the differential response of each cultivar for the same sort of traits. Among cultivars, "Arbosana" (V4) manifested a response of traits in close proximity followed by "Moraiolo" (V3), "Gemlik" (V2), and "Leccino" (V1).

Table 7. Correlation table describing the significance of association between different metabolic, physiological, and growth traits of different cultivars of olives due to different regimes of pre-cooling and BAP levels.

Starch	0.91 ***	0.95 ***	0.94 ***	0.94 ***	0.93 ***	0.92 ***	0.88 ***	0.95 ***	0.94 ***	0.91 ***	0.92 ***	0.94 ***	0.93 ***	0.92 ***	0.94 ***
	Glucose	0.92 ***	0.97 ***	0.97 ***	0.98 ***	0.96 ***	0.97 ***	0.90 ***	0.85 ***	0.88 ***	0.91 ***	0.91 ***	0.94 ***	0.93 ***	0.90 ***
		Fructose	0.93 ***	0.96 ***	0.92 ***	0.95 ***	0.90 ***	0.92 ***	0.92 ***	0.89 ***	0.92 ***	0.93 ***	0.92 ***	0.93 ***	0.94 ***
			Sucrose	0.98 ***	0.98 ***	0.97 ***	0.97 ***	0.95 ***	0.91 ***	0.92 ***	0.95 ***	0.95 ***	0.96 ***	0.93 ***	0.92 ***
				Chlorophyll	0.96 ***	0.98 ***	0.96 ***	0.94 ***	0.92 ***	0.92 ***	0.96 ***	0.95 ***	0.95 ***	0.95 ***	0.93 ***
					A_{CO2}	0.96 ***	0.96 ***	0.93 ***	0.89 ***	0.92 ***	0.93 ***	0.95 ***	0.97 ***	0.94 ***	0.93 ***
						Carotenoids	0.96 ***	0.94 ***	0.91 ***	0.91 ***	0.94 ***	0.95 ***	0.96 ***	0.95 ***	0.94 ***
							Proline	0.90 ***	0.85 ***	0.87 ***	0.91 ***	0.91 ***	0.93 ***	0.91 ***	0.90 ***
								Alkaloids	0.97 ***	0.93 ***	0.95 ***	0.97 ***	0.95 ***	0.92 ***	0.93 ***
									Flavonoids	0.94 ***	0.96 ***	0.94 ***	0.90 ***	0.89 ***	0.91 ***
										Phenols	0.96 ***	0.93 ***	0.90 ***	0.89 ***	0.87 ***
											Tannins	0.95 ***	0.91 ***	0.90 ***	0.89 ***
												PIS	0.97 ***	0.93 ***	0.96 ***
													LPS	0.95 ***	0.96 ***
														NLPS	0.94 ***
															NSPE

*** Indicates significant extent of paired association among traits at $p \leq 0.001$.

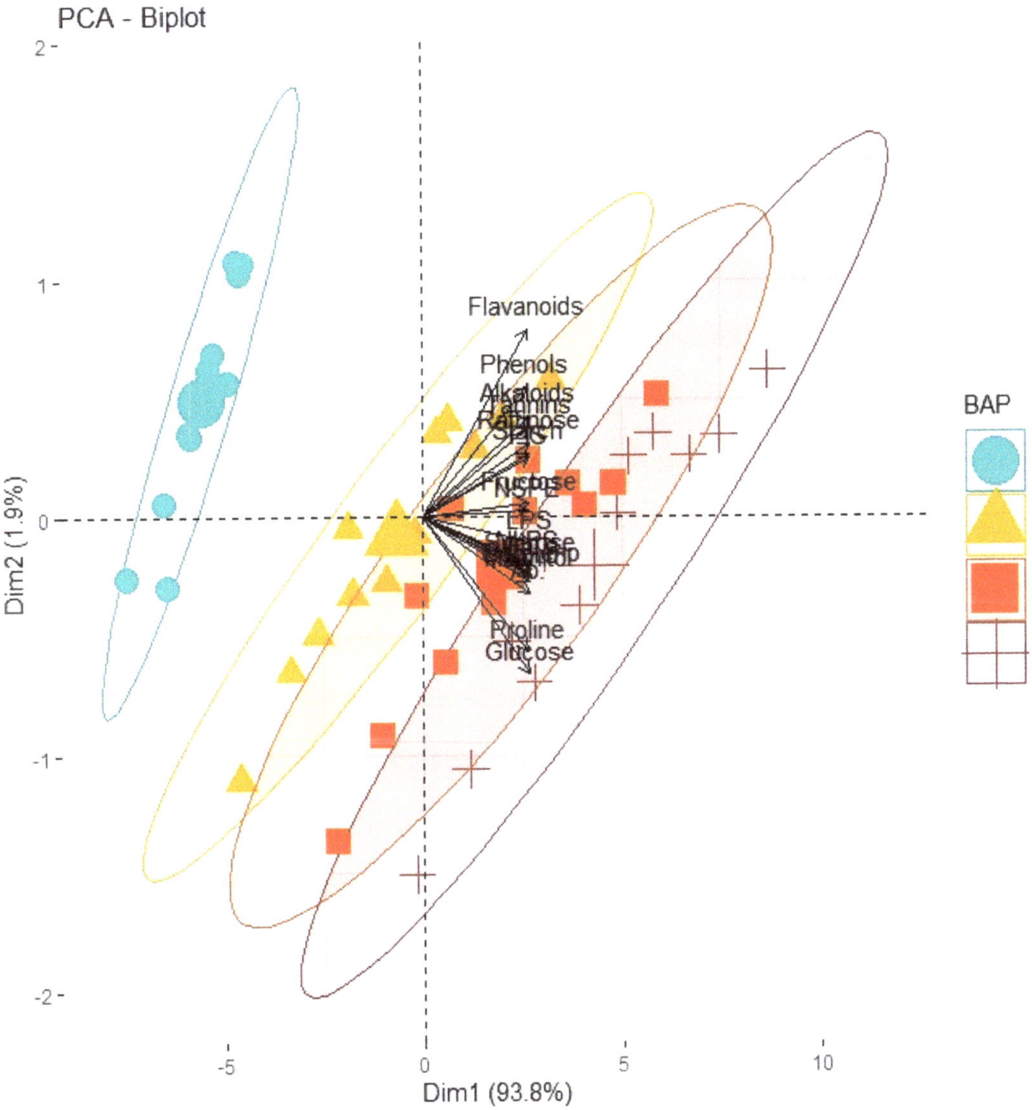

Figure 1. PCA graph demonstrating the extent of association and divergence of physiological, biochemical, and growth-related traits of different olive genotypes from origin under varying concentrations (L1, control; L2, 0.5 mgL^{-1}; L3, 1.5 mgL^{-1}; L4, 2.5 mgL^{-1}) of BAP with respect to control.

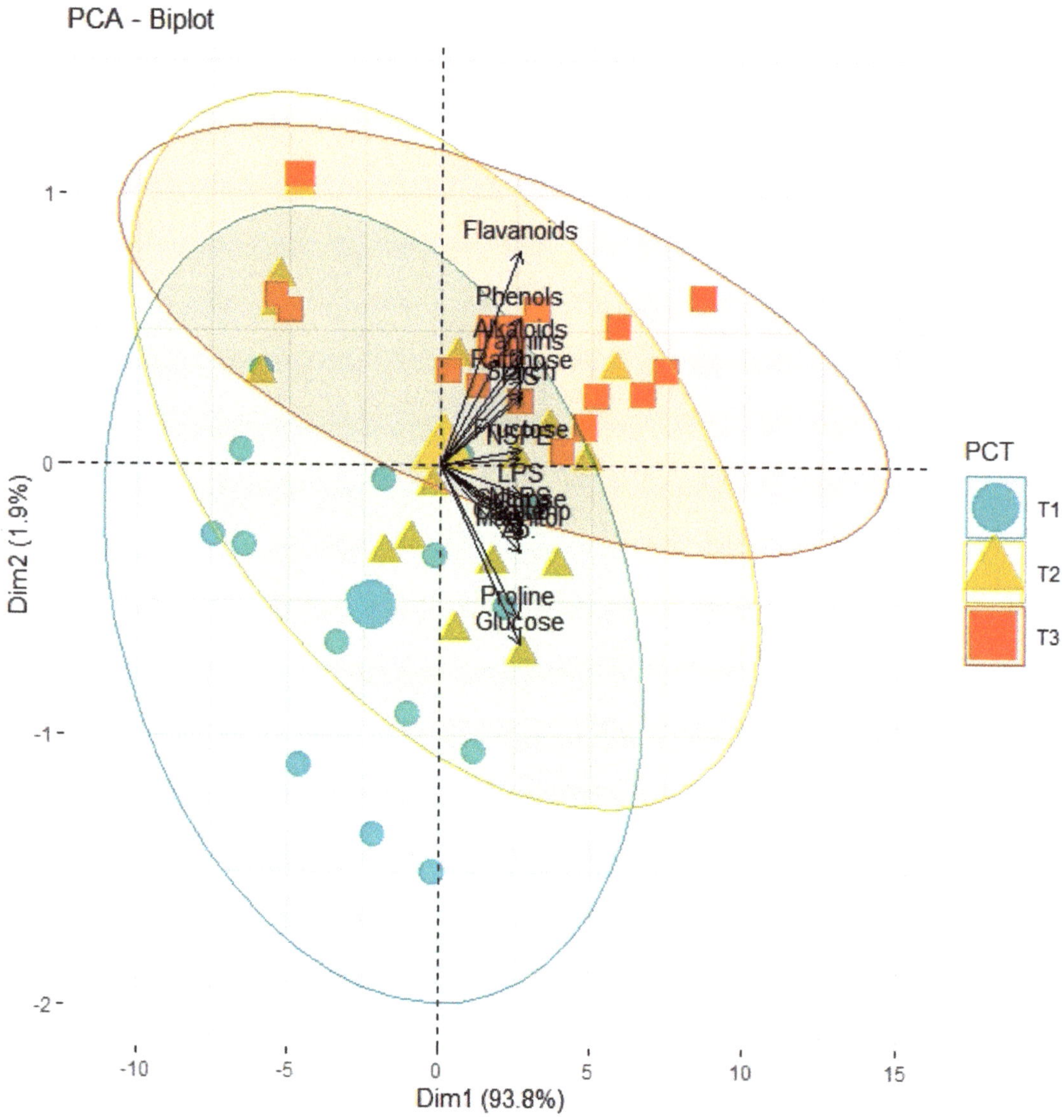

Figure 2. PCA graph demonstrating the extent of association and divergence of physiological, biochemical, and growth-related traits of different olive genotypes from origin under varying pre-cooling treatments (T1, control; T2, 24 h; T3, 48 h).

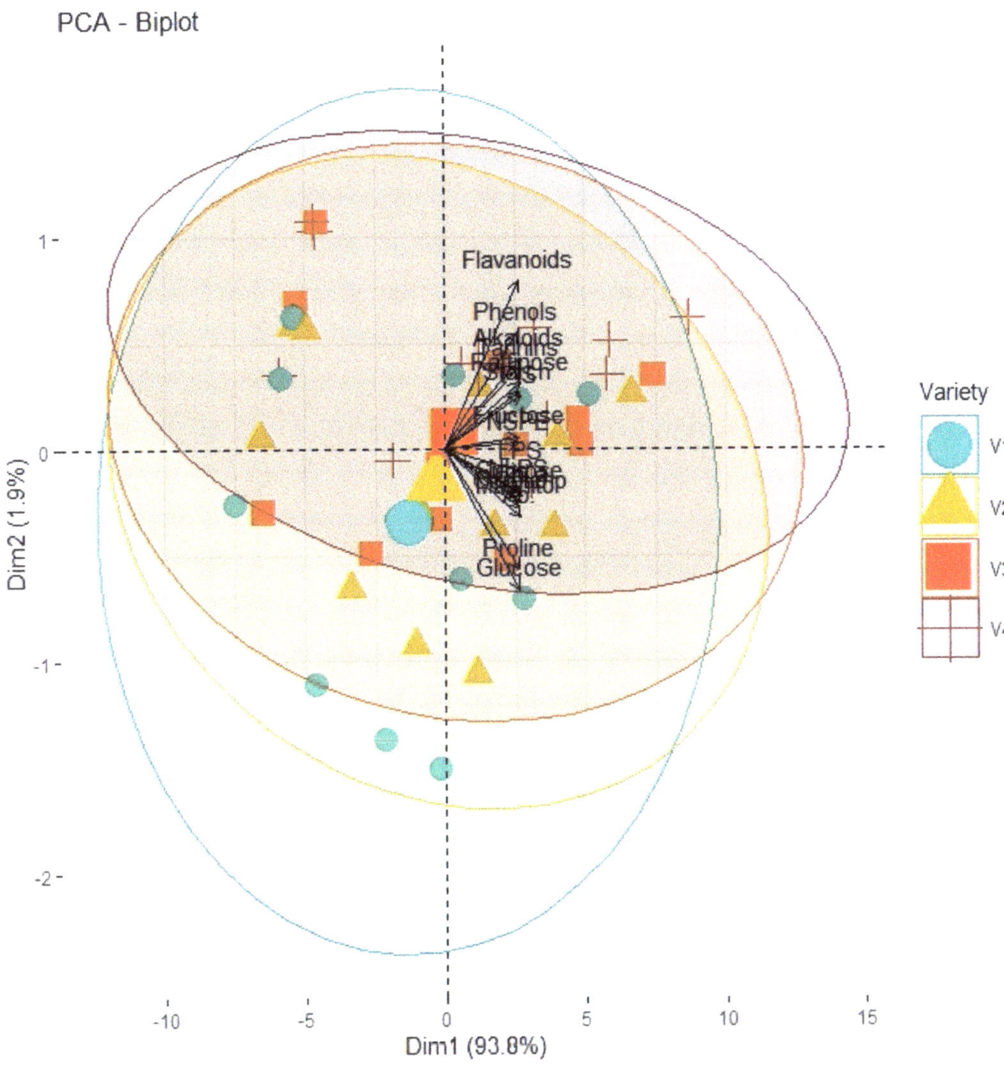

Figure 3. PCA graph explicating the varying extent of association and divergence of physiological, biochemical, and growth-related traits from origin with respect to type of genotypes (V1, Leccino; V2, Gemlik; V3, Moraiolo; V4, Arbosana).

4. Discussion

The current study intended to elucidate the impacts of various combinations of pre-cooling treatments and BAP concentrations on morphological, physiological, and biochemical traits of different olive genotypes cultured on OM during in vitro conditions. Despite incessant efforts, the delineation of optimized protocols for shoot induction in olive always remained a case of potential concern. In this context, the current study deeply assessed the impacts of different pre-cooling and BAP treatments on different olive cultivars for setting appropriate shoot induction protocol. For BAP, the present study noticed that for the induction of shooting and proliferation activity, at least its supplementation is mandatory at a concentration of 0.5 mgL^{-1}. Likewise, in the past, various studies proved that the addition of zeatin in OM is mandatory for inducing shooting in axillary buds of olive, however, its concentration at 2.5 mgL^{-1} depicted comparatively better results [1,6]. However, the supplementation of zeatin was variable in terms of effectivity for different olive cultivars [9]. Therefore, BAP supplementation in combination with pre-cooling treatments of explants was used as an alternative strategy. Apart from this pre-cooling treatment of explants such as axillary buds and nodal segments is an effective tool to increase their shoot induction percentage, as it has a tendency to break their dormancy by triggering the activity of vernalin [26]. In fact, vernalin has a tendency to reciprocate the effect of ABA whose deposition results in the synthesis of callose that blocks intercellular transport occurring through plasmodesmata [21]. Likewise, the present study recorded that integration of an extended span of pre-cooling and high concentration of BAP made significant improvements in growth traits such as LPS, SIP, NLPS, and NSPE of all olive cultivars (Table 1). These findings were inconsistent with the findings of Peixe et al. [1] Chaari-Rkhis et al. [6] and Zuccherelli and Zuccherelli [35] who reported a significant increase in LPS, SIP, NLPS, and NSPE under increased concentrations of growth regulators within OM. Furthermore, internodal segments cultured after 24 and 48 h of pre-cooling also recorded consistent increases in physiological processes and metabolic activities with increasing concentrations of BAP as evident from high accumulation pigments and primary metabolites (Tables 2, 3, 5 and 6). Perhaps, plant growth is strongly associated with the synthesis of structural and regulatory enzymes in addition to the synthesis of enzymes participating in photosynthesis [36]. Proline has a tendency to protect the essential enzymes and to balance the subcellular machinery that facilitates the specific function of carbon and nitrogen [37]. Proline serves as an osmoprotectant in the cell; therefore, it prevents plantlets from dehydration within the micropropagation system when they sense any sort of abiotic stress [36]. With increasing BAP concentrations, proline content depicted a significant reduction in all genotypes that is because of potential role of BAP in cohering metabolic activities as reported by Hamooh and Shah [9]. Furthermore, the increase in chlorophyll is also associated with dramatic increase in the levels of various metabolites including both primary and secondary [38]. High chlorophyll content is directly connected with enhanced CO_2 assimilation due to increase in quantity of photosynthetic pigments such as chlorophyll and carotenoids [39]. Perhaps due to these reasons all genotypes of olive depicted strong positive correlation between physiological and growth traits within micropropagation system due to increasing BAP concentrations (Table 7). In addition, high content of carbohydrate may increase the production of substrates involved in shikimic acid pathway that ultimately trigger the production of secondary metabolites [40]. Therefore, increased production of secondary metabolites is directly associated with the high production of carbohydrates as reported by Ghasemzadeh et al. [41]. Correspondingly, the current study recorded complete parallelism between the increment of primary and secondary metabolic contents (Tables 3, 4 and 7). Moreover, the current study noticed a significant change in the levels of both primary (fructose, glucose, sucrose, and starch) and secondary (alkaloids, phenols, tannins, and flavonoids) metabolites due to optimization of BAP concentrations with pre-cooling treatments within in vitro micropropagation system (Tables 3–6). Various protocols are optimized for different olive cultivars using different concentrations of growth regulators such as BAP and zeatin within OM by testing their impacts on growth traits. However, the current study recorded dynamic impacts of changing hormonal concentrations on physiological and metabolic activities in addition to growth traits. Hormones, being signaling entities, are potentiate enough to establish the coherence between physiological and biochemical activities for triggering the growth and differentiation process at the cellular level [42]. The current study validated these views by recording strong paired association between physiological, biochemical, and growth traits through correlation analysis (Table 7). Although all cultivars under study responded in an analogous way to the varying conditions of the micropropagation system, the extent of response was different with respect to pre-cooling treatments, BAP concentration, and cultivars (Figures 1–3). This could be attributed to the different genetic makeup of all genotypes due to which they responded differently to treatments. As a whole, on olive media, during both optimizations, the performance of "Arbosana" was exceptionally different at the morphological, physiological, and biochemical levels.

5. Conclusions

Overall, those cultivars whose explants were pre-treated at 48 h cooling treatments before culturing depicted noteworthy performance at a BAP concentration of 2.5 mgL^{-1}. The current study remained fruitful in optimizing shoot induction and proliferation protocol for olive cultivars by comprehensively evaluating the effects of all treatments at physiological, biochemical, and morphological levels. Moreover, this study proved that BAP is a good hormonal supplement to OM that can increase the effectiveness of the culture medium for olive propagation within in vitro conditions.

Author Contributions: W.L., C.-b.D., and X.L. came up with the idea and supervised the study; S.K. and Z.H.S. conducted experiments and wrote the manuscript; M.R. and Z.H.S. analyzed the data statistically; Y.Z. (Yinan Zheng), Y.Z. (Yue Zhang), F.A., and X.C. critically reviewed the manuscript. All authors have read and agreed to the published version of the manuscript.

Funding: This research received no external funding.

Institutional Review Board Statement: Not applicable.

Informed Consent Statement: Not applicable.

Data Availability Statement: Not applicable.

Conflicts of Interest: The authors declare no conflict of interest.

References

1. Peixe, A.; Raposo, A.; Lourenço, R.; Cardoso, H.; Macedo, E. Coconut water and BAP successfully replaced zeatin in olive (*Olea europaea* L.) micropropagation. *Sci. Hortic.* **2007**, *113*, 1–7. [CrossRef]
2. Espinosa-Leal, C.A.; Puente-Garza, C.A.; García-Lara, S. In vitro plant tissue culture: Means for production of biological active compounds. *Planta* **2018**, *248*, 1–18. [CrossRef] [PubMed]
3. Lambardi, M.; Ozudogru, E.A.; Roncasaglia, R. In vitro propagation of olive (*Olea europaea* L.) by nodal segmentation of elongated shoots. In *Protocols for Micropropagation of Selected Economically-Important Horticultural Plants*; Humana Press: Totowa, NJ, USA, 2012; pp. 33–44.
4. Benelli, C.; De Carlo, A. In vitro multiplication and growth improvement of *Olea europaea* L. cv Canino with temporary immersion system (Plantform™). *3 Biotech* **2018**, *8*, 371. [CrossRef] [PubMed]
5. Ali, A.; Ahmad, T.; Abbasi, N.A.; Hafiz, I.A. Effect of different media and growth regulators on in vitro shoot proliferation of olive cultivar 'Moraiolo'. *Pak. J. Bot.* **2009**, *41*, 783–795.
6. Chaari-Rkhis, A.; Maalej, M.; Drira, N.; Standardi, A. Micropropagation of olive tree *Olea europaea* L. 'Oueslati'. *Turk. J. Agric. For.* **2011**, *35*, 403–412.
7. Revilla, M.A.; Pacheco, J.; Casares, A.; Rodriguez, R. In vitro reinvigoration of mature olive trees (*Olea europaea* L.) through micrografting. *In Vitro Cell. Dev. Biol.-Plant* **1996**, *32*, 257. [CrossRef]
8. Rugini, E. In vitro propagation of some olive (*Olea europaea* L.) cultivars with different root ability, and medium development using analytical data from developing shoots and embryos. *Sci. Hortic.* **1984**, *24*, 123–134. [CrossRef]
9. Hamooh, B.T.; Shah, Z.H. In vitro evaluation of shoot induction and proliferation protocol for olive cultivars by assessing morpho-physiologic effects of pre-cooling and growth regulators. *Int. J. Biol.* **2017**, *11*, 126–139.
10. Ahmadi-Lahijani, M.J.; Kafi, M.; Nezami, A.; Nabati, J.; Erwin, J.E. ABA and BAP improve the accumulation of carbohydrates and alter carbon allocation in potato plants at elevated CO_2. *Physiol. Mol. Biol. Plants* **2021**, *27*, 313–325. [CrossRef]
11. Tahiri, A.; Destain, J.; Thonart, P.; Druart, P. In vitro model to study the biological properties of humic fractions from landfill leachate and leonardite during root elongation of *Alnus glutinosa* L. Gaertn and *Betula pendula* Roth. *Plant Cell Tiss. Organ Cult.* **2015**, *122*, 739–749. [CrossRef]
12. Ferrara, g.; Loffredo, E.; Simeone, R.; Senesi, N. Evaluation of antimutagenic and desmutagenic effects of humic and fulvic acids on root tips of *Vicia faba*. *Environ. Toxicol.* **2000**, *15*, 513–517. [CrossRef]
13. Ferrara, G.; Loffredo, E.; Senesi, N.; Marcos, R. Humic acids reduce the genotoxicity of mitomycin C in human lymphoblastoid cell line TK6. *Mutat. Res. Genet. Toxicol. Environ. Mutagen.* **2006**, *603*, 27–32. [CrossRef] [PubMed]

14. Ferrara, G.; Loffredo, E.; Senesi, N. Anticlastogenic, antitoxic and sorption effects of humic substances on the mutagen maleic hydrazide tested in leguminous plants. *Eur. J. Soil Sci.* **2004**, *55*, 449–458. [CrossRef]
15. Salem, J.; Hassanein, A.; El-Wakil, D.A.; Loutfy, N. Interaction between Growth Regulators Controls In Vitro Shoot Multiplication in Paulownia and Selection of NaCl-Tolerant Variants. *Plants* **2022**, *11*, 498. [CrossRef]
16. Lebedev, V.; Arkaev, M.; Dremova, M.; Pozdniakov, I.; Shestibratov, K. Effects of Growth Regulators and Gelling Agents on Ex Vitro Rooting of Raspberry. *Plants* **2018**, *8*, 3. [CrossRef]
17. Leva, A.; Petruccelli, R.; Rinaldi, L. Somaclonal Variation in Tissue Culture: A Case Study with Olive. In *Recent Advances in Plant in vitro Culture*; Leva, A., Rinaldi, L.M.R., Eds.; InTech: Rijeka, Croatia, 2012.
18. Antonopoulou, C.; Dimassi, K.D.; Therios, I.T.; Hatzissavvidis, C. Does dikegulac affect in vitro shoot proliferation and hyperhydricity incidence in olive explants? *Hortic. Sci.* **2018**, *45*, 125–130. [CrossRef]
19. Brunner, A.M.; Evans, L.M.; Hsu, C.Y.; Sheng, X. Vernalization and the chilling requirement to exit bud dormancy: Shared or separate regulation? *Front. Plant Sci.* **2014**, *5*, 732. [CrossRef]
20. Pan, W.; Liang, J.; Sui, J.; Li, J.; Liu, C.; Xin, Y.; Zhang, Y.; Wang, S.; Zhao, Y.; Zhang, J.; et al. ABA and Bud Dormancy in Perennials: Current Knowledge and Future Perspective. *Genes* **2021**, *12*, 1635. [CrossRef]
21. Liu, J.; Sherif, S.M. Hormonal orchestration of bud dormancy cycle in deciduous woody perennials. *Front. Plant Sci.* **2019**, *10*, 1136. [CrossRef]
22. Alexandre, C.; Hennig, L. FLC-independent vernalization responses. *Int. J. Plant Dev. Biol.* **2007**, *1*, 202–211.
23. Leida, C.; Conejero, A.; Arbona, V.; Gómez-Cadenas, A.; Llácer, G.; Badenes, M.L.; Ríos, G. Chilling-dependent release of seed and bud dormancy in peach associates to common changes in gene expression. *PLoS ONE* **2012**, *10*, 35777. [CrossRef] [PubMed]
24. Dhami, N.; Cazzonelli, C.I. Prolonged cold exposure to Arabidopsis juvenile seedlings extends vegetative growth and increases the number of shoot branches. *Plant Signal. Behav.* **2020**, *15*, 1789320. [CrossRef] [PubMed]
25. Jones, A.M.; Saxena, P.K. Inhibition of phenylpropanoid biosynthesis in *Artemisia annua* L.: A novel approach to reduce oxidative browning in plant tissue culture. *PLoS ONE* **2013**, *8*, 76802. [CrossRef]
26. Martini, A.N.; Papafotiou, M.; Vemmos, S.N. Season and explant origin affect phenolic content, browning of explants, and micropropagation of × *Malosorbus florentina* (Zucc.) Browicz. *HortScience* **2013**, *48*, 102–107. [CrossRef]
27. Gao, J.; Zhang, P.; Xue, J.; Xue, Y.; Wang, S.; Zhang, X. Advances in phenolic substances and their effects on browning in woody plant tissue culture. *Acta Hortic. Sin.* **2019**, *46*, 1645–1654.
28. Mahmood, S.; Daur, I.; Al-Solaimani, S.G.; Ahmad, S.; Madkour, M.H.; Yasir, M.; Hirt, H.; Ali, S.; Ali, Z. Plant Growth Promoting Rhizobacteria and Silicon Synergistically Enhance Salinity Tolerance of Mung Bean. *Front. Plant Sci.* **2016**, *7*, 876. [CrossRef] [PubMed]
29. Boussadia, O.; Steppe, K.; Zgallai, H.S.; Ben El-Hadj, S.; Braham, M.; Lemeur, R.; Van-Labeke, M.C. Effects of nitrogen deficiency on leaf photosynthesis, carbohydrate status and biomass production in two olive cultivars 'Meski' and 'Koroneiki'. *Sci. Hortic.* **2010**, *123*, 336–342. [CrossRef]
30. Chow, P.S.; Landhäusser, S.M. A method for routine measurements of total sugar and starch content in woody plant tissues. *Tree Physiol.* **2004**, *24*, 1129–1136. [CrossRef]
31. Siddiqui, N.; Rauf, A.; Latif, A.; Mahmood, Z. Spectrophotometric determination of the total phenolic content, spectral and fluorescence study of the herbal Unani drug Gul-e-Zoofa (*Nepeta bracteate* Benth). *J. Taibah Univ. Med. Sci.* **2017**, *12*, 360–363. [CrossRef]
32. Zhao, L.J.; Liu, W.; Xiong, S.H.; Tang, J.; Lou, Z.H.; Xie, M.X.; Xia, B.H.; Lin, L.M.; Liao, D.F. Determination of total flavonoids contents and antioxidant activity of Ginkgo biloba leaf by near-infrared reflectance method. *Int. J. Anal. Chem.* **2018**, *2018*, 8195784. [CrossRef]
33. Fadda, A.; Mulas, M. Chemical changes during myrtle (*Myrtus communis* L.) fruit development and ripening. *Sci. Hortic.* **2010**, *125*, 477–485. [CrossRef]
34. Li, L.; Long, W.; Wan, X.; Ding, Q.; Zhang, F.; Wan, D. Studies on quantitative determination of total alkaloids and berberine in five origins of crude medicine "Sankezhen". *J. Chromatogr. Sci.* **2015**, *53*, 307–311. [CrossRef] [PubMed]
35. Zuccherelli, G.; Zuccherelli, S. In vitro propagation of fifty olive cultivars. *Acta Hortic.* **2002**, *586*, 931–934. [CrossRef]
36. Guo, R.; Yuan, G.; Wang, Q. Effect of sucrose and mannitol on the accumulation of health-promoting compounds and the activity of metabolic enzymes in broccoli sprouts. *Sci. Hortic.* **2011**, *128*, 159–165. [CrossRef]
37. Ejaz, B.; Sajid, Z.A.; Aftab, F. Effect of exogenous application of ascorbic acid on antioxidant enzyme activities, proline contents, and growth parameters of *Saccharum* spp., hybrid cv. HSF-under salt stress. *Turk. J. Biol.* **2012**, *36*, 630–640. [CrossRef]
38. Aslam, M.; Sultana, B.; Anwar, F.; Munir, H. Foliar spray of selected plant growth regulators affected the biochemical and antioxidant attributes of spinach in a field experiment. *Turk. J. Agric. For.* **2016**, *40*, 136–145. [CrossRef]
39. Hayat, S.; Hayat, Q.; Alyemeni, M.N.; Wani, A.S.; Pichtel, J.; Ahmad, A. Role of proline under changing environments: A review. *Plant Signal. Behav.* **2012**, *7*, 1456–1466. [CrossRef]
40. Santos-Sánchez, N.F.; Salas-Coronado, R.; Hernández-Carlos, B.; Villanueva-Cañongo, C. Shikimic Acid Pathway in Biosynthesis of Phenolic Compounds. In *Plant Physiological Aspects of Phenolic Compounds*; Soto-Hernández, M., García-Mateos, R., Palma-Tenango, M., Eds.; IntechOpen: London, UK, 2019.

41. Ghasemzadeh, A.; Jaafar, H.Z.E.; Rahmat, A. Elevated carbon dioxide increases contents of flavonoids and phenolic compounds, and antioxidant activities in Malaysian young ginger (*Zingiber officinale* Roscoe.) varieties. *Molecules* **2010**, *15*, 7907–7922. [CrossRef]
42. El Sabagh, A.; Islam, M.S.; Hossain, A.; Mubeen, M.; Iqbal, M.A.; Waleed, M.; Reginato, M.; Battaglia, M.; Ahmed, S.; Rehman, A.; et al. Phytohormones as growth regulators during abiotic stress tolerance in plants. *Front. Agron.* **2022**, *8*, 68. [CrossRef]

horticulturae

MDPI

Article

Rapid In Vitro Propagation of Fig (*Ficus carica* L.) 'Violette de Solliès' Supported by Molecular and Microscopy Analyses

Wan Ting Ling, Li Vern Tan, Soo Ping Khor, Dahmendra Sriskanda, Sreeramanan Subramaniam and Bee Lynn Chew *

School of Biological Sciences, Universiti Sains Malaysia, George Town 11800, Penang, Malaysia
* Correspondence: beelynnchew@usm.my

Abstract: *Ficus carica* L. is a common fig that is an incredibly nutritional fruit, well-known for its medicinal and economic values. This study aims to establish an efficient protocol for the mass propagation of fig plantlets (*Ficus carica* L.) for the cultivar 'Violette de Solliès'. Surface-sterilized shoot-tip explants were cultured on Murashige and Skoog (MS) medium supplemented with different concentrations of cytokinins (6-benzylaminopurine, BAP; thidiazuron, TDZ; kinetin, Kn; and zeatin, Zea). Induced shoots were rooted on Woody Plant Medium (WPM) with different concentrations of auxins (naphthalene-acetic acid, NAA; indole-3-acetic acid, IAA; and indole-3-butyric acid, IBA). Rooted explants were acclimatized in eight different soil substrates prior to cultivation in a commercial plot. The propagated plantlets were analyzed for genetic stability and clonal fidelity using RAPD and SCoT molecular markers, whereas scanning electron microscopy (SEM) was performed to observe the stomata morphology of post-acclimatized plants. MS media supplemented with 5.0 mg/L BAP was the optimal treatment for multiple shoot induction (15.20 ± 1.03 shoots), whereas the highest percentage of rooting (93.33%) was achieved in WPM supplemented with 3.0 mg/L IBA. Plantlets were successfully acclimatized in biochar soil with a survival rate of 100%. RAPD and SCoT analysis showed no polymorphism occurrences across six subculture cycles, whereas observations via SEM indicated normal stomata structures on the leaves of acclimatized plantlets. This study documents an efficient micropropagation protocol for *Ficus carica* cv. 'Violette de Solliès' for the production of uniformed and true-to-type plant stocks suitable for commercial propagation.

Keywords: *Ficus carica*; polymorphism analysis; RAPD markers; SCoT markers; stomata; scanning electron microscopy

Citation: Ling, W.T.; Tan, L.V.; Khor, S.P.; Sriskanda, D.; Subramaniam, S.; Chew, B.L. Rapid In Vitro Propagation of Fig (*Ficus carica* L.) 'Violette de Solliès' Supported by Molecular and Microscopy Analyses. *Horticulturae* **2022**, *8*, 1025. https://doi.org/10.3390/horticulturae8111025

Academic Editors: Alicja Tymoszuk and Dariusz Kulus

Received: 12 October 2022
Accepted: 29 October 2022
Published: 2 November 2022

1. Introduction

Ficus is a genus from the family of Moraceae, consisting of over 800 species including *Ficus carica* L., the common edible fig [1]. The *Ficus carica* plant is characterized by its smooth grey bark, lobed leaves, and fibrous roots. The skin color of the fig fruits is different between cultivars, varying from yellowish green to copper, bronze, or dark purple [2]. This deciduous fruit tree is native to the Southwest regions of Asia and the eastern Mediterranean [3], which was later introduced and cultivated abundantly in Turkey, Egypt, Greece, Iran, and Morocco [4]. As a top producer and exporter of figs, Turkey contributed up to 27% of fig production worldwide in 2018, producing a total of 306,000 tons of figs and generating approximately $286 million in exports for both dried and fresh figs [5]. *Ficus carica* is not native to the tropical regions of the world, but has recently been introduced as a superfruit in Malaysia, where most of its cultivation is still at a small scale of 50–1500 plants per farm [6]. The price of cuttings range between RM 30 and RM 100 (USD 6 and USD 22) per plant, depending on the cultivar type, and is in high demand for commercial farm establishments and home gardening [6]. Fig cultivation in Malaysia is still new and is currently expanding under the Indonesia, Malaysia, and Thailand Growth Triangle (IMT-GT) agriculture program, which comprises an area of approximately 10 hectares in the northern region of the country [7].

The fig fruit has been a common food source and folk remedy since ancient times due to its high nutritional and medicinal values. The fruit consists of mainly fiber, potassium, calcium, and iron, which is higher than in a banana, orange, or apple [8]. The fig plant also produces a variety of secondary metabolites that are known for their potential health effects, namely, being antibacterial, antioxidant, antidiabetic, anti-inflammatory, antipyretic, and anticancer [3]. Studies have also shown that both ethanolic or methanolic extracts of the fruits, leaves, and stem barks possess antidiabetic activity [9–11]. Previous studies also reported that diabetic complications such as cholesterolemia and hypoglycemia were successfully controlled in streptozotocin-induced diabetic rats via feeding with extracts of fruits and leaves of fig, indicating the strong inhibitory activity against α-amylase and α-glucosidase enzymes in mammals [11–14]. Furthermore, the efficiency of ethyl acetate extract of fig leaves exhibited significant antidiabetic activity, successfully stimulating the production of insulin from regenerated pancreatic beta cells that resulted in the reduction in blood glucose levels of diabetic rats [15]. Similarly, the blood glucose level of alloxan-induced diabetic rats was reduced after feeding on *Ficus carica* leaf extract [10,16].

To date, there are over 750 varieties or more of figs in the world, mostly thriving in the Western Asia and Mediterranean region, as well as in countries with a mild climate [17]. One of the highly sought-after fig cultivars is 'Violette de Solliès', which is also deemed as a premium fig, from the Solliès region of France, as it is recognized for being the sweetest fig available. 'Violette de Solliès' or VDS produces relatively large, dark, purple-colored fruits with a soft, pinkish-red pulp and are sweet with a slight hint of cherry. However, its availability in the market is still relatively low as there are shortages in planting materials due to the production of nonviable seeds and the inefficiency of conventional propagation methods. Fig fruits are not dependent on pollination to fully set and develop; thus, the probability of obtaining viable seeds is very low, limiting the propagation via seed. The other alternative is propagation through cuttings, which is commonly marred by poor rooting efficiency, limiting the chance of survival to only 20 to 30% [18]. Additionally, cuttings are more sensitive to ecological changes such as drastic shift in temperature and moisture levels, further limiting their survival rate [19].

Plant tissue culture is an alternative propagation method whereby plant cells, tissues, or organs are cultured under aseptic controlled conditions. It is an efficient and widely adopted tool for the propagation of many fruit crops of economic importance, such as apples, bananas, pineapples, and citrus, in commercial nursery production [20]. This alternative method can be manipulated to continuously and consistently produce good-quality pathogen-free planting materials at a competitive price for large-scale production. Previous studies on the tissue culture and micropropagation of figs have reported the use of single shoot-tip explants [21], nodal explants [22], leaves [1,23,24], apical meristems, and axillary buds [24–26], which have successfully propagated figs from different cultivars. Up to now, there have been no tissue culture studies and micropropagation methods being reported for the cultivar of 'Violette de Solliès', albeit on its significance as one of the sweetest fig cultivars in the world. The present study aims to establish an efficient micropropagation protocol for the cultivar of 'Violette de Solliès' for the large-scale production of high-quality planting materials suitable for commercial field establishments. This study also looks to confirm if plantlets generated are true-to-type via molecular markers used to identify polymorphism, and microscopy analysis is used to observe the morphological stomata structures between the plantlets and the mother plant.

2. Materials and Methods

2.1. Establishment of In Vitro Cultures

The mother plants were obtained from the Malaysian Superfruits Valley, Perlis at an age of approximately one year, after being grown from air layering, with a height of approximately 1.5 m above the ground level; they were maintained at the Herbarium Unit of Universiti Sains Malaysia. Young shoot tips of 'Violette de Solliès' (VDS), approximately 3 cm in length, were collected fresh from healthy growing mother plants prior to the surface sterilization step. Shoot-tip explants were washed with 2% Sunlight® dishwashing liquid, which was followed by rinsing under running tap water for ten minutes. The explants were then surface sterilized with 70% ethanol for 5 min, which was followed by a gentle agitation with 25% Clorox® containing 2 to 3 drops of Tween 20 for 5 min. Explants were then rinsed with sterile distilled water eight times before being dried on sterile filter paper prior to culture.

2.2. Induction of Multiple Shoots

Sterile shoot-tip explants were cultured on MS [27] basal medium with 2% (w/v) sucrose solidified with 0.8% (w/v) plant agar (Duchefa, Haarlem, The Netherlands) and supplemented with different plant growth regulators, namely, 6-benzylaminopurine (BAP), kinetin (Kn), thidiazuron (TDZ), and zeatin (Zea) (Duchefa Biochemie, Haarlem, The Netherlands), at various concentrations of 1.0, 2.0, 3.0, 4.0, 5.0, and 6.0 mg/L. Shoot-tip explants cultured in MS medium without the supplementation of cytokinin were used as the control treatment. Cultures were maintained under white LED (light-emitting diode) light (Philips TLD 36 W/865–6500 K, 3070 lm) with a 16/8 h photoperiod at a temperature of 25 ± 2 °C and relative humidity of 50 ± 10%. A total of three experimental replications were carried out with each replication consisting of five explant replicates. Parameters such as the percentage of shoot induction, average number, and length of adventitious shoots per explant were measured after 6 weeks of culture.

2.3. Root Induction from Shoot Explants

Shoots generated in vitro (from the treatment of 6.0 mg/L BAP) at the length of approximately 2 cm were excised and cultured in Woody Plant Medium (WPM) [28] supplemented with auxins, namely, indole-3-acetic acid (IAA), indole-3-butric acid (IBA), and naphthalene-acetic acid (NAA), at the concentrations of 0.0, 1.0, 2.0, 3.0, and 4.0 mg/L. Parameters such as the percentage of root induction, and average root number and length per explant were measured after 6 weeks of culture. The experiment was repeated thrice with five replicates for each replication.

2.4. Acclimatization on Different Soil Substrates

Shoot explants with well-grown roots (approximately 9 weeks in WPM supplemented with 3.0 mg/L IBA) were gently removed from the culture media, with the remaining agar rinsed off with tap water. Plantlets were transferred to germination pots (5 cm in diameter) filled with different soil substrates, namely, Jiffy pellet (Jiffy International, Stange, Norway), peat, perlite, vermiculite, a peat and perlite mixture (2:1), a peat and vermiculite mixture (2:1), and biochar soil (Serbajadi, Malaysia), with a garden soil mixture (black compost, coco peat, and red soil mixture in the ratio of 3:1:1) as the control. The plantlets were irrigated with tap water once every 3 days and kept in closed transparent containers under 25 ± 2 °C with a total of 15 replicates per soil substrate evaluated. The cover of the container was gradually removed within 30 days to decrease the humidity inside the container. The survival rate of the plantlets (%) was recorded after 6 weeks. Successfully acclimatized plantlets were transferred to polybags (10 cm × 12 cm) containing the garden soil mixture and maintained under a shaded roof with normal cultivation practices for approximately 8 weeks prior to planting in the commercial plot.

2.5. RAPD- and SCoT-PCR Molecular Analysis

The genomic DNA was isolated from leaf samples of three randomly selected regenerated tissue-cultured plantlets from the first up to the sixth subculture cycle, following the protocols of the Wizard® Genomic DNA purification kit (Promega, Madison, WI, USA). DNA quality and concentrations were measured via the NanoDrop Spectrophotometer (ASP-2680, ACTGene Inc., Piscataway, NJ, USA) and DNA samples were diluted to a fixed concentration of 20 ng/µL prior to storage at −20 °C for the subsequent steps. The reaction mixtures were prepared by mixing 50 ng of genomic DNA (20 ng/µL), 1.0 µM of primer

(Operon Technologies Inc., Alameda, CA, USA), and 10 μL of 2X GoTaq® Green Master Mix (Promega, Madison, WI, USA), with a total reaction volume of 20 μL. The PCR reaction was performed in the T100™ Thermal Cycler (Bio-Rad, Hercules, CA, USA) and initiated with a 4 min denaturation step at 94 °C; subsequently, another 40 cycles of the 1 min denaturation step at 94 °C were set up before the annealing step. An annealing step was performed at approximately 5 °C below the derived melting temperature of the primers for 1 min, followed by 1 min of the extension step at 72 °C. Lastly, a final extension of 10 min at 72 °C was carried out at the end of the PCR reaction. The amplified DNA samples were loaded into a 1.0% agarose gel with 1x TBE (Tris-Borate-EDTA) and RedSafe™ (iNtRON Biotechnology, Seongnam-si, South Korea) as the buffer and nucleic acid staining reagent, respectively. The voltage for agarose gel electrophoresis was set at 70 volts for 1.5 h, and the results were visualized and analyzed using a gel documentation system, UVIdoc HD5 Gel Imaging System (UVITec Limited, Cambridge, UK). A total of 60 RAPD primers from OPC, OPK, and OPU series and 36 SCoT primers [29] were tested. Seven primers which produced clear and reproducible bands from each marker were selected and used in the assessment of genetic stability. Both RAPD and SCoT reactions were repeated for a minimum of three replications to ensure reproducible bands and consistency of the band patterns.

2.6. Scanning Electron Microscopy (SEM) Analysis

Leaf samples were collected from the in vitro cultures, acclimatized plantlets (20 and 60 days old), and field-grown plants (control), and excised into pieces 1 cm^2 in size. The leaf samples were fixed in McDowell–Trump fixative and prepared in a 0.1 M phosphate buffer for 2 to 24 h at 4 °C, which was followed by rinsing thrice under the same buffer for 10 min. Next, the leaves were kept in 1% osmium tetroxide (OsO4) that was prepared in a 0.1 M phosphate buffer for 2 h at room temperature. All samples were then washed twice with distilled water for 10 min. Following this, the leaves were dehydrated with ethanol at increasing concentrations (50, 75, 95, and 99.5%). The dehydrated samples were then immersed in 2 mL of hexamethyldisilazane (HMDS) for 10 min and air-dried in a desiccator. The dried leaves were adhered to a SEM sample stub and sputtered with gold prior to viewing in the SUPRA® 55VP field-emission scanning electron micrograph (FE-SEM, Carl Zeiss, Jena, Germany).

2.7. Statistical Analysis

The experimental design applied in this study was a completely randomized design with three replications containing five explants per replicate. The data were analyzed using a one-way analysis of variance (ANOVA) via SPSS statistical software version 26, followed by a post hoc test, Duncan's multiple range test, where $p < 0.05$.

3. Results and Discussion

3.1. Multiple Shoot Induction

In the current study, MS media supplemented with 2.0, 4.0, and 6.0 mg/L BAP, 5.0 and 6.0 mg/L Kn, and all concentrations of TDZ and Zea (except Zea at 2.0 mg/L) generated the maximum shoot induction with a percentage of 100%, which was higher than the control at only 40% (Table 1). It was also evident that the number of adventitious shoots induced was directly proportional to the concentration of plant growth regulators used. In the present study, BAP was identified as the most potent plant growth regulator in the induction of multiple shoots, followed by Zea, TDZ, and Kn. The treatment of 6.0 mg/L BAP generated the highest number of adventitious shoots with the value of 16.40 ± 2.11 shoots per explant (Figure 1a), indicating that BAP alone is efficient enough for rapid shoot proliferation in this cultivar. However, this value is not significant for the treatment of 5.0 mg/L BAP, which also resulted in a higher number of induced shoots (15.20 ± 1.03). On the other hand, treatments with Zea were found to stimulate shoot elongation. The optimal value in the increment of shoot length was observed in explants treated with 1.0 mg/L Zea, recording an average shoot length of 1.53 ± 0.21 cm, regardless of this value not being significant for the results of Zea treatments of 2.0, 3.0, and 4.0 mg/L. In this study, short and abundant shoots with small defined leaves were observed in treatments with BAP. It was also evident that relatively low concentrations of Kn (0 to 4.0 mg/L) induced very few to no multiple shoots, as illustrated in Figure 1b. Nevertheless, the production of multiple shoots increased as Kn concentration increased. Abnormal shoot formation (short and fasciated shoots) was noticed in all the TDZ treatments, particularly for high TDZ concentrations (5.0 and 6.0 mg/L), as shown in Figure 1c.

Table 1. The effects of different concentrations of plant growth regulators on shoot regeneration after 6 weeks of culture.

Plant Growth Regulators (PGRs)	Concentrations (mg/L)	Percentage of Shoot Induction (%)	Number of Induced Shoots per Explant (n)	Length of Induced Shoots (cm)
Control	0.0	40	0.95 ± 0.33^{h}	0.25 ± 0.10^{xyz}
6-Benzylaminopurine (BAP)	1.0	93.33	3.80 ± 0.67^{fg}	0.81 ± 0.12^{s}
	2.0	100	4.67 ± 0.57^{fg}	0.61 ± 0.11^{stuv}
	3.0	93.33	4.33 ± 0.62^{fg}	0.73 ± 0.16^{st}
	4.0	100	5.47 ± 0.93^{efg}	0.59 ± 0.09^{stuvw}
	5.0	100	15.20 ± 1.03^{a}	0.44 ± 0.03^{tuvwxy}
	6.0	100	16.40 ± 2.11^{a}	0.38 ± 0.04^{uvwxyz}
Kinetin (Kn)	1.0	20	0.27 ± 0.15^{h}	0.09 ± 0.05^{z}
	2.0	33.33	0.67 ± 0.40^{h}	0.15 ± 0.07^{yz}
	3.0	33.33	0.73 ± 0.41^{h}	0.22 ± 0.13^{xyz}
	4.0	46.67	0.80 ± 0.24^{h}	0.17 ± 0.06^{yz}
	5.0	100	8.20 ± 1.12^{bcd}	0.49 ± 0.05^{tuvwx}
	6.0	100	7.13 ± 0.53^{cde}	0.36 ± 0.05^{uvwxyz}
Thidiazuron (TDZ)	1.0	100	3.53 ± 0.50^{g}	0.31 ± 0.03^{vwxyz}
	2.0	100	4.53 ± 0.58^{fg}	0.29 ± 0.02^{wxyz}
	3.0	100	4.87 ± 0.44^{efg}	0.30 ± 0.02^{vwxyz}
	4.0	100	4.33 ± 0.30^{fg}	0.31 ± 0.03^{vwxyz}
	5.0	100	9.33 ± 0.87^{bc}	0.30 ± 0.03^{vwxyz}
	6.0	100	8.87 ± 0.61^{bc}	0.24 ± 0.02^{xyz}
Zeatin (Zea)	1.0	100	3.60 ± 0.32^{g}	1.53 ± 0.21^{r}
	2.0	86.67	3.07 ± 0.82^{g}	1.27 ± 0.19^{r}
	3.0	100	3.73 ± 0.37^{fg}	1.24 ± 0.14^{r}
	4.0	100	6.13 ± 1.06^{def}	1.47 ± 0.11^{r}
	5.0	100	8.67 ± 0.51^{bc}	0.64 ± 0.05^{stu}
	6.0	100	10.33 ± 0.70^{b}	0.66 ± 0.06^{stu}

Means of different values shown by different letters in the same column are statistically different using Duncan's multiple range test ($p < 0.05$). Mean and standard error (±SE) were calculated from three experimental replicates.

Cytokinin is known for its promotive effects on shoot multiplication, as it functions as a signaling molecule involved directly in the cell division of meristematic cells [30,31]. It controls the size of the shoot meristem, the number of leaf primordia, shoot growth, leaf production, and the development of axillary buds, as well as mediating apical dominance [32]. However, the response of tissues towards exogenous plant growth regulators can be due to the physiological status of the plant and/or the interaction between the endogenous regulators, resulting in a different response depending on the growth regulator supplemented. In this study, BAP was found to be the most effective cytokinin in stimulating multiple shoot production of 'Violette de Solliès'. The effectiveness of BAP in inducting multiple shoots is credited to its ability in reducing apical dominance in explants, leading to lateral shoot production [33]. In addition, BAP was also reported to be more stable in tissue culture when compared to other cytokinins due to its slower rate of metabolism. The results in this study corroborate the previous finding, whereby the maximum number of shoots (4.43 shoots per explant) were produced by explants of the "Bursa Siyahi" fig cultivar when supplemented with 1.0 mg/L BAP [34]. Additionally, the treatment with BAP at 2.5 mg/L was effective in inducing multiple shoots (19.7 shoots) on fig cultivars "Sultany", "Aboudi", and "White Adcy" [35]. Furthermore, the comparison of the effectiveness of different cytokinins on the shoot multiplication of the fig cultivar "Salti Kodari" found BAP to be the most effective cytokinin in comparison to Kn and Zea [36]. The highest number of multiple shoots (4.15 $\pm$ 0.43) was induced in the treatment of 0.8 mg/L BAP, which was higher than other cytokinin treatments of zeatin and thidiazuron for the Golden Orphan cultivar [22]. On the other hand, the treatments of BAP resulted in a higher number of multiple shoots in comparison to Zea induced in shoot-tip explants for the cultivar of Japanese BTM 6 [26]. In the current study, the 5.0 mg/L BAP was selected as the optimal concentration of BAP for shoot multiplication based on the high response in shoot number and shoot length for micropropagation purposes.

Figure 1. Shoot explant of *Ficus carica* cv. 'Violette de Solliès' at different stages of micropropagation: (**a**) multiple shoot induction in MS medium supplemented with 6.0 mg/L BAP; (**b**) multiple shoot induction in MS medium supplemented with 5.0 mg/L Kn; (**c**) multiple shoot induction in MS medium supplemented with 5.0 mg/L TDZ; (**d**) multiple shoot induction in MS medium supplemented with 6.0 mg/L Zea; (**e**) root induction in WPM supplemented with 3.0 mg/L IBA; (**f**) root morphology prior to acclimatization in biochar soil. Scale bars = 1 cm.

The results in the current study highlighted the potential of Zea in the elongation of shoots for the explants of the fig cultivar 'Violette de Solliès'. A previous study via transcriptome analysis mentioned that the expression of genes linked to zeatin biosynthesis in sugarcane is upregulated and is associated with the elongation of internodes [37]. Shoot clusters of lingonberry (*Vaccinium vitis-idaea* L.) were also successfully elongated when supplemented with Zea [38]. Besides, Zea at 0.5 mg/L effectively promoted shoot elongation in *Magnolia* sp. var. "Vulcan", as explants recorded a maximum average shoot length of 4.9 $\pm$ 0.2 cm [39]. These studies further indicated the efficiency of BAP in shoot proliferation and the potential elongation effects of Zeatin for the species of *Ficus carica*. On the contrary, treatments with Kn and TDZ were observed to be less effective in promoting multiple shoots and shoot elongation for the current cultivar of 'Violette de Solliès'. Another study reported similar results with the treatment of Kn producing only 1.415 shoots per explant in the micropropagation of figs native to Kurdistan, Iraq [2]. Kn (0.8 and 1.2 mg/L) had no effect on the shoot multiplication for the fig cultivar of "Salti Kodari", whereby the highest number of shoots induced was only at an average of 1.2 shoots [36]. However, a different observation was reported for the fig cultivar "Roxo de Valinhos'", whereby Kn at 2.0 mg/L was found to induce the highest number of shoots (4.25 shoots per explant) with no signs of vitrification [40]. The difference in the response to supplementation of exogenous plant growth regulators could be potentially influenced by the genotype of the plant itself. This was previously proven as figs from the cultivar "Deanna" achieved a shoot multiplication rate of 60% when treated with 3.5 mg/L BAP in combination with 0.5 mg/L gibberellic acid, whereas no shoots were induced by explants from fig cultivars "Poona Fig" and "Brown Turkey" [41]. In the current study, TDZ was discovered to exhibit abnormal effects on 'Violette de Solliès' explants as the shoots induced were found to be fasciated, preventing healthy explant growth. TDZ-induced deformed shoots with hyperhydration is possible, especially when TDZ is supplemented at excessive concentrations or under a prolonged culture duration [42,43]. The regulation of endogenous hormone levels could be the reason for abnormal shoot formation [44]. Additionally, the hyperhydricity and fasciation of shoots might be due to the latent consequences of TDZ on endogenous auxin such as IAA, and certain cytokinins, particularly BAP [45]. In the micropropagation of Chinese skullcap (*Scutellaria baicalensis* Georgi), the concentration of BAP in the MS medium decreased drastically when supplemented with TDZ at high concentrations (5.0 mg/L), which led to the increase in the IAA:BAP ratio [45]. High TDZ concentrations, therefore, caused an imbalance in the auxin/cytokinin ratio, which is essential for effective shoot proliferation. In the current experiment, it was evident that Kn and TDZ were not favorable for the induction of multiple shoots for the cultivar of 'Violette de Solliès'.

3.2. *Root Induction via the Supplementation of Auxins*

Of the three auxins evaluated in this study, IBA was observed to be the most effective in inducing roots for the in vitro explants of *Ficus carica* cv. 'Violette de Solliès'. With reference to Table 2, the highest rooting percentage and number of induced roots was observed in the treatment of 3.0 mg/L IBA with the values of 93.33% and 6.80 ± 1.43, respectively. Even though the treatment of 4.0 mg/L IAA resulted in the highest number of induced roots (7.27 ± 1.56 roots per shoot), this value is not significant in comparison to the treatment of 3.0 mg/L IBA. Similarly, all IBA treatments (1.0 mg/L up to 4.0 mg/L) also resulted in the stimulation of root elongation, as the values recorded were not significantly different from the values of the highest elongation observed in the treatment of 2.0 mg/L IAA. There were no morphological abnormalities or differences observed in the roots induced for all auxin treatments. The roots induced exhibited two different root types; aerial roots that were thin and hairy formed above the culture medium, and thick, brittle roots were embedded in the culture medium (Figure 1e). Figure 1f demonstrates the root morphology after 9 weeks of culture prior to acclimatization.

The efficiency of IBA in stimulating in vitro root formation was also reported for other fig cultivars. The addition of IBA increased the root number and length of "Sultany", "Gizy", and "Aboudi" fig cultivars [46]. The percentage of rooting in the cultivars "Brown Turkey" and "Brunswick", as well as "Zidi", were improved with IBA supplementation [47,48]. The treatment of 0.4 mg/L IBA resulted in the optimal percentage of rooting (83.33%) in the micropropagation of the yellow fig of the Golden Orphan cultivar [22]. Exogenous IBA was the main regulator of adventitious root induction, particularly in plant species with a very low content of endogenous auxin, such as *Arabidopsis thaliana* [49]. The supplementation of exogenous IBA is crucial as it leads to the conversion of IBA into IAA, which is necessary for the formation and stimulation of adventitious shoots [49]. The ability of IBA to promote roots could be credited to its relatively high stability [50]. IBA was also found to be more stable than IAA following exposure to high pressure and temperature during autoclave sterilization [51]. Moreover, IBA also acts as a slow-releasing hormone, enabling its effects to remain longer in the culture medium and resulting in the constant promotion of roots [50]. IBA can also be converted to free IAA at a continuous rate in the media, further enabling the constant presence of IAA rather than direct supplementation of IAA for plant regeneration [52]. These studies further indicated the efficiency of IBA in root induction, as observed in the current study for *Ficus carica*.

Table 2. The effects of IAA, IBA, and NAA on root regeneration after 6 weeks of culture.

Plant Growth Regulators (PGRs)	Concentrations (mg/L)	Rooting Percentage (%)	Number of Roots Induced per Explant	Length of Induced Roots (cm)
Control	0.0	46.67	2.20 ± 0.97 ^cd^	0.35 ± 0.12 ^rstu^
IAA	1.0	73.33	2.13 ± 0.52 ^cd^	0.30 ± 0.07 ^stu^
	2.0	86.67	4.47 ± 0.83 ^abc^	0.63 ± 0.14 ^r^
	3.0	73.33	3.33 ± 0.82 ^bcd^	0.39 ± 0.10 ^rstu^
	4.0	80.00	7.27 ± 1.56 ^a^	0.51 ± 0.09 ^rst^
IBA	1.0	73.33	5.07 ± 1.27 ^abc^	0.46 ± 0.10 ^rstu^
	2.0	86.67	6.73 ± 1.78 ^ab^	0.51 ± 0.08 ^rst^
	3.0	93.33	6.80 ± 1.43 ^ab^	0.59 ± 0.11 ^rs^
	4.0	73.33	3.67 ± 0.83 ^bcd^	0.48 ± 0.12 ^rstu^
NAA	1.0	66.66	3.27 ± 0.92 ^bcd^	0.19 ± 0.04 ^u^
	2.0	46.67	1.00 ± 0.31 ^d^	0.28 ± 0.09 ^tu^
	3.0	73.33	3.60 ± 0.85 ^bcd^	0.19 ± 0.04 ^u^
	4.0	73.33	5.40 ± 1.26 ^abc^	0.18 ± 0.04 ^u^

Means of different values shown by different letters in the same column are statistically different using Duncan's multiple range test ($p < 0.05$). Mean and standard error (±SE) were calculated from three experimental replications.

In the current study, auxin IAA was also observed to support the induction of in vitro roots for 'Violette de Solliès' as the supplementation at 2.0 and 4.0 mg/L recorded the highest average root number and length. In the rooting of fig from the cultivar "Conadria", the addition of IAA (0.1 mg/L) alone or in combination with IBA (1.0 mg/L) generated the maximum rooting percentage (100%). IAA was also found to be efficient at promoting root formation for other plant species. *Orthosiphon stamineus* explants produced the highest number of roots (11.9 ± 4.1 roots per explants) with the treatment of 3.0 mg/L IAA [53]. Auxin IAA can stimulate cell differentiation in the section responsible for rooting in cultures prior to root formation, further indicating its essential role in the aspect of the development of primary roots and lateral roots, as well as for the root apical meristem (RAM) and root vascular differentiation [53,54]. Moreover, the accumulation of IAA in the phloem was discovered to also induce the formation of adventitious roots in plants [55].

The treatment of auxin NAA, however, did not result in much significance in terms of root induction for the in vitro cultures of 'Violette de Solliès', as the roots formed rapidly grew in the treatments of IBA and IAA. NAA is another type of auxin commonly used in the rooting of different plant species. Rooting studies on *Ficus carica* of the Golden Orphan cultivar revealed that treatments of NAA in the rooting media resulted in callus formation, indirectly inhibiting the formation of roots for micropropagation [22]. NAA was observed to be inferior to IBA with regard to the rooting of fig, as the highest rooting percentage recorded in the treatment with NAA was 76.66%, whereas with IBA, the highest rate recorded was 91.11% [56]. NAA was also less efficient in rooting when compared to IBA for *Ficus carica* explants, recording a rooting rate as low as 8.34% [48].

Auxin NAA was also found to be unsuitable for the regeneration of certain plant tissues due to its low absorbance efficiency that could potentially reduce the rooting efficiency in culture [57]. However, NAA was effective at inducing roots from in vitro clones of *Ficus carica* [46]. Additionally, NAA was found to have successfully induced the highest average number of roots (4.87) and length (4.07 cm) for *Ficus carica* cv. Sultani explants [24]. This further indicated that the reactivity towards different types of auxins is species or cultivar specific, thus cementing the importance of an optimal rooting evaluation to ensure the production of healthy plantlets with higher survival rates.

3.3. Acclimatization and Effect of Different Soil Substrates on the Survival of Plantlets

In the present study, Jiffy pellets and biochar soil were the most effective soil substrates for the acclimatization of 'Violette de Solliès' plantlets, with both substrates recording the highest plantlet survival rate of 100% (Table 3). Vermiculite was the least suitable soil substrate as plantlets recorded the lowest survival rate (66.66%), which was even lower than the control (garden soil mix) at 73.33%. There were no morphological abnormalities observed in acclimatized plantlets (Figure 2a,b). Healthy acclimatized *Ficus carica* cv. 'Violette de Solliès' plantlets were then potted and supplemented with fertilizers of equal NPK ratios for eight weeks before being planted in the greenhouse. The images for the acclimatized plantlets, acclimatized plants grown in the commercial greenhouse, and fruits produced from the tissue-cultured plants after a year of being established in the greenhouse are shown in Figure 2.

Table 3. Percentage of survival rates of rooted plantlets of *Ficus carica* cv. 'Violette de Solliès' after 6 weeks of acclimatization in garden soil mixture (control) and different types of soil substrates.

Types of Soil Substrates	Survival Rate (%)
Garden soil mixture	73.33
Jiffy pellet	100.00
Peat moss	73.33
Perlite	80.00
Vermiculite	66.66
Peat moss and perlite mixture (1:1)	93.33
Peat moss and vermiculite (2:1)	93.33
Biochar soil	100.00

Figure 2. Ex vitro plantlets and fruits of *Ficus carica* cv. 'Violette de Solliès': (**a**) acclimatized plants in biochar soil after 2 weeks (scale bar = 1 cm); (**b**) acclimatized plants in polybags after 8 weeks (scale bar = 5 cm); (**c**) acclimatized plants grown in commercial greenhouse (scale bar = 2 cm); (**d**) fruit produced from tissue-cultured plants 10 months after acclimatization (scale bar = 1 cm).

The efficiency of the Jiffy pellets in aiding the acclimatization step in this study was attributed to its high porosity that enables rapid root penetration, allowing better nutrient transport and uptake from the soil [58]. Jiffy pellets are made of *Sphagnum* peat moss or coco fibers with high water retention capacities, which provides an ongoing supply of moisture essential for continuous plant growth and survival. The results from this study are in agreement with the findings from Mengesha et al. (2013), where pineapple plantlets were reported to show improved growth when acclimatized in Jiffy pellets, reporting a survival rate as high as 98%. Similarly, Jiffy pellets were also successfully utilized for the acclimatization of hazelnut plantlets, as plantlets had developed well-grown roots and recorded the highest survival rate of 97% [59]. As for the cultivar of 'Violette de Solliès', plantlets also recorded the maximum survival (100%) when acclimatized in biochar soil. Biochar soil is made up of carbon-rich organic materials, such as tree bark, wood chips, rice husk, and other agricultural wastes through pyrolysis, and its effectiveness is credited to its ability for improving soil structure as well as increasing pH and water retention capacities, leading to optimal plant growth [60,61]. Biochar soil was also used to improve the growth of maize [62] and mung bean [63], leading to healthy crop production. Both soil substrates were found most suitable for the acclimatization of 'Violette de Solliès' plantlets, but biochar soil is preferred due to its lower cost and easy market availability.

Based on current studies, it was also observed that 'Violette de Solliès' plantlets generated from the optimum tissue culture protocol were vigorous and fast-growing compared to conventionally grown figs. Fig fruits were observed to successfully produce from tissue-cultured plants approximately 10 months after acclimatization, which is approximately 2 to 3 times faster than the conventionally grown figs in the same cultivation plot. The tissue culture protocol generated from this study produced healthy growing plantlets that are disease-free and had a complete whole-plant system prior to acclimatization, which consequently, promotes vigorous plant growth when transferred to the commercial cultivation plot.

3.4. Polymorphism Analysis via RAPD and SCoT Molecular Markers

The seven RAPD primers generated a total of 47 bands ranging from 140 bp (OPU_20) to 1250 bp (OPU_12). In this study, there were no polymorphic bands detected via RAPD analysis through the comparison of banding patterns between the mother plant and micropropagated plant DNA samples, ranging from the first up to the sixth subculture cycle (Table 4). As for the SCoT analysis, 51 bands ranging from 200 bp (SCoT_22) to 2100 bp (SCoT_29) were obtained with an average of 7.29 bands per primer. Similar to the observation through RAPD analysis, no polymorphism was observed from the first up to the sixth subculture cycle of micropropagated plants (Table 4). The banding results revealed a total of 98 scorable bands from the RAPD and SCoT primers, where all the bands were monomorphic amongst the micropropagated plants and the mother plant (control) (Figure 3).

Somaclonal variation is a genetic variation associated with plant tissue culture caused by changes either at the genotypic or phenotypic level. The occurrence of mutations in plants generated via tissue culture can be triggered by numerous stress factors such as wounding and the exposure of explants to sterilizing agents, media components (plant growth regulators and salts), and in vitro culture environments (different lighting conditions, humidity, and temperature) [64]. Others stress factors such as the rapid multiplication of cultures [65] or frequent subculturing may lead to mutations and somaclonal variations. Therefore, it is recommended that the developed micropropagation protocol should ensure the number of subcultures required is kept to a minimum [66]. For instance, a frequency increase in somaclonal variation from 1.3% to 3.8% between the 5th and 11th subculture of the Brazilian banana "Nanicão" could negatively impact the quality of plantlets produced [67].

Table 4. The nucleotide sequences of primers used for RAPD and SCoT molecular marker analysis.

Primers	Primer Sequence (5′-3′)	Total No. of Bands	Number of Monomorphic Bands	Number of Polymorphic Bands	Percentage of Polymorphic Bands (%)	Size Range (bp)
RAPD						
OPC_02	GTGAGGCGTC	6	6	0	0	300–900
OPC_09	CTCACCGTCC	9	9	0	0	320–1100
OPC_20	ACTTCGCCAC	5	5	0	0	350–1100
OPK_04	CCGCCCAAAC	5	5	0	0	200–900
OPU_05	TTGGCGGCCT	8	8	0	0	250–1200
OPU_12	TCACCAGCCA	4	4	0	0	650–1250
OPU_20	ACAGCCCCCA	10	10	0	0	140–1200
SCoT						
SCoT_07	CAACA**ATG**GCTACCACGG	6	6	0	0	600–2000
SCoT_21	ACCAC**ATG**GCGACCCACA	7	7	0	0	260–900
SCoT_22	AACC**ATG**GCTACCACCAC	10	10	0	0	200–2000
SCoT_24	CACC**ATG**GCTACCACCAT	8	8	0	0	550–1500
SCoT_26	CACC**ATG**GCTACCACCAT	6	6	0	0	400–900
SCoT_29	ACC**ATG**GCTACCACCGTC	10	10	0	0	350–2100
SCoT_35	C**ATG**GCTACCACCGGCCC	4	4	0	0	400–550

In this study, no polymorphism was observed from the first up to the sixth subculture cycle of micropropagated plants, as confirmed via RAPD and SCoT molecular marker analysis. A previous study utilized inter-simple sequence repeat (ISSR) markers for the identification of the genetic alteration of the fig cultivar "Black fig" exposed to different plant growth regulators, growth retardants, photoperiods, and cold acclimatization [68]. In addition, homogeneity between the fig mother plant and the micropropagated plantlets was successfully evaluated by El-Dessoky et al. (2016), using RAPD and ISSR markers with 45 scorable bands produced from six RAPD primers showing 6.6% polymorphism [69]. As opposed to the findings by El-Dessoky et al. (2016), the current study reported 100% monomorphic bands based on RAPD analysis of the total 60 RAPD primers screened, compared to only 16 RAPD primers screened by El-Dessoky et al. (2016). RAPD markers could only amplify a small region of the plant genome; therefore, the use of other molecular markers to compliment and validate the other regions of the DNA is most preferred, particularly to ensure higher reliability in the detection of polymorphism [70]. The current study includes the application of SCoT molecular markers which target the short, conserved regions of plant genes surrounding the ATG start codon [29] to further confirm the genetic stability of micropropagated *Ficus carica* plantlets. SCoT is a reliable gene-targeted marker, where the molecular profiles generated by SCoT primers can be correlated to the plant's functional genes, and thus, correspond to its respective traits [29,71,72]. The application of SCoT molecular markers for genetic variation detection in micropropagated plantlets have been previously utilized for the species of *Pittosporum eriocarpum* Royle [73], *Ansellia africana* [72], and *Bauhinia racemosa* Lam [74]. However, up to now, the application of SCoT molecular markers in polymorphism detection for the *Ficus* genus has not been reported. The current study is the first to assess the suitability of SCoT molecular markers for polymorphism detection of micropropagated *Ficus carica* plantlets. In the present study, the generated SCoT molecular profile has further confirmed the clonal uniformity of the micropropagated fig plantlets up to the sixth number of the subculture cycle. This has further validated the reliability of the developed micropropagation protocol in this study for the mass production of plantlets which maintain the novel genetic make-up of the parent plant, making it suitable for commercial planting.

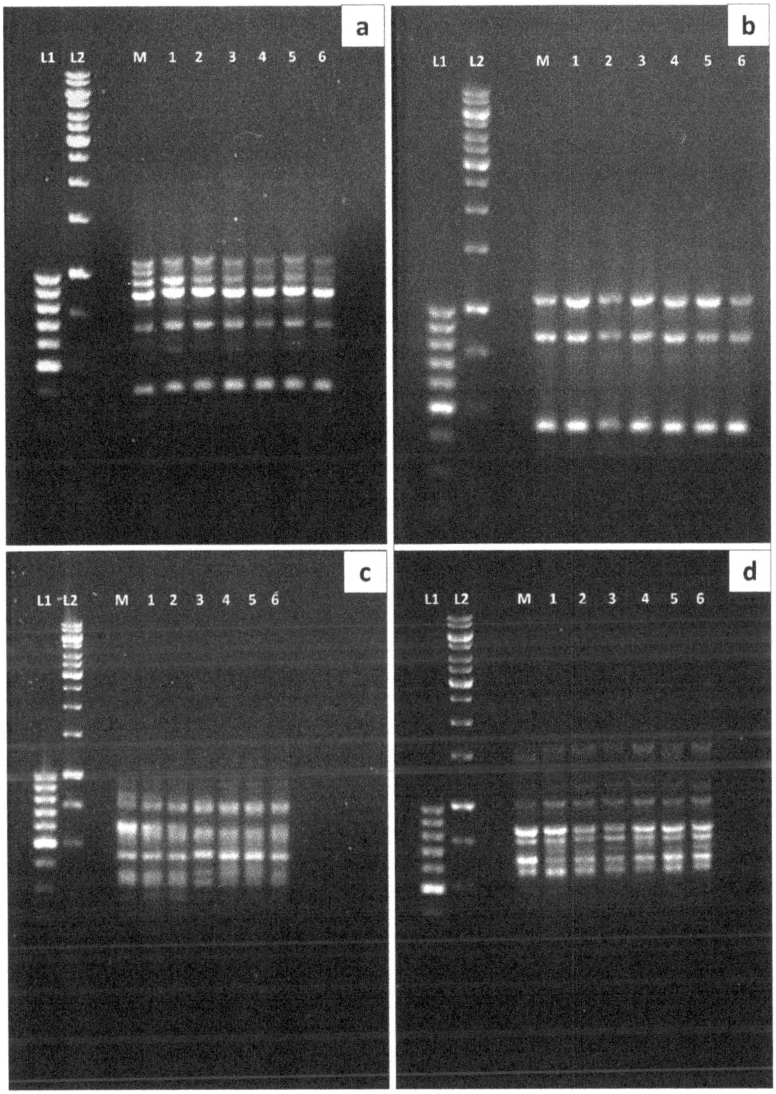

Figure 3. RAPD and SCoT marker profiles of mother plant and micropropagated plants (first to sixth subculture cycles) of *Ficus carica* cv. 'Violette de Solliès': (**a**) OPU_U20; (**b**) OPC_20; (**c**) SCoT_21; (**d**) SCoT_24. Lane L1 and L2 represent the 100 bp and 1 kb ladder DNA, respectively; lane M represents mother plant and lanes 1–6 represent the micropropagated plants from first to sixth subculture (n = 3).

3.5. Observation of Stomatal Structures via SEM

In the present study, SEM was utilized to assist with further structural observations on the stomata of the in vitro regenerated plantlets. From the analysis, it was evident that there were no significant differences observed in the stomatal structures from the abaxial epidermis of fig leaves obtained from plantlets after 20 and 60 days of acclimatization in comparison to the field-grown plants. However, a crystal-like formation was prominent in the stomata of in vitro explants, as illustrated in Figure 4a. These crystalline structures completely disappeared post-acclimatization (Figure 4b,c) and were also absent in field-grown plants (Figure 4d). This observation further validates that the exposure to in vitro culture conditions and subculture durations did not result in alterations in the morphological characteristics of the fig plantlets grown post-acclimatization, regardless of being previously subjected to culture conditions. The stomata of the leaf samples that were 20 and 60 days old and from acclimatized plantlets when observed under SEM were deemed to be normal with kidney-shaped guard cells and external periclinal walls that were well formed without signs of deformations; this was the same as for field-grown plants. Normal stomata formation is crucial to prevent disruption of the gaseous exchange channels that facilitate the plant respiration, photosynthesis, and transpiration required for optimal plant growth. Besides, the stomata also play a key role in inducing plant responses and their ability to adapt, as it is highly sensitive to environmental changes [75]. However, structural abnormalities in the stomata may be environmentally induced, resulting in the breakdown of stomatal functions which could negatively impact plant growth and development, leading to eventual death. Based on our observations and screening of the leaf samples under SEM in this study (Figure 4), the in vitro and acclimatized plant stomatal complexes were of the actinocytic type. The actinocytic type of stomata was observed in all plants of the *Ficus* genus and is found only on the abaxial surface (hypostomatic) of the leaf [76].

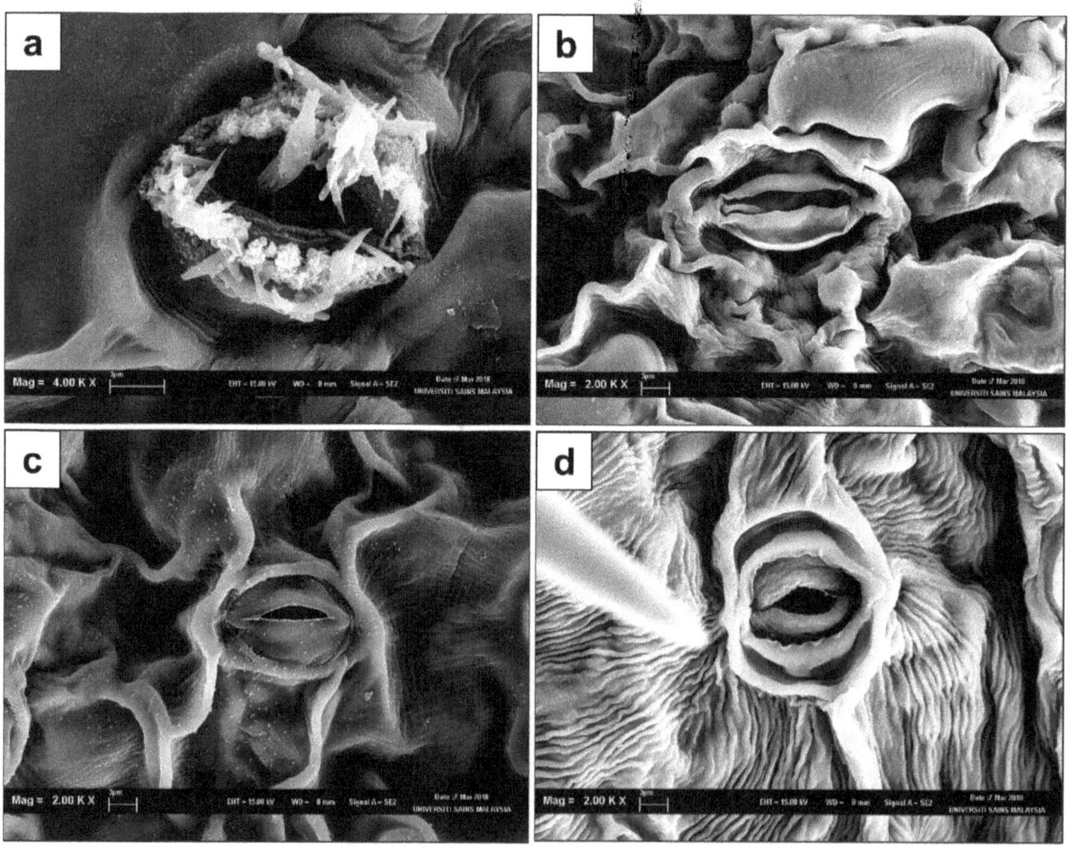

Figure 4. Scanning electron microscopy images of leaves of *Ficus carica* cv. 'Violette de Solliès' pre- and post-acclimatization: (**a**) in vitro leaf; (**b**) 20 days after acclimatization; (**c**) 40 days after acclimatization; (**d**) field-grown plant.

On the other hand, crystalline-like structures were also observed on the leaves of in vitro explants in this study. It was hypothesized that these structures are cell-mediated calcium oxalate (CaOx) crystals. Calcium is known to be abundantly found in the environment and is a vital element for plant growth and development, mainly as a structural component of the cell wall that functions as an osmoticum, as well as playing a pivotal role in the signaling of developmental pathways [77]. However, calcium at high cytosolic free levels with concentrations exceeding 10^{-7} M could result in detrimental effects on the calcium signaling pathway, energy metabolism, and microskeletal dynamics [78]. Therefore, the uptake, distribution, and storage of calcium within the cell wall and the vacuole needs to be carefully regulated in plants [79]. Calcium oxalate crystals are often formed in plant cells (bundle sheath cells), tissues (epidermal, ground, and vascular tissues), or organs (leaves, roots, and stems) when the levels of calcium within the cell is present in excess (Webb, 1999). The leaves of plants from the *Ficus* genus were reported to accumulate minerals in the form of amorphous calcium carbonate cystoliths, calcium oxalates, and silica phytoliths [80]. This report potentially indicates that the observed crystalline-like structures of the in vitro leaf samples were associated with the deposition of these minerals.

4. Conclusions

The current study is the first to report on an efficient and effective micropropagation protocol of *Ficus carica* cv. 'Violette de Solliès'. The MS medium supplemented with 5.0 mg/L BAP was identified to be the optimal media for multiple shoot formation, whereas the WPM supplemented with 3.0 mg/L IBA induced the highest percentage of rooting. On the other hand, biochar soil was found to be the optimal soil substrate for fig acclimatization. The polymorphism analysis via RAPD and SCoT molecular techniques indicated no genetic variation occurrence between plantlets up to the sixth subculture cycle, whereas the normal structural development of stomata was observed in acclimatized plantlets as confirmed by SEM analysis, further indicating that the micropropagated plants are genetically uniform and true-to-type.

Author Contributions: Conceptualization, B.L.C., W.T.L. and S.S.; methodology, W.T.L. and L.V.T.; validation, S.P.K., L.V.T. and D.S.; resources, B.L.C.; data curation, S.P.K.; writing—original W.T.L. and L.V.T.; writing—B.L.C., D.S. and S.P.K.; visualization, B.L.C.; supervision, B.L.C.; project administration, B.L.C., W.T.L. and L.V.T.; funding acquisition, B.L.C. All authors have read and agreed to the published version of the manuscript.

Funding: The Economic Planning Unit of the Prime Minister's Department, Malaysia (Grant code: 304/PBIOLOGI/6501099/U120).

Acknowledgments: The authors would like to acknowledge the Economic Planning Unit of the Prime Minister's Department, Malaysia for funding the project (Grant code: 304/PBIOLOGI/6501099/U120). They also thank Universiti Sains Malaysia, FigDirect Sdn. Bhd, and the Agricultural Crop Trust (Malaysia) for their support of this project.

Conflicts of Interest: The authors have no conflict of interest to declare.

References

1. Dhage, S.; Pawar, B.; Chimote, V.; Jadhav, S.; Kale, A.A. In vitro callus induction and plantlet regeneration in Fig (*Ficus carica* L.). *J. Cell Tissue Res.* **2012**, *12*, 3395–3400.
2. Danial, G.H.; Ibrahim, D.A.; Brkat, S.A.; Khalil, B.M. Multiple shoots production from shoot tips of Fig tree (*Ficus carica* L.) and callus induction from leaf segments. *Int. J. Pure Appl. Sci. Technol.* **2014**, *9*, 117–124.
3. Chawla, A.; Kaur, R.; Sharma, A.K. *Ficus carica* Linn.: A review on its pharmacognostic, phytochemical and pharmacological aspects. *Int. J. Pharmacol. Res.* **2012**, *2012*, 215–232.
4. Flaishman, M.A.; Rodov, V.; Stover, E. The Fig: Botany, Horticulture, and Breeding. In *Horticultural Reviews*; Janick, J., Ed.; John Wiley & Sons, Inc.: Hoboken, NJ, USA, 2008; pp. 113–196. ISBN 978-0-470-38014-7.
5. Daily Sabah with AA Turkey Top Producer, Exporter of Hazelnuts, Cherries, Figs, Apricots. *Dly Sabah.* 19 August 2019. Available online: https://www.dailysabah.com/business/2019/08/19/turkey-top-producer-exporter-of-hazelnuts-cherries-figs-apricots (accessed on 1 November 2022).

6. Moniruzzaman, M.; Yaakob, Z.; Khatun, R.; Awang, N. Mealybug (Pseudococcidae) Infestation and organic control in Fig (*Ficus carica*) orchards of Malaysia. *Biol. Environ. Proc. R. Ir. Acad.* **2017**, *117*, 25–32. [CrossRef]
7. Kamarubahrin, A.; Haris, A.; Abdul Shukor, S.; Mohd Daud, S.N.; Ahmad, N.; Zulkefli, Z.; Muhamed, N.A.; Makmun, A.H. An overview Malaysia as a hub of planting prophetic fruits. *Malays. J. Sustain. Agric.* **2019**, *3*, 13–19. [CrossRef]
8. Crisosto, H.; Ferguson, L.; Bremer, V.; Stover, E.; Colelli, G. Fig (*Ficus carica* L.). In *Postharvest Biology and Technology of Tropical and Subtropical Fruits*; Elsevier: Amsterdam, The Netherlands, 2011; pp. 134–160. ISBN 978-1-84569-735-8.
9. Stalin, C.; Dineshkumar, P.; Nithiyananthan, K. Evaluation of antidiabetic activity of methanolic leaf extract of *Ficus carica* in alloxan - induced diabetic rats. *Asian J. Pharm. Clin. Res.* **2012**, *5*, 85–87.
10. Ahmad, M.Z.; Ali, M.; Mir, S.R. Anti-diabetic activity of *Ficus carica* L. stem barks and isolation of two new flavonol esters from the plant by using spectroscopical techniques. *Asian J. Biomed. Pharm. Sci.* **2013**, *3*, 7.
11. Mopuri, R.; Islam, M.S. Antidiabetic and anti-obesity activity of *Ficus carica*: In vitro experimental studies. *Diabetes Metab.* **2016**, *42*, 300. [CrossRef]
12. Canal, J.; Torres, M.; Romero, A.; Perez, C. Chloroform extract obtained from a decoction of *Ficus carica* leaves improves the cholesterolaemic status of rats with streptozotocin-induced diabetes. *Acta Physiol. Hung.* **2000**, *87*, 71–76. [CrossRef]
13. Rashidi, A.A.; Noureddini, M. Hypoglycemic effect of the aromatic water of leaves of *Ficus carica* in normal and streptozotocin induced diabetic rats. *Pharmacologyonline* **2011**, *1*, 372–379.
14. Sheikh, Y.; Maibam, B.C.; Biswas, D.; Laisharm, S.; Deb, L.; Talukdar, N.C.; Borah, J.C. Anti-diabetic potential of selected ethno-medicinal plants of North East India. *J. Ethnopharmacol.* **2015**, *171*, 37–41. [CrossRef] [PubMed]
15. Stephen Irudayaraj, S.; Christudas, S.; Antony, S.; Duraipandiyan, V.; Naif Abdullah, A.-D.; Ignacimuthu, S. Protective effects of *Ficus carica* leaves on glucose and lipids levels, carbohydrate metabolism enzymes and β-cells in type 2 diabetic rats. *Pharm. Biol.* **2017**, *55*, 1074–1081. [CrossRef] [PubMed]
16. El-Shobaki, F.; El-Bahay, A.; Esmail, R.; El-Megeid, A.A.; Esmail, N. Effect of figs fruit (*Ficus carica* L.) and its leaves on hyperglycemia in alloxan diabetic rats. *World J. Dairy Food Sci.* **2010**, *5*, 47–57.
17. Guvenc, M.; Tuzcu, M.; Yilmaz, O. Analysis of fatty acid and some lipophilic vitamins found in the fruits of the *Ficus carica* variety picked from the Adiyaman district. *Res. J. Biol. Sci.* **2009**, *4*, 320–323.
18. Kumar, V.; Radha, A.; Kumar Chitta, S. In vitro plant regeneration of Fig (*Ficus carica* L. cv. Gular) using apical buds from mature trees. *Plant Cell Rep.* **1998**, *17*, 717–720. [CrossRef]
19. Dolgun, O.; Tekintas, F. Production of Fig (*Ficus carica* L.) nursery plants by stem layering method. *Agric. Conspec. Sci. ACS* **2008**, *73*, 157–160.
20. Kajla, S.; Poonia, A.K.; Kharb, P.; Duhan, J.S. Role of biotechnology for commercial production of fruit crops. In *Biotechnology: Prospects and Applications*; Salar, R.K., Gahlawat, S.K., Siwach, P., Duhan, J.S., Eds.; Springer India: New Delhi, India, 2013; pp. 27–37. ISBN 978-81-322-1682-7.
21. Parab, A.; Chew, B.; Lit chow, Y.; Subramaniam, S. Organogenesis on apical buds in common Fig (*Ficus carica*) var. Black Jack. *Electron. J. Biotechnol.* **2021**, *54*, 69–76. [CrossRef]
22. Sriskanda, D.; Liew, Y.X.; Khor, S.P.; Merican, F.; Subramaniam, S.; Chew, B.L. An efficient micropropagation protocol for *Ficus carica* cv. Golden Orphan suitable for mass propagation. *Biocatal. Agric. Biotechnol.* **2021**, *38*, 102225. [CrossRef]
23. Kim, K.-M.; Kim, M.Y.; Yun, P.Y.; Chandrasekhar, T.; Lee, H.-Y.; Song, P.-S. Production of multiple shoots and plant regeneration from leaf segments of Fig tree (*Ficus carica* L.). *J. Plant Biol.* **2007**, *50*, 440–446. [CrossRef]
24. Soliman, H.I.; Gabr, M.; Abdallah, N.A. Efficient transformation and regeneration of Fig (*Ficus carica* L.) via somatic embryogenesis. *GM Crops* **2010**, *1*, 40–51. [CrossRef]
25. Al-Khaybari, A. *Propagation of Fig Trees (cv. Brown Turkey) by Tissue Culture Technique*; King Saud University: Riyadh, Saudi Arabia, 2008.
26. Ling, W.; Liew, F.; Lim, W.; Subramaniam, S.; Chew, B.L. Shoot induction from axillary shoot tip explants of Fig (*Ficus carica*) cv. Japanese BTM 6. *Trop. Life Sci. Res.* **2018**, *29*, 165–174. [CrossRef] [PubMed]
27. Murashige, T.; Skoog, F. A revised medium for rapid growth and bio assays with tobacco tissue cultures. *Physiol. Plant.* **1962**, *15*, 473–497. [CrossRef]
28. Lloyd, G.B.; McCown, B.H. Commercially-feasible micropropagation of mountain laurel, *Kalmia latifolia*, by use of shoot-tip culture. *Comb. Proc. Int. Plant Propagators' Soc. (USA)* **1980**, *30*, 421–427.
29. Collard, B.C.Y.; Mackill, D.J. Start Codon Targeted (SCoT) polymorphism: A simple, novel dna marker technique for generating gene-targeted markers in plants. *Plant Mol. Biol. Rep.* **2009**, *27*, 86–93. [CrossRef]
30. Košir, P.; Škof, S.; Luthar, Z. Direct shoot regeneration from nodes of *Phalaenopsis* orchids. *Acta Agric. Slov.* **2004**, *83*, 233–242.
31. Zanello, C.A.; Duarte, W.N.; Gomes, D.M.; Cardoso, J.C. Micropropagation from inflorescence nodal segments of *Phalaenopsis* and acclimatization of plantlets using different substrates. *Horticulturae* **2022**, *8*, 340. [CrossRef]
32. Le Bris, M. Hormones in Growth and Development. In *Reference Module in Life Sciences*; Elsevier: Amsterdam, The Netherlands, 2017; p. B9780128096338051000; ISBN 978-0-12-809633-8.
33. Seyyedyousefi, S.R.; Kaviani, B.; Dehkaei, N.P. The effect of different concentrations of NAA and BAP on micropropagation of *Alstroemeria*. *Eur. J. Exp. Biol.* **2013**, *3*, 133–136.
34. Demiralay, A.; Yalçin-Mendi, Y.; Aka-Kaçar, Y.; Çetiner, S. In vitro Propagation of *Ficus carica* L. var. Bursa Siyahi through meristem culture. *Acta Hortic.* **1998**, *480*, 165–168. [CrossRef]

35. Mustafa, N.; Taha, R. Influence of plant growth regulators and subculturing on *in vitro* multiplication of some Fig (*Ficus carica*) Cultivars. *J. Appl. Sci. Res.* **2012**, *8*, 4038–4044.
36. Shatnawi, M.; Shibli, R.A.; Shahrour, W.G.; Al-Qudah, T.S.; Taleb, A.-Z. Micropropagation and conservation of Fig (*Ficus carica* L.). *J. Adv. Agric.* **2019**, *10*, 1669–1679. [CrossRef]
37. Qiu, L.; Chen, R.; Fan, Y.; Huang, X.; Luo, H.; Xiong, F.; Liu, J.; Zhang, R.; Lei, J.; Zhou, H.; et al. Integrated MRNA and small RNA sequencing reveals microRNA regulatory network associated with internode elongation in sugarcane (*Saccharum officinarum* L.). *BMC Genom.* **2019**, *20*, 817. [CrossRef] [PubMed]
38. Debnath, S.C. Micropropagation of Lingonberry: Influence of genotype, explant orientation, and overcoming TDZ-Induced inhibition of shoot elongation using zeatin. *HortScience* **2005**, *40*, 185–188. [CrossRef]
39. Kim, T.-D.; Kim, J.-A.; Lee, N.-N.; Choi, C.-H. Multiple shoot induction and plant regeneration from axillary buds of *Magnolia* 'Vulcan. ' *J. Plant Biotechnol.* **2020**, *47*, 40–45. [CrossRef]
40. Fráguas, C.B.; Pasqual, M.; Dutra, L.F.; Cazetta, J.O. Micropropagation of Fig (*Ficus carica* L.) "Roxo de Valinhos" plants. *In Vitro Cell. Dev. Biol.* **2013**, *40*, 471–474. [CrossRef]
41. Dhage, S.S.; Chimote, V.P.; Pawar, B.D.; Kale, A.A.; Pawar, S.V.; Jadhav, A.S. Development of an efficient in vitro regeneration protocol in Fig (*Ficus carica* L.). *J. Appl. Hortic.* **2015**, *17*, 160–164. [CrossRef]
42. Sivanesan, I.; Song, J.Y.; Hwang, S.J.; Jeong, B.R. Micropropagation of *Cotoneaster Wilsonii* Nakai—A rare endemic ornamental plant. *Plant Cell Tissue Organ Cult. PCTOC* **2011**, *105*, 55–63. [CrossRef]
43. Feng, J.-C.; Yu, X.M.; Shang, X.L.; Li, J.D.; Wu, Y.X. Factors influencing efficiency of shoot regeneration in *Ziziphus jujuba* Mill. *'Huizao.' Plant Cell Tissue Organ Cult. PCTOC* **2010**, *101*, 111–117. [CrossRef]
44. Siwach, P.; Gill, A.R. Enhanced shoot multiplication in *Ficus religiosa* L. in the presence of adenine sulphate, glutamine and phloroglucinol. *Physiol. Mol. Biol. Plants* **2011**, *17*, 271–280. [CrossRef]
45. Zhang, C.G.; Li, W.; Mao, Y.F.; Zhao, D.L.; Dong, W.; Guo, G.Q. Endogenous hormonal levels in *Scutellaria baicalensis* calli induced by thidiazuron. *Russ. J. Plant Physiol.* **2005**, *52*, 345–351. [CrossRef]
46. Hepaksoy, S.; Aksoy, U. Propagation of *Ficus carica* L. clones by *in vitro* culture. *Biol. Plant.* **2006**, *50*, 433–436. [CrossRef]
47. Metwali, E.; Soliman, H.; Al-Zahrani, H.; Howladar, S.; Fuller, M. Influence of different concentrations of salt stress on in vitro multiplication of some Fig (*Ficus carica* L.) cultivars. *Life Sci. J.* **2014**, *11*.
48. Bayoudh, C.; Labidi, R.; Majdoub, A.; Mars, M. In vitro propagation of caprifig and female fig varieties (*Ficus carica* L.) from shoot-tips. *J. Agric. Sci. Technol.* **2015**, *17*, 1597–1608.
49. Fattorini, L.; Veloccia, A.; Della Rovere, F.; D'Angeli, S.; Falasca, G.; Altamura, M.M. Indole-3-butyric acid promotes adventitious rooting in *Arabidopsis thaliana* thin cell layers by conversion into indole-3-acetic acid and stimulation of anthranilate synthase activity. *BMC Plant Biol.* **2017**, *17*, 121. [CrossRef]
50. Epstein, E.; Ludwig-Muller, J. Indole-3-butyric acid in plants: Occurrence, synthesis, metabolism and transport. *Physiol. Plant.* **1993**, *88*, 382–389. [CrossRef]
51. Nissen, S.; Sutter, E. Stability of IAA and IBA in nutrient medium to several tissue culture procedures. *HortScience* **1990**, *25*, 800–802. [CrossRef]
52. Epstein, E.; Sagee, O.; Zelcer, A. Uptake and metabolism of indole-3-butyric acid and indole-3-acetic acid by *Petunia* cell suspension culture. *Plant Growth Regul.* **1993**, *13*, 31–40. [CrossRef]
53. Pick, A.; Ling, A.; Kok, K.; Hussein, S.; Ling, S.; Ong, S.L. Effects of plant growth regulators on adventitious roots induction from different explants of *Orthosiphon stamineus. Am. -Eurasian J. Sustain. Agric.* **2009**, *3*, 493–501.
54. Ponce, G.; Barlow, P.W.; Feldman, L.J.; Cassab, G.I. Auxin and ethylene interactions control mitotic activity of the quiescent centre, root cap size, and pattern of cap cell differentiation in maize. *Plant Cell Environ.* **2005**, *28*, 719–732. [CrossRef]
55. Aloni, R.; Aloni, E.; Langhans, M.; Ullrich, C.I. Role of cytokinin and auxin in shaping root architecture: Regulating vascular differentiation, lateral root initiation, root apical dominance and root gravitropism. *Ann. Bot.* **2006**, *97*, 883–893. [CrossRef]
56. Wakle, A.G.; Shinde, S.J.; Jadhav, S.D.; Gharate, P.S. Influence of different levels of IAA and NAA growth regulators on air layers in Fig (*Ficus carica* L.) cv. Dinkar. *J. Pharm. Innov.* **2021**, *10*, 119–124.
57. Hausman, J.F. Changes in peroxidase activity, auxin level and ethylene production during root formation by poplar shoots raised in vitro. *Plant Growth Regul.* **1993**, *13*, 263–268. [CrossRef]
58. Mengesha, A.; Ayenew, B.; Tadesse, T. Acclimatization of *in vitro* propagated pineapple (*Ananas comosus* (L.), var. Smooth cayenne) plantlets to ex vitro condition in Ethiopia. *Am. J. Plant Sci.* **2013**, *4*, 317–323. [CrossRef]
59. Nas, M.; Read, P. Improved rooting and acclimatization of micropropagated hazelnut shoots. *HortScience Publ. Am. Soc. Hortic. Sci.* **2004**, *39*, 1688–1690. [CrossRef]
60. Chan, K.Y.; Van Zwieten, L.; Meszaros, I.; Downie, A.; Joseph, S. Using poultry litter biochars as soil amendments. *Soil Res.* **2008**, *46*, 437. [CrossRef]
61. Ronsse, F.; van Hecke, S.; Dickinson, D.; Prins, W. Production and characterization of slow pyrolysis biochar: Influence of feedstock type and pyrolysis conditions. *GCB Bioenergy* **2013**, *5*, 104–115. [CrossRef]
62. Uzoma, K.C.; Inoue, M.; Andry, H.; Fujimaki, H.; Zahoor, A.; Nishihara, E. Effect of cow manure biochar on maize productivity under sandy soil condition: Cow manure biochar agronomic effects in sandy soil. *Soil Use Manag.* **2011**, *27*, 205–212. [CrossRef]
63. Saxena, J.; Rawat, J.; Kumar, R. Conversion of biomass waste into biochar and the effect on mung bean crop production. *CLEAN—Soil Air Water* **2017**, *45*, 1501020. [CrossRef]

64. Krishna, H.; Alizadeh, M.; Singh, D.; Singh, U.; Chauhan, N.; Eftekhari, M.; Sadh, R.K. Somaclonal variations and their applications in horticultural crops improvement. *3 Biotech* **2016**, *6*, 54. [CrossRef]
65. Israeli, Y.; Lahav, E.; Reuveni, O. In Vitro Culture of Bananas. In *Bananas and Plantains*; Gowen, S., Ed.; Springer Netherlands: Dordrecht, The Netherlands, 1995; pp. 147–178. ISBN 978-94-011-0737-2.
66. Bhatia, S.; Sharma, K. Chapter 13—Technical Glitches in Micropropagation. In *Modern Applications of Plant Biotechnology in Pharmaceutical Sciences*; Bhatia, S., Sharma, K., Dahiya, R., Bera, T., Eds.; Academic Press: Boston, MA, USA, 2015; pp. 393–404. ISBN 978-0-12-802221-4.
67. Rodrigues, P.; Tulmann Neto, A.; Cassieri Neto, P.; Mendes, B. Influence of the Number of Subcultures on Somaclonal Variation in Micropropagated Nanicã O (Musa spp., Aaa Group). *Acta Hortic.* **1997**, *490*, 469–474.
68. Fatah, N.; Abouarab, M.; Amin, A.; Diab, A. Short term preservation for fig (*Ficus carica* cv. Black Fig) by different osmotic stabilizers. *Egypt. J. Genet. Cytol.* **2016**, *45*, 47–61. [CrossRef]
69. Dessoky, E.S.; Attia, A.O.; Mohamed, E.A.M. An efficient protocol for *in vitro* propagation of fig (*Ficus carica* sp.) and evaluation of genetic fidelity using RAPD and ISSR markers. *J. Appl. Biol. Biotechnol.* **2016**, *4*, 057–063.
70. Khor, S.P.; Yeow, L.C.; Poobathy, R.; Zakaria, R.; Chew, B.L.; Subramaniam, S. Droplet-vitrification of Aranda Broga Blue orchid: Role of ascorbic acid on the antioxidant system and genetic fidelity assessments via RAPD and SCoT markers. *Biotechnol. Rep.* **2020**, *26*, e00448. [CrossRef] [PubMed]
71. Bhattacharyya, P.; Kumaria, S.; Tandon, P. High Frequency regeneration protocol for *Dendrobium nobile*: A model tissue culture approach for propagation of medicinally important orchid species. *S. Afr. J. Bot.* **2016**, *104*, 232–243. [CrossRef]
72. Bhattacharyya, P.; Kumar, V.; Van Staden, J. Assessment of genetic stability amongst micropropagated *Ansellia africana*, a vulnerable medicinal orchid species of Africa using SCoT markers. *S. Afr. J. Bot.* **2017**, *108*, 294–302. [CrossRef]
73. Thakur, J.; Dwivedi, M.D.; Sourabh, P.; Uniyal, P.L.; Pandey, A.K. Genetic Homogeneity revealed using SCoT, ISSR and RAPD markers in micropropagated *Pittosporum eriocarpum* Royle-An endemic and endangered medicinal plant. *PLoS ONE* **2016**, *11*, e0159050. [CrossRef]
74. Sharma, U.; Rai, M.K.; Shekhawat, N.S.; Kataria, V. Genetic homogeneity revealed in micropropagated *Bauhinia racemosa* Lam. using gene targeted markers CBDP and SCoT. *Physiol. Mol. Biol. Plants* **2019**, *25*, 581–588. [CrossRef]
75. Zhu, J.; Yu, Q.; Xu, C.; Li, J.; Qin, G. Rapid estimation of stomatal density and stomatal area of plant leaves based on object-oriented classification and its ecological trade-off strategy analysis. *Forests* **2018**, *9*, 616. [CrossRef]
76. Shakir, H.M.; Baji, S.H. Anatomical study of some characters in certain species of genus *Ficus L.* growing in Iraq. *J. Biol. Agric. Healthc.* **2016**, *6*, 98–105.
77. Webb, M.A. Cell-mediated crystallization of calcium oxalate in plants. *Plant Cell* **1999**, *11*, 751–761. [CrossRef]
78. Kretsinger, R.H. Evolution of the Informational Role of Calcium in Eukaryotes. In *Calcium-Binding Proteins and Calcium Function*; North Holland Publishing: New York, NY, USA, 1977; pp. 63–72.
79. Leigh, R.A.; Tomos, A.D.; Mansfield, T.A.; Davies, W.J.; Leigh, R.A. Ion distribution in cereal leaves: Pathways and mechanisms. *Philos. Trans. R. Soc. Lond. B. Biol. Sci.* **1993**, *341*, 75–86.
80. Pierantoni, M.; Tenne, R.; Rephael, B.; Brumfeld, V.; van Casteren, A.; Kupczik, K.; Oron, D.; Addadi, L.; Weiner, S. Mineral deposits in *Ficus* leaves: Morphologies and locations in relation to function. *Plant Physiol.* **2018**, *176*, 1751–1763. [CrossRef] [PubMed]

horticulturae

Article

Partial Elimination of Viruses from Traditional Potato Cultivar 'Brinjak' by Chemotherapy and Its Impact on Physiology and Yield Components

Snježana Kereša [1,*], Darko Vončina [2], Boris Lazarević [3], Anita Bošnjak Mihovilović [1], Milan Pospišil [4], Marina Brčić [4], Ana Matković Stanković [1] and Ivanka Habuš Jerčić [1]

[1] Department of Plant Breeding, Genetics and Biometrics, Faculty of Agriculture, University of Zagreb, Svetošimunska cesta 25, 10000 Zagreb, Croatia
[2] Department of Plant Pathology, Faculty of Agriculture, University of Zagreb, Svetošimunska cesta 25, 10000 Zagreb, Croatia
[3] Department of Plant Nutrition, Faculty of Agriculture, University of Zagreb, Svetošimunska cesta 25, 10000 Zagreb, Croatia
[4] Department of Field Crops, Forage and Grassland, Faculty of Agriculture, University of Zagreb, Svetošimunska cesta 25, 10000 Zagreb, Croatia
* Correspondence: skeresa@agr.hr; Tel.: +385-1-239-3801

Abstract: Viruses are responsible for more than 50% of annual potato tuber yield losses and cause great economic damage. The traditional Croatian potato cultivar 'Brinjak' is important for local growers because of its economically profitable production and as a gene pool for future breeding programs. However, the full genetic potential of the cultivar cannot be exploited due to virus infection. In this study, we attempted to eliminate potato virus M (PVM) and potato virus S (PVS) from potato cultivar 'Brinjak' and to evaluate the effects on physiological parameters and yield. Shoot apices were isolated from PVM + PVS-infected sprouts and cultivated for six weeks on MS medium with the addition of 50 or 100 mg L^{-1} ribavirin. The surviving shoot apices were micropropagated. The in vitro post-eradication period lasted 200 days. DAS-ELISA and RT-PCR were performed on R0 and R1 plants 90 days after acclimatization to determine the sanitary status of the plants. Chlorophyll fluorescence and multispectral imaging were performed on the R0 plants at the same time. The success of PVS elimination was 33% at both ribavirin concentrations. However, neither concentration was successful in eliminating PVM. Plants with mixed infection (PVM + PVS) had more severe disease symptoms compared to PVM-infected plants, affecting photochemistry and multispectral parameters and, consequently, yield. PVM + PVS plants had significantly lower number and weight of tubers per plant and lower average tuber weight than plants with single PVM infection in most of the generations studied. The results indicate a strong negative impact of PVS in mixed infections with PVM and show the importance of its elimination from potato plants.

Keywords: *Solanum tuberosum*; ribavirin; virus elimination; potato virus M; potato virus S; multispectral parameters; yield

Citation: Kereša, S.; Vončina, D.; Lazarević, B.; Bošnjak Mihovilović, A.; Pospišil, M.; Brčić, M.; Matković Stanković, A.; Habuš Jerčić, I. Partial Elimination of Viruses from Traditional Potato Cultivar 'Brinjak' by Chemotherapy and Its Impact on Physiology and Yield Components. *Horticulturae* **2022**, *8*, 1013. https://doi.org/10.3390/horticulturae8111013

Academic Editors: Alicja Tymoszuk and Dariusz Kulus

Received: 29 September 2022
Accepted: 28 October 2022
Published: 1 November 2022

1. Introduction

Potato (*Solanum tuberosum* L.) is one of the five most important crops in the world today and is currently grown on over 20.7 million hectares across all continents with an estimated global production of 437 million tons [1]. In 2020, potato production in Croatia amounted to 174,280 tons of tubers from approximately 9330 ha [2]. Commercial production is entirely based on a limited number of foreign varieties that have replaced traditional cultivars/ecotypes. The intensive substitution of traditional by modern high-yielding varieties in agricultural production during the last century resulted in some erosion of genes [3], as the traditional cultivars are a source of genetic variability for many traits [4].

This is also the case with the traditional cultivar 'Brinjak', which is resistant to *Phytophthora infestans* and *Streptomyces scabies*, according to reports from local producers in Lika (a region in Croatia), where the cultivar 'Brinjak' has been grown for many generations. Therefore, traditional varieties are of great importance for biodiversity conservation, as a gene pool for future plant breeding programs [5], but also for economically profitable production once the variety is included in the List of Varieties of the Republic of Croatia as a conservation variety. It is expected that the cultivar 'Brinjak' will soon be included in that list, which will enable its commercial production and sale in the Republic of Croatia. Viruses are important agricultural pathogens with an estimated economic loss of >USD 30 billion annually [6] and are responsible for more than 50% of yield losses of potato tubers [7]. While there are no effective measures for controlling viruses and curing plants once they are infected [8,9], virus-free plants are required for commercial potato production and maintenance of potato germplasm [10]. Commercially important potato viruses cause a reduction in the starch content of tubers [11] and other biochemical and physiological changes manifested by a reduction in dry matter and vitamin content, starch granules, a decrease in amylase content, and starch acidity [11,12]. In addition to poor vegetative growth (stunting, dwarfing) and significant yield losses in tubers, viruses in potato plants can cause symptoms such as leaf curling, mosaic, leaf drop, chlorosis, necrosis, reduced product quality, undersized tubers, and sometimes death of the entire plant [11–15]. Potato leafroll virus (PLRV), potato virus A, M, S, X, and Y (PVA, PVM, PVS, PVX, and PVY) are usually the most important viral pathogens associated with significant economic impact and production loss [9,15,16]. PVM and PVS can be transmitted by aphids, especially the green peach aphid (*Myzus persicae*), by vegetative propagation (via tubers), or mechanically (e.g., through contaminated tools and wounds) [17,18]. Plants infected with PVM and PVS do not always show symptoms, depending on the cultivar and virus isolates [9,15,19]. However, PVM can reduce tuber yield by 11–45% [11,20], and PVS by 10–20% [21,22]. Mixed infection results in a yield reduction of 20–30% [22]. Symptoms of virus infections leading to changes in plant morphology, anatomy, physiology, and biochemistry can be quantified using spectral reflectance analysis. Since plant diseases interact with their host and lead to disruption of metabolic processes and development of specific symptoms, various optical sensors have been used for non-destructive diagnosis and detection of plant diseases [23,24]. Virus infection can reduce the photosynthetic efficiency of potato by decreasing the assimilation area, photosynthetic rate, ribulose-1,5-diphosphate carboxylase activity, chlorophyll and xanthophyll cycle pigment content, and PSII concentration and activity [25,26]. The sensitivity of PSII to biotic factors has made chlorophyll fluorescence one of the most widely used techniques in the study of plant disease interactions [27,28]. In addition, visible and multispectral imaging can assess the spectral information of infected leaves and can be used to calculate various vegetation indices [23,24,29].

The production of certified seed potatoes always starts with the micropropagation of virus-free stock plants, followed by the production of minitubers, and then tubers through several generations until the production of certified seed potatoes supplied to potato growers for commercial production [16]. In this process, potato growers receive healthy seed potatoes, or seed potatoes infected with viruses at a very low percentage, which is permissible. Various in vitro virus eradication methods such as meristem tip culture, thermotherapy, chemotherapy, cryotherapy, electrotherapy, and somatic embryogenesis [30–34] have been successfully used in different species. The efficiency of each method depends on the virus to be eliminated, the size of the explant, the duration of treatment, and also the variety [9,35,36]. Successful eradication of viruses from infected potato plants through chemotherapy or in combination with other methods, and the efficiency of ribavirin are well documented in the literature [9,10,32,35–47]. Ribavirin is a synthetic purine nucleoside analog with a structure closely related to guanosine [48]. It is a broad-spectrum antiviral nucleoside that is active against a variety of RNA and DNA viruses. Several possible mechanisms have been proposed for the antiviral activity of ribavirin, including the possibility that ribavirin negatively affects the synthesis of the RNA cap structure of viral RNA transcripts [49]. Exogenous applications of the antivirus agent ribavirin in chemotherapy inhibits viral RNA synthesis [35,50,51].

The aim of this study was to eliminate potato viruses M (PVM) and S (PVS) in the traditional potato cultivar 'Brinjak' by applying ribavirin in a culture medium and to quantify the effects of single (PVM) and mixed (PVM + PVS) infection on potato photochemical and multispectral parameters and yield.

2. Materials and Methods

2.1. Plant Material

The tubers of the traditional potato cultivar 'Brinjak' were obtained from the field collection of the University of Zagreb, Faculty of Agriculture, which is maintained within the framework of the Croatian National Program for Conservation and Sustainable Use of Plant Genetic Resources Important for Food and Agriculture. Plants of the cultivar 'Brinjak' are short, upright, stem-type, and with late maturity. The plants produce mostly smaller tubers that are round-shaped with a light beige color skin. Flesh has a medium yellow color. At maturity, the tubers have a high dry matter content.

2.2. Culture Establishment on Media Containing Ribavirin and Micropropagation

The tubers of the two mother potato plants of the cultivar 'Brinjak' were kept in the dark at a temperature of approximately 18 °C for five months. The 2–6 cm long sprouts grown on the tubers were used as the starting material for setting up the experiments. The sprouts were washed under running tap water for 20 minutes and then disinfected in 70% ethanol for 1 minute. They were then shaken for 15 minutes in a 4% solution of sodium dichloroisocyanurate dihydrate (Izosan® G, Pliva Hrvatska, Zagreb, Croatia) supplemented with 3 drops 100 mL^{-1} of surfactant Tween-20. The sprouts were then washed three times for 3 min in sterile distilled water.

Shoot apices > 0.5–1 mm in size were aseptically isolated from the terminal and lateral buds of the elongated sprouts under a stereomicroscope. Shoot apices were placed in 12-cm-high tubes on the surface of 10 mL of the Murashige and Skoog medium [52] containing vitamins (Duchefa Biochemie, Haarlem, The Netherlands) supplemented with 0.04 mg L^{-1} Kinetin (Kin), 0.5 mg L^{-1} Gibberellic acid (GA), 3% sucrose, 0.8% BD bacto agar (Becton, Dickinson and Company, NJ, USA), and 50 or 100 mg L^{-1} ribavirin (Duchefa Biochemie, Haarlem, The Netherlands). Ribavirin was prepared as a stock solution of 50 mg mL^{-1}, filter-sterilized, and added to the autoclaved medium before solidification. The pH of the medium was adjusted to 5.8. For each ribavirin treatment (50 or 100 mg L^{-1}), 24 shoot apices were placed. After 6 weeks, the number of surviving shoot apices was determined, and the live shoot apices were subcultured on hormone-free MS medium (HFMS), which was the same medium used to establish the culture, except that it did not contain plant growth regulators and ribavirin. Shoots formed from shoot apices were cut into single-nodal segments without leaves and subcultured in Magenta boxes on the same medium in several cycles until the production of a sufficient number of shoots from the surviving explants. Therefore, each surviving shoot apex was micropropagated to produce a clone. Post-eradication period in vitro lasted 200 days, and shoots that passed this stage were used for all subsequent steps. Culture establishment and micropropagation were performed in a growth chamber at 23 °C with a 16 h photoperiod of cool white light (40 µmol m^{-2} s^{-1}).

2.3. Microtuberization

Twenty single-nodal segments with a leaf of each micropropagated potato clone were placed vertically on the medium for microtuberization in two Sterivent high containers (Duchefa Biochemie, The Netherlands). The medium for microtuberization consisted of MS medium with vitamins (Duchefa Biochemie, Haarlem, The Netherlands) supplemented with 4 mg L^{-1} Kin, 1 mg L^{-1} 6-benzylaminopurine (BAP), 8% sucrose, 0.7% plant agar (Duchefa Biochemie, Haarlem, The Netherlands). The pH was adjusted to 5.8. Explants were grown for 20 days at a photoperiod of 16/8 day/night with cool white light

(40 μmol m^{-2} s^{-1}) at a temperature of 22 °C. Then, they were moved to short-day conditions, i.e., a photoperiod of 8/16 day/night at the same light intensity and temperature, and grown for four months. To determine the effect of viruses on microtuberization, the percentage of shoots with microtubers, the number of microtubers, and the weight of microtubers were determined.

2.4. Acclimatization and Minituber Production of R0 Plants

Three-week-old rooted shoots were planted in sterile Kekkila TSM 2 substrate (Kekkilä Professional, Vantaa, Finland) in small pots (3 cm × 3 cm × 7 cm) and covered by a transparent plastic cover for 7 days for acclimatization. After 10 days, the plants were transplanted into larger pots (10 cm × 10 cm × 20 cm) containing sterile Kekkila TSM 3 substrate, in which they were grown to maturity. The number of transplanted plants per clone varied from 6 to 10. The substrate was sterilized by gamma irradiation (Ruđer Bošković Institute, Zagreb, Croatia). Irradiation was performed with a ^{60}Co gamma ray source and an absorbed dose of 50 kGy. R0 plants were grown for 150 days in a growth chamber at a photoperiod of 16/8 day/night of LED tubes (Valoya L35 NS12) and a light intensity of 300 μmol m^{-2} s^{-1} at a temperature of 22 °C. Plants were watered twice a week and fertilized once a month with organic liquid fertilizer "Organomex 6-2-4" (Omex Agrifluids Limited, King's Lynn, UK).

2.5. Production of R1 Clones from Microtubers and Minitubers

Three and a half months after microtuberization and storage at room temperature, two microtubers from each clone were planted in plastic pots (10 cm × 10 cm × 20 cm) in sterile Kekkila TSM 3 substrate, in which they were grown to maturity. Similarly, minitubers, the largest from each R0 plant, were planted in the same way. The plants were grown to maturity under the same conditions as the R0 clones. At harvest, the number and weight of tubers per plant were determined for all three generations of plants.

2.6. Verification of Virus Status by DAS-ELISA and RT-PCR

To gain insight into the virus sanitary status of the mother plants and, later, their progeny, DAS-ELISA was performed on six different viruses: PLRV, PVA, PVM, PVS, PVX, and PVY. Four randomly selected plants per clone in the R0 generation, one plant per clone in the R1 generation produced from microtubers, and two randomly selected plants per clone in the R1 generation produced from minitubers were analyzed. Lower leaves from different sides of each plant studied were collected 90 days after acclimatization. The collected leaves from each plant were mixed and used to prepare an average sample of 0.1 g, which served as a potential antigen source. Commercially available ELISA kits from Bioreba (Reinach, Switzerland) were used for virus detection according to the manufacturer's instructions. Viruses detected by DAS-ELISA in the mother plants that were not detected by the same method in the progeny (R0 micropropagated plants, R1 from microtubers, and R1 from minitubers) were verified by reverse transcription-polymerase chain reaction (RT-PCR). For this purpose, RNA was isolated from leaf samples collected 90 days after acclimatization using the RNeasy Plant Mini Kit (Qiagen, Hilden, Germany). Leaves taken from different sides of each plant were mixed into an average sample consisted of 0.1 g of plant tissue that was minced in MiniG 1600, SpexSample Prep (Metuchen, NJ, USA) using grinding balls. The quality and quantity of the isolated RNA were checked spectrophotometrically (A260/A280) using a NanoPhotometer P330 (Implen, Munich, Germany). One-step RT-PCR reactions were performed in the Mastercycler (Eppendorf, Hamburg, Germany) under the following conditions: reverse transcription at 50 °C for 30 min, initial activation step at 95 °C for 15 min, followed by 35 cycles of denaturation at 94 °C for 30 s, annealing at 53 °C for 30 s, and elongation at 72 °C for 30 s. The final elongation step was performed at 72 °C for 10 min. For virus detection, the primers described in [7] were combined with the one-step RT-PCR kit (Qiagen, Germany) in a reaction volume of 10 μL consisting of: 0.5 μL of each primer (10 μM), 0.4 μL dNTP mix

(10 μM), 2 μL Qiagen OneStep RT-PCR enzyme mix, 2 μL Qiagen OneStep RT-PCR buffer, 2 μL Q-Solution, 3.2 μL RNase-free water, and 1 μL template RNA. To visualize the RT-PCR products, a 1.5% agarose gel was prepared in a 1X TBE buffer with the addition of GelRed (CareDx AB, Stockholm, Sweden) and subjected to horizontal gel electrophoresis (Bio-Rad, Hercules, CA, USA).

2.7. Chlorophyll Fluorescence and Multispectral Analysis

Chlorophyll fluorescence and multispectral imaging were conducted on all R0 plants 90 days after planting using CropReporter^{TM} (PhenoVation B.V., Wageningen, The Netherlands), while image processing and analysis were performed using DA software (PhenoVation B.V., Wageningen, The Netherlands). Plants were imaged from a distance of 70 cm, and measurements were made according to the protocol described in [53]. For the chlorophyll fluorescence imaging, plants were first dark-adapted for 30 min. On dark-adapted plants, photosynthesis was excited using 4000 μmol m^{-2} s^{-1} red LED light flash. Minimum chlorophyll fluorescence (F_0) was measured after ten (10) μs and maximum chlorophyll fluorescence (F_m) after 800 ms. Following these measurements, plants were relaxed in the dark for 15 s and then were light-adapted for 5 min using actinic light of 250 μmol m^{-2} s^{-1}. Saturating pulse (4000 μmol m^{-2} s^{-1}) was again applied for photosynthetic excitation of the light-adapted plants. The steady-state fluorescence yield (F_s') was measured before the onset of the saturation pulse, and the maximum chlorophyll fluorescence (F_m') of the light-adapted leaves was measured at saturation. After the measurement, the actinic light was turned off, and in the presence of far-red light, the minimal fluorescence yield of the illuminated plant (F_0') was estimated. These measured parameters were used for the calculation of different chlorophyll fluorescence parameters, which are shown in Table 1.

Table 1. List of calculated chlorophyll fluorescence parameters with abbreviations, the equation for calculation, and the reference.

Abbrev	Trait	Wavelength/Equation
F_v/F_m	Maximum Efficiency of Photosystem Two	$F_v/F_m = (F_m - F_0)/F_m$ [54]
F_q'/F_m'	Effective Quantum Yield of Photosystem Two	$F_q'/F_m' = (F_m' - F_s')/F_m'$ [55]
ETR	Electron Transport Rate	$ETR = F_q'/F_m' \times PPFD \times (0.5)$ [55]
NPQ	Non-Photochemical Quenching	$NPQ = (F_m - F_m')/F_m'$ [56]

Following the chlorophyll fluorescence imaging, actinic light 250 μmol m^{-2} s^{-1} was again switched on and multispectral imaging was performed. A list of all measured and multispectral parameters and calculated vegetation indices are given in Table 2.

Table 2. List of analyzed multispectral parameters with abbreviations, wavelength for measurement or equation for calculation, and the reference if appropriate.

Abbrev.	Trait	Wavelength/Equation
$R_{Red'}$	Reflectance in Red	640 nm
R_{Green}	Reflectance in Green	550 nm
R_{Blue}	Reflectance in Blue	475 nm
R_{SpcGrn}	Reflectance in Specific Green	510–590 nm
R_{FarRed}	Reflectance in Far-Red	710 nm

Table 2. *Cont.*

Abbrev.	Trait	Wavelength/Equation
R_{NIR}	Reflectance in Near Infra-Red	769 nm
R_{Chl}	Reflectance Specific to Chlorophyll	730 nm
HUE	Hue (0–360°)	HUE = 60 × (0 + (R_{Green}—R_{Blue})/(max-min)), if max = R_{Red}; HUE = 60 × (2 + (R_{Blue}—R_{Red})/(max-min)), if max = R_{Green}; HUE = 60 × (4 + (R_{Red}—R_{Green})/(max-min)) if max = R_{Blue}; 360 was added in case HUE < 0
SAT	Saturation (0–1)	SAT = (max—min)/(max + min) if VAL > 0.5, or SAT = (max—min)/(2.0—max—min) if VAL < 0.5, where max and min are selected from the R_{Red}, RG$_{reen}$, R_{Blue}
VAL	Value (0–1)	VAL = (max + min)/2; where max and min are selected from the R_{Red}, R_{Green}, R_{Blue}
ARI	Anthocyanin Index	ARI = $(R_{550})^{-1}$—$(R_{700})^{-1}$ [57]
CHI	Chlorophyll Index	CHI = $(R_{700})^{-1}$—$(R_{769})^{-1}$ [58]
NDVI	Normalized Differential Vegetation Index	NDVI = $(R_{NIR} - R_{Red})/(R_{NIR} + R_{Red})$ [59]

2.8. Data Analysis

All experiments were set up in a completely randomized design. Data from all measurements were subjected to ANOVA. Bonferroni post hoc test at $p \leq 0.05$ was used for means comparison. The statistical analysis of the data was carried out by the SAS/STAT® [60] program package.

3. Results and Discussion

3.1. Culture Establishment and Micropropagation

Out of 24 shoot apices placed on medium supplemented with 50 mg L^{-1} of ribavirin, 6 (25%) survived, whereas only 3 (12.5%) survived on medium supplemented with 100 mg L^{-1} of ribavirin. There was no contamination. The cause of the low percentage of surviving shoot apices may be ribavirin, but also their physiological condition, because the sprouts for shoot apices isolation were used after the tubers had been kept in the dark for five months. Most of the surviving explants (Figure 1a,b) grew very slowly or were even inhibited in growth 100 days after placement in in vitro culture, i.e., two months after subcultivation on HFMS medium without ribavirin (Figure 1c,d). The strength of ribavirin treatment in the previous medium did not affect inhibition; surviving shoots derived from both ribavirin concentrations, whether 50 or 100 mg L^{-1}, were equally inhibited. Phytotoxicity of ribavirin to potato regeneration and plantlet growth was also reported by [44], who found that plant height and fresh weight generally showed a decreasing tendency with increasing ribavirin concentration in a culture medium. Singh [61] reported that the percentage of shoot regeneration and plantlet development decreased when the concentration of antiviral chemicals, including ribavirin, was increased from 5 to 30 mg L^{-1}.

Figure 1. Surviving explants after 6 weeks of growth on ribavirin (**a**,**b**); inhibited growth on HFMS medium 100 days from the start of the experiment—shoot originating from RIB 50 (**c**) and RIB 100 (**d**); micropropagation by single node culture (**e**). Bars: (**a**,**b**) = 2 mm; (**c**,**d**) = 5 mm; (**e**) = 10 mm.

Eight months after the start of the experiment, all nine surviving shoot apices that developed into shoots were micropropagated, but with different efficiency. Some shoots were still partially growth-inhibited and micropropagation resulted in a small number of plants per clone, while other shoots grew well after initial inhibition (Figure 1e) and were micropropagated into a large number of plants by single node culture. The slow growth of individual shoots that we observed according to ELISA results not only did not depend on ribavirin concentration (50 or 100 mg L^{-1}), but also did not depend on virus infection (single or mixed). Micropropagated plants rooted spontaneously on HFMS medium in a very high percentage, and no special step was required for rooting.

3.2. The Influence of The Sanitary Status of Plants on Microtuberization

The sanitary status of the plants affected the number of microtubers per plant, which was higher on average in plants infected with PVM only (Table 3, Figure 2a,b). The data show that mixed infection (PVM + PVS) reduced the number of microtubers. Bettoni et al. [9] reported a significantly lower number of microtubers in plants infected with PVS compared to virus-free plants in the cultivar 'Duncle', but mixed infection with PVS and PVM in other cultivars or PVS and PVA did not affect the number of microtubers. We could not assess the effect of PVM on microtuberization or mixed infections with PVM + PVS compared to healthy plants because none of the plants were virus-free. The percentage of plants that formed microtubers under short-day conditions (Figure 2c) was similar and was 82% in plants infected with both viruses and 78% in plants infected with PVM only. The average microtuber weight was not significantly different between PVM- and PVM- + PVS-infected plants (Table 3). This is in agreement with the results of Zhang et al. [10], who found that infection with viruses had little effect on microtuber production compared to healthy plants, except in the case of triple infection with PVX, PVS, and PVY viruses.

Table 3. Microtuber yield components, as affected by the sanitary status of the plants.

Virus Infection	Number of Microtubers Per Plant	Average Microtuber Weight (mg)
PVM	1.3 a	207 a
PVM + PVS	1.0 b	199 a

Values within the column followed by the same letter are not significantly different at $p < 0.05$.

Figure 2. Microtubers produced in one container of a clone infected with PVM (**a**) or with both viruses, PVM + PVS (**b**); microtuberization under short-day conditions (**c**); bar = 1 cm.

3.3. Acclimatization and Plant Growth

Micropropagated plants planted in a sterile substrate acclimatized with 100% success, and after transplanting into larger pots, they developed quickly under optimal conditions. The difference in symptoms between plants infected with both (PVM + PVS) or only one virus (PVM) in the R0 generation of plants produced by micropropagation, although noticeable, was not very pronounced. However, differences in viral disease symptoms were evident in R1 plants produced from microtubers and minitubers. In contrast to the pronounced symptoms in PVS + PVM-infected plants (reduced growth, wrinkling, twisting, and mosaicism), plants infected only with PVM had, on some leaves, very small lesions resembling a hypersensitivity reaction (Figure 3a,b).

3.4. Virus Detection and Efficiency of Virus Elimination

Initial virus screening of the mother plants for the presence of six viruses confirmed infection with only two viruses: PVM and PVS (Supplementary Table S1). This finding was interesting because the traditional potato cultivar Brinjak has long been grown with tubers without any sanitary selection. Despite that, other common viruses frequently detected in imported cultivars (e.g., PLRV, PVX, and PVY) were not confirmed. According to the ELISA results, PVS was successfully eliminated by six weeks of chemotherapy with ribavirin in three (14-1, 14-3, and 14-4) of nine R0 clones, and PVS-free status was also confirmed in the progeny (R1 from microtubers and minitubers). However, the same treatments were ineffective in eliminating PVM (Supplementary Table S1). Successful elimination of PVS was also confirmed by RT-PCR, where only the mother plants reacted positively, resulting in an amplicon of 885 base pairs, while selected three ELISA-negative PVS plants were also negative in molecular tests (Figure 4). These results are in partial agreement with the results of [36], who found that ribavirin at a concentration of 100 mg L^{-1} was not successful in eliminating either PVM or PVS after one subculture of 45 days, but only after the second or third subculture on ribavirin medium, and even then, elimination was genotype-dependent. The combination of ribavirin in one subculture with cryotherapy eliminated PVS in a certain percentage of plants, but not PVM. However, two subcultures with ribavirin + cryotherapy achieved 71–100% success in eliminating virus S and virus M, depending on the genotype [36].

Figure 3. Symptoms in mixed infection with PVM + PVS compared to a plant infected only with PVM; view on the whole plant (**a**) and details on the leaf (**b**). Arrows show small lesions.

Bettoni et al. [9] found that the effectiveness of virus eradication varied depending on the type of virus; PVS and PVM were more difficult to eliminate than PVA regardless of the method used. Using chemotherapy alone, Bettoni et al. [9] obtained an eradication success of 20–50% for PVS and 20% for PVM after a four-week treatment of shoot segments with ribavirin at a concentration of 100 mg L^{-1}. Even with a combination of different consecutive and simultaneous treatments, e.g., chemotherapy + (chemotherapy + thermotherapy) + cryotherapy, PVM was eliminated in 70% of the plants, while the elimination of PVS with this combination of treatments was 100% [9]. Yang et al. [44] achieved 17% successful elimination of PVS and PVM with ribavirin (75 and 100 mg L^{-1}) with one subcultivation of 45 days. At the second 45-day subcultivation, the percentage of virus elimination was higher, and after the third subcultivation, i.e., after 135 days, the plants were free from both PVS and PVM.

Figure 4. RT-PCR results for potato virus S (PVS) with an expected amplicon size of 885 base pairs (bps) performed on two PVS-infected mother plants (P1 and P2) and PVS-free plants obtained by ribavirin treatment B 14-1 RIB 50, B 14-3 RIB 100, and B14-4 RIB 50 in the following order: lines 1–3—R0 plants, lines 4–6—R1 plants from microtubers, lines 7–9—R1 plants form minitubers; N—negative control, M—marker (GelPilot 100 bp Plus Ladder, Qiagen, Hilden, Germany). For further details, please refer to Supplementary Table S1.

There was no difference in the efficiency of PVS elimination between the different ribavirin concentrations (Table 4). Ribavirin at a lower concentration (50 mg L^{-1}) eliminated PVS from 33% of the surviving shoot apices, as did ribavirin at a concentration of 100 mg L^{-1}. Oana et al. [35] also found no differences between the results obtained when 35 or 50 mg L^{-1} ribavirin was added to the culture medium and claimed that even 35 mg L^{-1} of ribavirin was sufficient for potato virus eradication, but with longer treatment. They also noted that successful eradication seemed to depend on the duration of treatment and the cultivar rather than the virus. Therefore, for complete elimination of the viruses in the traditional cultivar 'Brinjak', additional cycles of ribavirin chemotherapy or another method or combination of methods should be used.

Table 4. The efficiency of virus elimination by ribavirin from potato cultivar 'Brinjak'.

Ribavirin Concentration (mg L^{-1})	Plants Free of Virus after Chemotherapy	
	PVM	PVS
50	0/6 (0%)	2/6 (33%)
100	0/3 (0%)	1/3 (33%)

Bougie and Bisaillon [49] found that ribavirin acts as a substrate for a viral RNA capping enzyme; however, such RNA transcripts blocked with ribavirin are not efficiently translated. Consequently, ribavirin prevents replication of a large number of DNA and RNA viruses [62].

Sometimes, as reported by [9], a plant tested for a particular virus may be negative (e.g., when a sample of plant tissue is tested from in vitro conditions), and upon subsequent inspection, e.g., after growing in a greenhouse, the same plant is found to be positive for the tested virus. In this study, except for the very long post-eradication period of 200 days + 90 days of cultivation in the growth chamber, three generations of clones obtained by treatment with ribavirin were used for virus detection by the methods DAS-ELISA and RT-PCR, eliminating the possibility of false-negative results. Accordingly, Yang et al. [44], who examined the plants by qPCR after a six-month post-eradication period, proved that ribavirin can thoroughly remove the viruses from the infected plants and not only suppress their replication. The same authors, using SSR markers, showed that ribavirin treatment did not cause genetic variation in plants.

3.5. The Influence of Sanitary Status of Plants on Chlorophyll Fluorescence and Multispectral Parameters

Mixed infection with both viruses (PVM + PVS) significantly decreased the effective quantum yield of PSII (F_q'/F_m') and increased non-photochemical quenching (NPQ) (Table 5), which indicates that less light is used for photochemistry and more energy is lost through heat dissipation in plants with mixed infection. Zhou et al. [26] hypothesized that virus infection affects photosynthesis mainly by interfering with the Calvin cycle, leading to down-regulation of the efficiency of excitation energy capture by open PSII reaction centers (F_q'/F_m'). Although F_v/F_m is the most common fluorescence parameter used in plant stress studies [27,28], the results of this study show that F_v/F_m ranged from 0.79 to 0.81 for both PVM and PVM + PVS plants, indicating good fitness of plants. As noted by [15,19], plants infected with PVM and PVS do not always show symptoms, which depends on the variety and virus isolates.

Table 5. Chlorophyll fluorescence parameters as affected by the sanitary status of the plants.

Virus Infection	F_v/F_m	F_q'/F_m'	ETR	NPQ
PVM	0.81 a	0.48 a	5581 a	0.38 b
PVM + PVS	0.79 a	0.42 b	5234 a	0.50 a

Values within the column followed by the same letter are not significantly different at $p < 0.05$.

Considering the multispectral parameters, mixed infection significantly decreases the chlorophyll index (CHI) and increases the reflectance in the far-red (R_{FarRed}) (Table 6). These results are consistent with [63,64], who found that mixed infection can increase symptoms severity and virus accumulation. Decreased chlorophyll content affects photosynthetic light absorption and increases its reflection, as evidenced by increased (although not significant) reflection of R_{Red}, R_{Green}, and R_{SpcGrn} in plants with mixed infection. Thus, besides decreased F_q'/F_m' and increased NPQ, plants with mixed infection will have lower light absorption, likely resulting in decreased sugar production in photosynthesis.

Table 6. Multispectral parameters as affected by the sanitary status of the plants.

Virus Infection	R_{Red}	R_{Green}	R_{Blue}	R_{FarRed}	R_{NIR}	R_{SpcGrn}	HUE	SAT	VAL	CHI	ARI	NDVI
PVM	1975 a	3019 a	1581 a	6046 b	2831 a	3394 a	105 a	0.05 a	0.46 a	3.8 a	3.9 a	0.85 a
PVM + PVS	2001 a	3148 a	1563 a	6403 a	2849 a	3558 a	104 a	0.05 a	0.49 a	3.5 b	3.8 a	0.85 a

Values within the column followed by the same letter are not significantly different at $p < 0.05$.

3.6. The Influence of the Sanitary Status of Plants on Yield Components

The sanitary status of R0 clones produced from micropropagated plants affected tuber weight per plant and average tuber weight (Table 7). Mixed infection with both viruses (PVM + PVS) reduced tuber weight per plant by 31% and average tuber weight by 64% compared to plants infected with PVM only. In R1 clones obtained from microtubers, mixed infection resulted in a significantly lower number of tubers per plant (59%), as well as lower tuber weight per plant (44%) (Table 8). An even greater difference was observed in R1 clones produced from minitubers, where the decrease in tuber weight per plant was 86% in plants with mixed infection compared to those infected with PVM alone (Table 9). The results obtained are in agreement with previous studies by [22,65], which found that mixed infection with two viruses resulted in a much greater yield loss than single infection. A statistically significant increase in the number of tubers per plant in PVS-free R1 clones obtained from microtubers and minitubers could be due to the higher number of stems per plant compared to plants with mixed infection. However, in the R0 clones obtained from micropropagated plants, each plant had only one main stem, and there was no significant difference in the number of tubers per plant between differently infected plants. These results are in agreement with those of [66], who reported that the number of main stems correlated positively with tuber number and tuber yield and negatively with average tuber weight.

Table 7. Minituber yield components of R_0 clones produced from micropropagated plants, as affected by the sanitary status of the plants.

Virus Infection	Number of Tubers per Plant	Tuber Weight per Plant (g)	Average Tuber Weight (g)
PVM	4.0 a	52.0 a	15.6 a
PVM + PVS	4.7 a	39.8 b	9.5 b

Values within the column followed by different letters are significantly different at $p < 0.05$.

Table 8. Mini tuber yield components of R1 clones produced from microtubers, as affected by the sanitary status of the plants.

Virus Infection	Number of Tubers per Plant	Tuber Weight per Plant (g)	Average Tuber Weight (g)
PVM	7.3 a	65.0 a	9.6 a
PVM + PVS	4.6 b	45.1 b	10.1 a

Values within the column followed by the same letter are not significantly different at $p < 0.05$.

Table 9. Tuber yield components of R1 clones produced from minitubers, as affected by the sanitary status of the plants.

Virus Infection	Number of Tubers per Plant	Tuber Weight per Plant (g)	Average Tuber Weight (g)
PVM	9.8 a	133.9 a	16.1 a
PVM + PVS	7.2 b	72.1 b	12.1 a

Values within the column followed by the same letter are not significantly different at $p < 0.05$.

Yield reduction in plants infected with both viruses (PVM + PVS) is the result of changes in metabolic processes in the affected tissues. Phytopathogenic viruses often cause physical deformation of leaves [11]. This leads to a decrease in the intensity of photosynthesis, which is very important for plant productivity.

Reduced growth, wrinkling, twisting, and mosaic structure of leaves caused by mixed infection were not observed in plants infected only with PVM, which had only small necrotic lesions. Severe phenotypic changes in plants with mixed infection are the cause of lower chlorophyll index (CHI) and lower effective PSII quantum yield (F_q'/F_m'), which ultimately resulted in large differences in potato yield between plants infected with two (PVM + PVS) compared to those infected with only one (PVM) virus.

The traditional potato cultivar 'Brinjak' is resistant to the fungi *Phytophthora infestans* and *Streptomyces scabies* (personal communication), thus providing valuable germplasm for breeding programs and commercial growing. Infection of plants that have never been subjected to virus elimination before with only two—PVM and PVS—of the six viruses tested (PVX, PVY, PVA, PLRV, PVM, and PVS) may indicate that this cultivar possesses resistance to these viruses, as suggested by [10] for resistance to PVY in Norwegian cultivars. Elimination of these two viruses and use of virus-free planting material would probably increase yield and bring out the full genetic potential of this cultivar, in which growers are interested, especially in the mountainous regions of Croatia, where it is adapted.

4. Conclusions

The goal we set at the beginning of the research was only partially achieved, as ribavirin treatment at both concentrations (50 or 100 mg L^{-1}) eliminated PVS from 33% of surviving shoot apices, but these treatments were ineffective in eliminating PVM. Mixed infection with PVM + PVS significantly decreased the effective quantum yield of PSII (F_q'/F_m') and increased non-photochemical quenching (NPQ) compared to single infection with PVM. The multispectral parameters show that the mixed infection affected photosynthesis also at the chlorophyll level by significantly decreasing the chlorophyll index (CHI) and increasing far-red reflection (R_{FarRed}). Plants infected only with PVM, compared to those with mixed infection, had 36–59% higher number of tubers per plant, 31–86% higher tuber weight per plant, and up to 64% higher average tuber weight, depending on the generation. The results indicate a strong negative impact of PVS in mixed infection with PVM and highlight the importance of removing PVS from potato plants.

Supplementary Materials: The following supporting information can be downloaded at: https://www.mdpi.com/article/10.3390/horticulturae8111013/s1, Table S1: DAS-ELISA and RT-PCR results for mother plants and their progeny tested on the presence of potato virus M (PVM) and potato virus S (PVS).

Author Contributions: Conceptualization, S.K. and I.H.J.; methodology, S.K., D.V. and B.L.; formal analysis, S.K, B.L.; investigation, S.K, I.H.J., D.V. and B.L.; resources, M.P, S.K. and B.L.; data curation, S.K.; writing—original draft preparation, S.K, D.V., B.L., I.H.J., M.B. and A.M.S.; writing—review and editing, M.P. and A.B.M.; visualization, A.M.S.; supervision, S.K. All authors have read and agreed to the published version of the manuscript.

Funding: This research was funded by the University of Zagreb's financial support for scientific research.

Data Availability Statement: Data are contained within the article or Supplementary Materials.

Conflicts of Interest: The authors declare no conflict of interest.

References

1. FAO. 2022. Available online: https://www.fao.org/3/nj015en/nj015en.pdf (accessed on 15 September 2022).
2. FAOSTAT. 2022. Available online: https://www.fao.org/faostat/en/#data/QCL (accessed on 15 September 2022).
3. Raggi, L.; Pacicco, L.C.; Caproni, L.; Álvarez-Muñiz, C.; Annamaa, K.; Barata, A.M.; Batir-Rusu, D.; Díez, M.J.; Heinonen, M.; Holubec, V.; et al. Analysis of landrace cultivation in Europe: A means to support in situ conservation of crop diversity. *Biol. Conserv.* **2022**, *267*, 109460. [CrossRef]
4. Jovović, Z.; Stešević, D.; Meglič, V.; Dolničar, P. *Old Potato Varieties in Montenegro*; University of Montenegro Biotechnical Faculty Podgorica: Podgorica, Montenegro, 2013.
5. Bradshaw, J.E. Review and Analysis of Limitations in Ways to Improve Conventional Potato Breeding. *Potato Res.* **2017**, *60*, 117–193. [CrossRef]
6. Jones, R.A.C.; Naidu, R.A. Global Dimensions of Plant Virus Diseases: Current Status and Future Perspectives. *Annu. Rev. Virol.* **2019**, *6*, 387–409. [CrossRef] [PubMed]
7. Rashid, M.-O.; Li, J.H.; Liu, Q.; Wang, Y.; Han, C.G. Molecular detection and identification of eight potato viruses in Gansu province of China. *Curr. Plant Biol.* **2021**, *25*, 100184. [CrossRef]
8. Rubio, L.; Galipienso, L.; Ferriol, I. Detection of plant viruses and disease management: Relevance of genetic diversity and evolution. *Front. Plant Sci.* **2020**, *11*, 1092. [CrossRef]
9. Bettoni, J.C.; Mathew, L.; Pathirana, R.; Wiedow, C.; Hunter, D.A.; McLachlan, A.; Khan, S.; Tang, J.; Nadarajan, J. Eradication of Potato Virus S, Potato Virus A, and Potato Virus M From Infected in vitro Grown Potato Shoots Using in vitro Therapies. *Front. Plant Sci.* **2022**, *13*, 878733. [CrossRef]
10. Zhang, Z.; Wang, Q.C.; Spetz, C.; Blystay, D.R. In vitro therapies for virus elimination of potato-valuable germplasm in Norway. *Sci. Hortic.* **2019**, *249*, 7–14. [CrossRef]
11. Kolychikhina, M.S.; Beloshapkina, O.O.; Phiri, C. Change in potato productivity under the impact of viral diseases. *Earth Environ. Sci.* **2021**, *663*, 012035. [CrossRef]
12. Oyesola, O.L.; Aworunse, O.S.; Oniha, M.I.; Obiazikwor, O.H.; Bello, O.; Atolagbe, O.M.; Sobowale, A.A.; Popoola, J.O.; Obembe, O.O. Impact and Management of Diseases of *Solanum tuberosum*. In *Solanum Tuberosum: A Promising Crop for Starvation Problem*; Yildiz, M., Ozgen, Y., Eds.; IntechOpen: London, UK, 2021.
13. Jayashige, U.; Chuquilllanqui, C.; Salazar, L.F. Modified expression of virus resistance in potato in mixed virus infections. *Am. J. Potato Res.* **1989**, *66*, 137–144.
14. Onditi, J.; Nyongesa, M.; Vlugt, R. Prevalence, distribution and control of six major potato viruses in Kenya. *Trop. Plant Pathol.* **2021**, *46*, 311–323. [CrossRef]
15. Kreuze, J.F.; Souza-Dias, J.A.C.; Jeevalatha, A.; Figueira, A.R.; Valkonen, J.P.T.; Jones, R.A.C. Viral diseases in potato. In *The Potato Crop*; Campos, H., Ortiz, O., Eds.; Springer: Cham, Switzerland, 2020; pp. 389–430.

16. Wang, B.; Ma, Y.; Zhang, Z.; Wu, Z.; Wu, Y.; Wang, Q.; Li, M. Potato viruses in China. *Crop Prot.* **2011**, *30*, 1117–1123. [CrossRef]
17. Hlaoui, A.; Boukhris-Bouhachem, S.; Sepúlveda, D.A.; Correa, M.C.G.; Briones, L.M.; Souissi, R.; Figueroa, C.C. Spatial and Temporal Genetic Diversity of the Peach Potato Aphid *Myzus persicae* (Sulzer) in Tunisia. *Insects* **2019**, *10*, 330. [CrossRef] [PubMed]
18. Bragard, C.; Dehnen-Schmutz, K.; Gonthier, P.; Jacques, M.; Miret, J.A.J.; Justesen, A.F.; MacLeod, A.; Magnusson, C.S.; Milonas, P.; Navas-Cortes, J.A.; et al. Pest categorisation of potato virus M (non-EU isolates). *EFSA J.* **2020**, *18*, e05854. [PubMed]
19. Bradshaw, J.E. Improving resistance to diseases and pests: A dynamic situation. In *Potato Breeding: Theory and Practice*; Bradshaw, J.E., Ed.; Springer Nature: Cham, Switzerland, 2021; pp. 247–337.
20. Mosahebi, G.; Koohi-Habibi, M.; Okhovvat, S.M. Study on potato virus M (PVM) occurrence in potato fields in Iran. *Commun. Agric. Appl. Biol. Sci.* **2005**, *70*, 441–443.
21. Wang, X.M.; Jin, L.P.; Yin, J. Progress in Breeding for Virus Resistance in Potatoes. *Chin. Potato J.* **2005**, *19*, 285–289.
22. Huang, P.; Yan, Q.; Ding, Y. Occurrence and Control of Potato Virus S in Guizhou. *Guizhou Agricul. Sci.* **2009**, *37*, 88–89.
23. Mutka, A.M.; Bart, R.S. Image-based phenotyping of plant disease symptoms. *Front. Plant Sci.* **2015**, *5*, 734. [CrossRef]
24. Kuska, M.T.; Mahlein, A.K. Aiming at decision making in plant disease protection and phenotyping by the use of optical sensors. *Eur. J. Plant. Pathol* **2018**, *152*, 987–992. [CrossRef]
25. Synkova, H.; Semoradova, S.; Schnablova, R.; Muller, K.; Pospísilova, J.; Ryslava, H.; Malbeck, J.; Serovska, N. Effects of biotic stress caused by *Potato virus Y* on photosynthesis in ipt transgenic and control *Nicotiana Tableacum* L. *Plant Sci.* **2006**, *171*, 607–616. [CrossRef]
26. Zhou, Y.H.; Peng, Y.H.; Lei, J.L.; Zou, L.Y.; Zheng, J.H.; Yu, J.Q. Effects of potato virus Y^{NTN} infection on gas exchange and Photosystem II function in leaves of *Solanum tuberosum* L. *Photosynthetica* **2004**, *42*, 417–423. [CrossRef]
27. Maxwell, M.; Johnson, G.N. Chlorophyll fluorescence—A practical guide. *J. Exp. Bot.* **2000**, *51*, 659–668. [CrossRef] [PubMed]
28. Murchie, E.H.; Lawson, T. Chlorophyll fluorescence analysis: A guide to good practice and understanding some new applications. *J. Exp. Bot.* **2013**, *64*, 3983–3998. [CrossRef] [PubMed]
29. Li, L.; Zhang, Q.; Huang, D. A Review of Imaging Techniques for Plant Phenotyping. *Sensors* **2014**, *14*, 20078–20111. [CrossRef] [PubMed]
30. Panattoni, A.; Luvisi, A.; Triolo, E. Review: Elimination of viruses in plants: Twenty years of progress. *Span. J. Agric. Res.* **2013**, *11*, 173–188. [CrossRef]
31. Wang, B.; Wang, R.R.; Li, J.W.; Ma, Y.L.; Sheng, W.M.; Li, M.F.; Wang, Q.C. Development of Three-vitrification-based cryopreservation procedures for shoot tips of China's potato. *Cryoletters* **2013**, *34*, 369–380.
32. Kidulile, C.E.; Ateka, E.M.; Alankonya, A.E.; Ndunguru, J.C. Efficacy of chemotherapy and thermotherapy in elimination of East African cassava mosaic virus from Tanzanian cassava landrace. *J. Phytopathol.* **2018**, *166*, 739–745. [CrossRef]
33. Gong, H.; Igiraneza, C.; Dusengemungu, L. Major in vitro techniques for potato virus elimination and post eradication detection methods. A review. *Am. J. Potato Res. Vol.* **2019**, *96*, 379–389. [CrossRef]
34. Kereša, S.; Kurtović, K.; Goreta Ban, S.; Vončina, D.; Habuš Jerčić, I.; Bolarić, S.; Lazarević, B.; Godena, S.; Ban, D.; Bošnjak Mihovilović, A. Production of Virus-Free Garlic Plants through Somatic Embryogenesis. *Agronomy* **2021**, *11*, 876. [CrossRef]
35. Danci, O.; Erdei, L.; Vidacs, L.; Danci, M.; Anca, B.; Baciu, A.; David, I.; Berbentea, F. Influence of ribavirin on potato plants regeneration and virus eradication. *J. Hortic. For. Biotechnol.* **2009**, *13*, 421–425.
36. Kushnarenko, S.; Romadanova, N.; Aralbayeva, M.; Zholamanova, S.; Alexandrova, A.; Karpova, O. Combined ribavirin treatment and cryotherapy for efficient Potato virus M and Potato virus S eradication in potato (*Solanum tuberosum* L.) in vitro shoots. *Vitr. Cell. Dev. Biol. Plant* **2017**, *53*, 425–432. [CrossRef]
37. Wambugu, F.M.; Secor, G.A.; Gudmestad, N.C. Eradication of potato virus Y and S from potato by chemotherapy of cultured axillary bud tips. *Am. Potato J. Vol.* **1985**, *62*, 667–672. [CrossRef]
38. Bittner, H.; Schenk, G.; Schuster, G.; Kluge, S. Elimination by chemotherapy of potato virus S from potato plants grown in vitro. *Potato Res. Vol.* **1989**, *32*, 175–179. [CrossRef]
39. Zapata, C.; Miller, J.C.; Smith, R.H. An In vitro Procedure to Eradicate Potato Viruses X, Y, and S from Russet Norkotah and Two of Its Strains. *Vitr. Cell. Dev. Biol. Plant* **1995**, *31*, 153–159. [CrossRef]
40. Faccioli, G. Control of Potato Viruses using Meristem and Stem-cutting Cultures, Thermotheraphy and Chemotherapy. In *Virus and Viruslike Diseases of Potatoes and Production of Seed-Potatoes*; Loebenstein, G., Berger, P.H., Brunt, A.A., Lawson, R.H., Eds.; Springer: Dordrecht, The Netherlands, 2001; pp. 365–390.
41. Faccioli, G.; Colalongo, M.C. Eradication of potato virus Y and potato leafroll virus by chemotherapy of infected potato stem cuttings. *Phytopathol. Mediterr.* **2002**, *41*, 76–78.
42. Nascimento, L.C.; Pio-Ribeiro, G.; Willadino, L.; Andrade, G.P. Stock indexing and potato virus y elimination from potato plants cultivated in vitro. *Sci. Agric.* **2003**, *60*, 525–530. [CrossRef]
43. Gopal, J.; Garg, I.D. An efficient protocol of chemo-cum-thermotherapy for elimination of potato (*Solanum tuberosum* L.) virus by meristem-tip culture. *Indian J. Agric. Sci.* **2011**, *81*, 544–549.
44. Yang, L.; Nie, B.; Liu, J.; Song, B. A Reexamination of the Effectiveness of Ribavirin on Eradication of Viruses in Potato Plantlets in vitro Using ELISA and Quantitative RT-PCR. *Am. J. Potato Res.* **2014**, *91*, 304–311. [CrossRef]
45. Smyda-Daymund, P. Virus elimination from in vitro potato plants. *Plant Breed. Seed Sci.* **2017**, *78*, 81–85. [CrossRef]

46. Shoala, T.; Eid, K.E.; El-Fiki, I.A.I. Impact of Chemotherapy and Thermotherapy Treatments on The Presence of Potato Viruses Pvy, Pvx and Plrv in Tissue-Cultured Shoot Tip Meristem. *J. Plant Prot. Pathol.* **2019**, *10*, 581–585. [CrossRef]
47. Karjadi, A.K.; Kardjadi; Gunaeni, N. The Effect Of Antiviral Ribavirin, Explant Size, Varieties On Growth And Development In Potato Meristematic. *IOP Conf. Ser. Earth Environ. Sci.* **2021**, *985*, 012022. [CrossRef]
48. Prusiner, P.; Sundaralingam, M. A New Class of Synthetic Nucleoside Analogues with Broad-spectrum Antiviral Properties. *Nat. New Biol.* **1973**, *244*, 116–118. [CrossRef] [PubMed]
49. Bougie, I.; Bisaillon, M. The Broad Spectrum Antiviral Nucleoside Ribavirin as a Substrate for a Viral RNA Capping Enzyme. *J. Biol. Chem.* **2004**, *279*, 22124–22130. [CrossRef] [PubMed]
50. Wang, M.-R.; Cui, Z.-H.; Li, J.-W.; Hao, X.-Y.; Zhao, L.; Wang, Q.-C. In vitro thermotherapy-based methods for plant virus eradication. *Plant Methods* **2019**, *14*, 87. [CrossRef]
51. Crotty, S.; Cameron, C.; Andino, R. Ribavirin's antiviral mechanism of action: Lethal mutagenesis? *J. Mol. Med.* **2002**, *80*, 86–95. [CrossRef]
52. Murashige, T.; Skoog, F. A revised medium for rapid growth and bioassay with tobacco tissue culture. *Physiol. Plant* **1962**, *15*, 473–497. [CrossRef]
53. Lazarević, B.; Šatović, Z.; Nimac, A.; Vidak, M.; Gunjača, J.; Politeo, O.; Carović-Stanko, K. Application of phenotyping methods in detection of drought and salinity stress in basil (*Ocimum basilicum* L.). *Front. Plant Sci.* **2021**, *12*, 174. [CrossRef]
54. Kitajima, M.; Butler, W. Quenching of chlorophyll fluorescence and primary photochemistry in chloroplasts by dibromothymoquinone. *Biochim. Biophys. Acta.* **1975**, *376*, 105–115. [CrossRef]
55. Genty, B.; Briantais, J.M.; Baker, N.R. The relationship between the quantum yield of photosynthetic electron transport and quenching of chlorophyll fluorescence. *Biochim. Biophys. Acta Gen. Subj.* **1989**, *990*, 87–92. [CrossRef]
56. Bilger, W.; Björkman, O. Role of the xanthophyll cycle in photoprotection elucidated by measurements of light-induced absorbance changes, fluorescence and photosynthesis in leaves of Hedera canariensis. *Photosynth. Res.* **1990**, *25*, 173–185. [CrossRef]
57. Gitelson, A.A.; Merzlyak, M.N.; Chivkunova, O.B. Optical Properties and Non-destructive Estimation of Anthocyanin Content in Plant Leaves. *Photochem. Photobiol.* **2001**, *74*, 38–45. [CrossRef]
58. Gitelson, A.A.; Gritz, Y.; Merzlyak, M.N. Relationships between leaf chlorophyll content and spectral reflectance and algorithms for non-destructive chlorophyll assessment in higher plant leaves. *J. Plant Physiol.* **2003**, *160*, 271–282. [CrossRef] [PubMed]
59. Rouse, J.W.; Haas, R.H.; Schell, J.A.; Deering, D.W. Monitoring vegetation systems in the Great Plains with ERTS. In *NASA. Goddard Space Flight Center 3d ERTS-1 Symp*; NASA: Washington, DC, USA, 1974; Volume 1, pp. 309–317.
60. *SAS/STAT®Software*, version [9.4] 2002–2012; SAS Institute Inc.: Cary, NC, USA, 2013.
61. Singh, B. Effect of antiviral chemicals on in vitro regeneration response and production of PLRV-free plants of potato. *J. Crop. Sci. Biotechnol.* **2015**, *18*, 341–348. [CrossRef]
62. De Clercq, E. Antiviral Agents: Characteristic Activity Spectrum Depending on the Molecular Target With Which They Interact. *Adv. Virus Res.* **1993**, *42*, 1–55. [PubMed]
63. Nyalugwe, E.P.; Wilson, C.R.; Coutts, B.A.; Jones, R.A.C. Biological properties of potato virus X in potato: Effects of mixed infection with potato virus S and resistance phenotypes in cultivars from three continents. *Plant Dis.* **2012**, *96*, 43–54. [CrossRef]
64. Moreno, A.B.; López-Moya, J.J. When viruses play team sports: Mixed infections in plants. *Phytopathology* **2020**, *110*, 29–48. [CrossRef] [PubMed]
65. Liu, W.P. Synergistic effect of potato virus Y (PVY) and potato spindle tuber viroid (PSTV) on tuber yield of potato. *J. Heilongjiang August First Land Reclam. Uni.* **2007**, *19*, 40–43.
66. Lemaga, B.; Caesar, K. Relationships between numbers of main stems and yield components of potato (*Solanum tuberosum* L. cv. Erntestolz) as influenced by different daylengths. *Potato Res.* **1990**, *33*, 257–267. [CrossRef]

horticulturae

Article

In Vitro Propagation of *Pyracantha angustifolia* (Franch.) C.K. Schneid.

Behzad Kaviani [1,*], Bahareh Deltalab [2], Dariusz Kulus [3,*], Alicja Tymoszuk [3], Hamideh Bagheri [1] and Taha Azarinejad [4]

1 Department of Horticultural Science, Rasht Branch, Islamic Azad University, Rasht 4147654919, Iran
2 Agroecology, Razi University, Kermanshah 6715685438, Iran
3 Laboratory of Ornamental Plants and Vegetable Crops, Faculty of Agriculture and Biotechnology, Bydgoszcz University of Science and Technology, Bernardyńska 6, 85-029 Bydgoszcz, Poland
4 Maragheh Branch, Islamic Azad University, Maragheh 5358197202, Iran
* Correspondence: kaviani@iaurasht.ac.ir (B.K.); dkulus@gmail.com (D.K.); Tel.: +98-9111777482 (B.K.); +48-523749536 (D.K.)

Abstract: Narrow-leaf firethorn or pyracantha (*Pyracantha angustifolia* (Franch.) C.K. Schneid.), from the family Rosaceae, is a species of large and thorny evergreen shrub. In this study, a procedure is presented for efficient axillary shoot multiplication and root induction in *P. angustifolia* using Murashige and Skoog (MS), woody plant (WPM), and Linsmaier and Skoog (LS) culture media supplanted with 6-benzylaminopurine (BAP) and indole-3-butyric acid (IBA). The disinfection of the axillary buds was performed with a 70.23% success rate on a basal MS medium augmented with 0.5 mg·L^{-1} gibberellic acid (GA_3). Uniform and axenic explants were then cultured on MS, WPM, and LS media enriched with different concentrations of BAP, 0.3 mg·L^{-1} GA_3, and 0.1 mg·L^{-1} IBA. The highest multiplication coefficient (2.389) was obtained for the MS medium supplemented with 2.5 mg·L^{-1} BAP. After one month, newly formed micro-shoots were transferred to rooting media (MS, WPM, and LS) containing different concentrations of IBA, together with a constant concentration of 0.1 mg·L^{-1} BAP. The micro-shoots were kept in the dark for one week and then cultured in a 16/8 h light/dark regime. The MS medium supplemented with 1 mg·L^{-1} IBA was the most effective in stimulating rooting (88.76% of micro-shoots). The highest number of roots (3.5 per micro-shoot) was produced in the MS medium enriched with 1.5 mg·L^{-1} IBA. The rooted plantlets were transferred into pots filled with perlite and peat moss in a 2:1 proportion and acclimatized to ambient greenhouse conditions, with a resultant mean 92.84% survival rate. Thus, this protocol can be successfully applied for the in vitro mass propagation of *P. angustifolia*.

Keywords: culture medium; micropropagation; ornamental trees and shrubs; plant growth regulators

Citation: Kaviani, B.; Deltalab, B.; Kulus, D.; Tymoszuk, A.; Bagheri, H.; Azarinejad, T. In Vitro Propagation of *Pyracantha angustifolia* (Franch.) C.K. Schneid. *Horticulturae* **2022**, *8*, 964. https://doi.org/10.3390/horticulturae8100964

Academic Editor: Luigi De Bellis

Received: 30 September 2022
Accepted: 16 October 2022
Published: 18 October 2022

1. Introduction

Pyracantha is a genus of large, thorny, and perennial evergreen shrubs in the family Rosaceae, subfamily Amygdaloideae, and order Rosales. They are native to an area extending from southwest Eastern Europe to Southeast Asia. The plants can reach up to 4.5 m in height. Their leaves are small and slender, with serrated margins and numerous thorns [1]. The name "pyracantha" is derived from the Greek words "pyr", meaning fire, and "akanthos", meaning "a thorn", hence "fire-thorn". The genus *Pyracantha* includes 11 species (*P. angustifolia*, *P. atalantioides*, *P. coccinea*, *P. crenatoserrata*, *P. crenulata*, *P. crenulata-serrata*, *P. densiflora*, *P. fortuneana*, *P. inermis*, *P. koidzumii*, and *P. rogersiana*) with small white flowers [2]. The flowers are produced during late spring and early summer. Fruit, i.e., red, orange, or yellow pomes, develop in late summer and mature in late autumn [1,3]. *Pyracantha* is commonly used in the food and medicinal industries, commercial landscapes, and as a hedge or barrier plant [4]. One of the known species of *Pyracantha* is *P. angustifolia*, which is distinguished from the other *Pyracantha* species by its morphological characteristics of leaves and flowers [4]. Their leaves are relatively narrow, with rounded tips and entire margins, and their undersides have a density of white hair [5]. The color of its mature fruit is yellow to dark orange, and it is usually covered with white hair. The antioxidant and anti-inflammatory effects of *Pyracantha angustifolia* fruit extracts have been reported [1]. This species is cultivated as an ornamental garden plant, especially in temperate regions, as the shrub is resistant to cold and seasonal drought. Many cultivars of *P. angustifolia* are the result of hybridization with other inbred cultivars [6]. Due to the high demand for this ornamental shrub, the slow propagation pace of *Pyracantha*, and relatively low and uneven production of valuable phytochemicals under natural conditions, micropropagation is a suitable method for the mass production of this genus.

In vitro propagation has been recognized as an important and efficient technique for the large-scale propagation of horticultural plants, overcoming the problems caused by heterogeneous seed production [7]. It also allows for the elicitation and more uniform production of secondary metabolites [8]. The commercial use of propagating woody plants through tissue culture is more difficult than with herbaceous plants due to the limited multiplication efficiency and plantlets' survival during acclimatization [9]. Nonetheless, several woody plant species are successfully micro-propagated in vitro [10]. The success of the in vitro propagation methods depends on several factors, such as the plant's genotype, medium type, plant growth regulators (PGRs), and explant parameters, which should be carefully optimized [11]. To date, there is no available information about the micropropagation of *Pyracantha angustifolia*. Only one paper has been published on the micropropagation of *P. coccinea* [12]. These researchers solely used the MS medium supplemented with 6-benzylaminopurine (BAP) or indole-3-butyric acid (IBA), and they found that the highest shoot proliferation rate (3.40 axillary buds) was obtained using 3/4 strong MS with 0.3 $mg{\cdot}L^{-1}$ IBA. As for the rooting step, 1/4 MS medium with 18.9 $mg{\cdot}L^{-1}$ IBA was most effective. The successful use of IBA and BAP for the shoot multiplication of some members of Rosaceae, such as *Prunus cerasifera* Ehrh. [13] and *Prunus persica* L. [14], has been reported. Likewise, studies have shown that the best auxin for rooting in several members of Rosaceae is IBA [14–16]. The results of various studies have shown that different PGR combinations and concentrations have significant roles in increasing the micropropagation efficiency of ornamental and fruit trees or shrubs belonging to the Rosaceae family [16–18]. The most frequently used explant type in these studies was the stem node.

The present study aimed to evaluate the effects of different concentrations of BAP and IBA, added individually or in combination, into the Murashige and Skoog (MS), woody plant (WPM), and Linsmaier and Skoog (LS) media on the in vitro propagation of *P. angustifolia* through direct organogenesis.

2. Materials and Methods

2.1. Plant Material

Plant material was obtained from the municipality of Tabriz city in East Azarbaijan province, Iran. East Azarbaijan province is located in the northwest of Iran at an altitude of 1800 m above sea level at 46° and 25′ east longitude and 38° and 2′ north latitude from the Greenwich meridian. This province is located in a mountainous region and has a climate with cold and long winters and mild summers. Plant samples were collected from the lateral buds of *Pyracantha angustifolia* shrubs in Tabriz city in September 2020. The samples were taken in the form of 10–15 cm-long cuttings and, to preserve moisture and prevent possible injuries, they were wrapped in a wet cloth and transported to the laboratory in an ice flask. The leaves of the cuttings were removed and then cut into approximately 2 cm-long pieces with 1 or 2 buds.

2.2. Explants Disinfection

In order to reduce surface contamination, the explants were washed carefully with dishwashing liquid and Captan 80 WDG fungicide (3 g per 500 mL water for 20 min).

Next, for better washing and removal of inhibitory substances, the explants were placed under tap water for an hour. After preliminary washing, all the explants were treated with 70% (*v*/*v*) ethyl alcohol for 1 min and then placed in sodium hypochlorite (NaOCl) at a concentration of 2% (*v*/*v*) for 5 min. All disinfection steps were performed under a laminar flow hood cabinet and 12 drops of dishwashing liquid (1 mL per 500 mL water for 20 min) were added to increase the contact surface of the disinfectant liquid with the explant. Then, the explants were washed three times in sterile distilled water, each time for 5 min. To remove the parts of the explants that were damaged in the disinfection treatments, the bases of the explants were cut off.

2.3. Establishment of In Vitro Culture

Each of the three disinfected explants were placed inside sterile Petri dishes filled with 50 mL of basal Murashige and Skoog [19] (MS) culture medium containing 3% (*w*/*v*) sucrose and 0.8% (*w*/*v*) agar (SIGMA Aldrich, Milwaukee, Brookfield, WI, USA) augmented with 0.5 mg·L^{-1} filtered sterilized gibberellic acid (GA_3) (SIGMA Aldrich, Milwaukee, WI, USA). The pH of the media was adjusted to 5.6–5.8 with 0.1 N NaOH or HCl prior to autoclaving. All media contained in the culture bottles were autoclaved at 105 kPa and 121 °C for 20 min. In the first week of culture, the explants were kept in the dark at 24 ± 2 °C and, thereafter, at the same temperature with a 16/8 h light/dark regime with a light intensity of 50–60 μmol·m^{-2}·s^{-1} provided by cool—white fluorescent tubes. After 30 days, the newly formed micro-shoots (Figure 1A) were used for the subsequent experiments.

Figure 1. In vitro propagation of *Pyracantha angustifolia*. (**A**) Shoots produced in the initiation step in the MS medium containing 0.5 mg·L^{-1} GA_3. (**B**) Preparation of the explants (nodal segments with

axillary buds) for the multiplication step. (**C**) Micro-shoots produced in the proliferation step (LS medium containing 3 mg·L^{-1} BAP (left), 2 mg·L^{-1} BAP (medial), and 2.5 mg·L^{-1} BAP (right)). (**D**) Micro-shoots produced in the proliferation step (MS medium containing 2 mg·L^{-1} BAP (left), 2.5 mg·L^{-1} BAP (medial), and 3 mg·L^{-1} BAP (right)). (**E**) Micro-shoots produced in the MS medium containing 2.5 mg·L^{-1} BAP. (**F**) Micro-shoots produced in the WPM containing 2.5 mg·L^{-1} BAP. (**G**) Rooted micro-shoots in the MS medium containing 1 mg·L^{-1} IBA. (**H**) Plantlets obtained from the MS medium containing 1 mg·L^{-1} IBA. (**I**) Acclimatization of the plantlets in plastic pots filled with perlite and peat moss (2:1) in the acclimatization room.

2.4. *Shoot Proliferation*

After the establishment and growth of the micro-shoots, a total of 108 nodal explants with axillary buds (Figure 1B) were used as secondary explants in the proliferation stage (Figure 1C–F). Three types of basal culture media–MS, woody plant medium (WPM) [20], and Linsmaier and Skoog (LS) [21]—containing 3% sucrose and 0.8% agar were prepared. The pH levels of the media were adjusted to 5.6–5.8 before autoclaving as described above. The media were augmented with different concentrations of BAP (0, 2, 2.5, and 3 mg·L^{-1}) for the induction of direct organogenesis. In all experimental objects, 0.3 mg·L^{-1} GA_3 and 0.1 mg·L^{-1} IBA were used for hormonal balance. Jars of jam were used as the culture vessels and 50 mL of medium was poured into each jar. All the cultures were incubated at 24 ± 2 °C and a 16 h photoperiod of 50–60 μmol·m^{-2}·s^{-1} irradiance was provided by cool—white fluorescent tubes. The produced micro-shoots were sub-cultured twice after 30 days.

2.5. *Root Induction on Microshoots*

The micro-shoots produced in each culture vessel, from all experimental objects, were transferred to the rooting media. Three culture media–MS, WPM, and LS—were prepared and autoclaved. The media were supplemented with different concentrations of IBA (0, 0.1, 0.5, 1, and 1.5 mg·L^{-1}) (Figure 1G). For hormonal balance, 0.1 mg·L^{-1} BAP was added to these root induction media. All the cultures were kept in the growth chamber with the same light and temperature conditions as described previously.

2.6. *Acclimatization Process*

The rooted micro-shoots—95 plantlets from the MS medium, 55 from the LS medium, and 45 from the WPM medium (Figure 1H)—were taken out from the culture jars and the remains of the culture medium were removed from the roots. Then, the lower leaves of the plantlets were dissected with scissors for easier cultivation of the plantlets in the pots. The plantlets were placed in the autoclaved perlite and peat moss substrate (in a ratio of 2:1) in plastic pots and watered with sterile water. A transparent plastic cup was also placed over the plantlets. The pots were placed in an acclimatization room with 70 ± 5% relative humidity under a 16/8 h light/dark regime (light intensity of 100 μmol·m^{-2}·s^{-1}) at a temperature of 25 ± 2 °C. The plantlets were fed by a syringe with nutrients containing 1 g·L^{-1} of N.P.K and 500 mg·L^{-1} of iron every 4 days from the second week of transferring to the acclimatization room. After 30 days, the plantlets were transferred to larger plastic pots containing garden soil and perlite in the proportion of 2:1 and transferred to the greenhouse (Figure 1I).

2.7. *Experimental Design and Data Analysis*

The factorial in vitro experiment was conducted in a completely randomized block design with 10 replications. Each experimental unit consisted of 5 jars, and 3 explants were cultured in each jar. The effect of the culture media on the growth of the explants and the percentage of the survival of the plantlets were evaluated. The first observation was completed 15 days after cultivation and the second one was repeated 15 days later. Initially, the number of jars in each experimental unit was higher, but 5 days after culture initiation, by removing the contaminated jars, it was reduced to 5 jars in each experimental object. This design was repeated in the multiplication (proliferation) and rooting experiments.

During the micro-shoot proliferation stage, the multiplication coefficient, micro-shoot length (cm), and share of hyperhydrated micro-shoots (%) were evaluated after 30 days of culture. The micro-shoots produced in the proliferation stage were usually suitable for rooting, but to produce more shoots, the explants were sub-cultured twice.

The rooting of micro-shoots was performed for 21 days, after which the rooting effectiveness (%), root number, and root length (mean length of all the regenerated roots per one micro-shoot; cm) were measured.

The greenhouse experiments were set in a completely randomized block design with three replicates. Data were subjected to analysis of variance (ANOVA) and the means were compared by Duncan's test at $p < 0.05$ using the SAS version 9.1 software [22].

3. Results

3.1. Micro-Shoot Establishment

After disinfection, where efficiency reached 70.23%, the explants were transferred to the initiation culture medium (MS with 0.5 mg·L^{-1} GA_3), on which narrow-leaf firethorn shoots were successfully established (Figure 1A). The share of shoot-producing nodal segments was 88.8, 83.33, and 72.23% in the MS, LS, and WPM media, respectively. Secondary explants (nodal segments) were obtained after 30 days of culture (Figure 1B). They were then transferred to the proliferation media (MS, WPM, and LS) augmented with different concentrations of BAP (Figure 1C–F).

3.2. Micro-Shoot Proliferation

The analysis of variance showed that in terms of all the studied characteristics, there was a significant effect of the tested BAP concentrations and basal culture media types, singularly and in combination (Table 1, Supplementary Figures S1 and S2). In general, the MS medium was more effective than the LS and WPM media (Supplementary Figure S1). The highest multiplication coefficient (2.389) was obtained in the MS medium augmented with 2.5 mg·L^{-1} BAP (Table 2). This coefficient was approximately two-fold higher than that of the other treatments. The lowest multiplication coefficients (0.960 and 1.021), on the other hand, were obtained in the WPM and LS media, both without BAP (control). Likewise, the highest concentration of BAP (3 mg·L^{-1}) had a negative effect on the multiplication coefficient in all three media types (MS, WPM, and LS) (Table 2).

Table 1. Analysis of variance of the effects of the different concentrations of BAP and the basal culture medium type on the measured characteristics of the *Pyracantha angustifolia* plants grown in vitro in the multiplication step.

Source of Variance	df	Multiplication Coefficient MS	Plantlet Length MS	Hyperhydricity MS
BAP	3	0.379 **	0.598 **	2.259 *
Culture medium	2	2.943 **	0.874 **	19.704 **
BAP × culture medium	6	0.173 *	0.019 **	0.148 *
Error	24	0.053	0.04	0.481
CV (%)	-	14.79	11.33	18.61

*, **: significant at the 0.05 and 0.01 probability level, respectively; CV: coefficient of variation; df: degrees of freedom; MS: mean square.

The longest micro-shoots (2.132 and 2.081 cm) were produced in the MS and LS media, both of which were supplemented with 2.5 mg·L^{-1} BAP, respectively (Table 2). The shortest lengths (1.513 and 1.523 cm) were obtained in the WPM medium containing 3 mg·L^{-1} BAP and in the control object without BAP, respectively. The evaluation of the individual effects of medium type and BAP concentration on the length of the micro-shoots also showed the superiority of the MS medium and the 2.5 mg·L^{-1} BAP concentration (Supplementary Figures S1 and S2).

Table 2. Effects of the different concentrations of BAP and the basal culture medium type on the measured characteristics of the *Pyracantha angustifolia* plants grown in vitro in the multiplication step.

Treatment		Multiplication Coefficient	Plantlet Length (cm)	Hyperhydricity (%)
Culture Medium	BAP ($mg \cdot L^{-1}$)			
MS	0	1.213 ^{d} ± 0.220	1.624 ^{cd} ± 0.316	3.351 ^{cd} ± 0.751
	2	1.516 ^{c} ± 0.075	1.645 ^{c} ± 0.119	3.269 ^{d} ± 0.635
	2.5	2.389 ^{a} ± 0.488	2.132 ^{a} ± 0.414	2.053 ^{f} ± 0.332
	3	1.061 ^{de} ± 0.193	1.614 $^{c–e}$ ± 0.329	3.422 $^{b–d}$ ± 0.564
WPM	0	0.960 ^{e} ± 0.174	1.523 ^{g} ± 0.296	3.541 $^{a–c}$ ± 0.489
	2	1.082 ^{de} ± 0.196	1.543 ^{fg} ± 0.342	3.716 ^{a} ± 0.722
	2.5	1.971 ^{b} ± 0.358	1.838 ^{b} ± 0.157	2.741 ^{e} ± 0.533
	3	1.041 ^{e} ± 0.189	1.513 ^{g} ± 0.294	3.615 ^{ab} ± 0.702
LS	0	1.021 ^{e} ± 0.185	1.553 $^{e–g}$ ± 0.302	3.532 $^{a–c}$ ± 0.687
	2	1.617 ^{c} ± 0.294	1.594 $^{c–f}$ ± 0.409	3.564 $^{a–c}$ ± 0.692
	2.5	2.002 ^{b} ± 0.364	2.081 ^{a} ± 0.391	2.132 ^{f} ± 0.414
	3	1.092 ^{de} ± 0.198	1.564 $^{d–g}$ ± 0.537	3.553 $^{a–c}$ ± 0.690

Means ± standard deviations with different letters in the same column are significantly different ($p < 0.05$) based on the Duncan's test.

3.3. Micro-Shoot Hyperhydricity

The highest shares of hyperhydrated shoots (3.716 and 3.615%) were reported in the WPM medium supplemented with 2 and 3 $mg \cdot L^{-1}$ BAP, respectively. On the other hand, the lowest rate of hyperhydricity (2.053%) was found in the MS medium supplemented with 2.5 $mg \cdot L^{-1}$ BAP. In general, the micro-shoots produced in all three culture media (MS, WPM, and LS) enriched with 2.5 $mg \cdot L^{-1}$ BAP showed the lowest rates of hyperhydricity (Table 2). These findings were also confirmed during the individual evaluations of each of the factors (Supplementary Figures S1 and S2).

3.4. Micro-Shoot Rooting

Statistical analysis of the data showed that there was a significant influence of the studied factors, singularly and in combination, regarding rooting effectiveness (Table 3, Supplementary Figures S3 and S4). There was also a significant difference between the various concentrations of IBA and the culture medium type regarding the root number. Moreover, the type of basal culture medium affected the elongation of the roots (Table 3).

Table 3. Analysis of variance of the effects of the different concentrations of IBA and the basal culture medium type on the measured characteristics of the *Pyracantha angustifolia* plants grown in vitro in the rooting step, as well as the acclimatization effectiveness.

Source of Variance	df	Rooting Effectiveness MS	Root Number MS	Root Length MS	Acclimatization MS
IBA	4	6066.93 **	325.30 *	0.490 ^{ns}	425.80 **
Culture medium	2	845.48 **	14.40 *	7.941 **	26.63 ^{ns}
IBA × culture medium	8	159.98 *	23.50 **	0.189 *	169.95 **
Error	30	38.96	15.8	0.191	15.7
CV (%)	-	15.67	16.2	13.11	8.34

*, **: significant at the 0.05 and 0.01 probability level, respectively; ^{ns}: not significant at $p < 0.05$; CV: coefficient of variation; df: degrees of freedom; MS: mean square.

The highest and the lowest production of roots were observed in the MS and WPM media, respectively (Supplementary Figure S3). The maximum rooting efficiency (88.76%) was found in the MS medium with 1 $mg \cdot L^{-1}$ IBA. On the other hand, the minimum rooting effectiveness levels of 32.71% and 35.59% were reported for the control WPM medium and that with 0.1 $mg \cdot L^{-1}$ IBA, respectively (Table 4).

Table 4. Effects of the different concentrations of IBA and the basal culture medium type on the measured characteristics of the *Pyracantha angustifolia* plants grown in vitro in the rooting step, as well as the acclimatization effectiveness.

Treatment		Rooting Effectiveness (%)	Root Number	Root Length (cm)	Acclimatization (%)
Culture medium	IBA (mg·L^{-1})				
MS	0	73.30 ^{d} ± 13.34	0.875 ^{e} ± 0.14	3.790 ^{b} ± 0.639	96.29 ^{ab} ± 4.17
	0.1	76.09 ^{cd} ± 13.85	1.75 ^{d} ± 0.34	3.840 ^{b} ± 0.447	96.29 ^{ab} ± 3.17
	0.5	80.10 ^{bc} ± 14.58	2.25 ^{bc} ± 0.36	3.952 ^{b} ± 0.367	97.2 ^{ab} ± 3.27
	1	88.76 ^{a} ± 16.16	2.625 ^{b} ± 0.57	4.297 ^{a} ± 0.725	97.2 ^{ab} ± 5.27
	1.5	81.21 ^{b} ± 11.78	3.5 ^{a} ± 0.39	4.155 ^{a} ± 0.704	100 ^{a} ± 2.56
WPM	0	32.71 ^{j} ± 5.95	0.375 ^{i} ± 0.11	1.817 ^{i} ± 0.307	84.08 ^{e} ± 1.88
	0.1	35.59 ^{ij} ± 6.48	0.5 ^{h} ± 0.12	1.962 ^{hi} ± 0.132	86.11 ^{de} ± 4.04
	0.5	39.69 ^{hi} ± 10.22	0.61 ^{g} ± 0.14	2.162 ^{gh} ± 0.364	86.11 ^{de} ± 4.04
	1	44.88 ^{g} ± 4.17	0.625 ^{fg} ± 0.11	2.397 ^{f} ± 0.405	88.10 $^{c–e}$ ± 2.14
	1.5	41.59 ^{gh} ± 7.39	0.68 ^{f} ± 0.13	2.225 ^{fg} ± 0.377	91.11 $^{b–d}$ ± 2.33
LS	0	41.76 ^{gh} ± 9.60	0.625 ^{fg} ± 0.12	3.172 ^{e} ± 0.596	93.46 ^{bc} ± 3.42
	0.1	42.70 ^{gh} ± 6.78	0.725 ^{ef} ± 0.12	3.215 ^{e} ± 0.440	93.46 ^{bc} ± 6.41
	0.5	49.91 ^{f} ± 9.09	1.75 ^{cd} ± 0.26	3.335 ^{de} ± 0.665	94.46 $^{a–c}$ ± 3.53
	1	55.03 ^{e} ± 10.01	1.80 ^{cd} ± 0.31	3.540 ^{c} ± 0.390	94.46 $^{a–c}$ ± 5.52
	1.5	50.82 ^{ef} ± 8.25	1.84 ^{c} ± 0.33	3.440 ^{cd} ± 0.481	95.82 ^{ab} ± 4.23

Means ± standard deviations with different letters in the same column are significantly different ($p < 0.05$) based on Duncan's test.

Likewise, the highest and lowest root numbers were produced in the MS and WPM media, respectively (Supplementary Figure S3). The highest number of roots (3.5 per micro-shoot) was produced in the MS medium supplemented with 1.5 mg·L^{-1} IBA. On the other hand, the lowest root number (0.375 per micro-shoot) was produced in the WPM medium without IBA (Table 4).

The longest and shortest roots were found in the MS and WPM media, respectively. The longest roots (4.297 and 4.155 cm) were produced in the MS medium with 1 and 1.5 mg·L^{-1} IBA, respectively. The shortest roots (1.817 and 1.962 cm) were regenerated in the control WPM medium and that with 0.1 mg·L^{-1} IBA, respectively (Table 4).

3.5. Acclimatization Efficiency

It was found that 84.08–100% of the plantlets survived transplantation and acclimation to the greenhouse conditions (Figure 1I). The composition of the culture medium, particularly the auxin concentration (Supplementary Figure S4), used during rooting affected the survival of the plantlets (Tables 3 and 4). The highest survival rate was reported in the 1.5 mg·L^{-1} IBA treatment in the MS culture medium and the lowest in the WPM control object (0 mg·L^{-1} IBA).

4. Discussion

Production of true-to-type plants within a short time is the main goal for successful in vitro propagation. Many woody ornamental species are recalcitrant to vegetative propagation by conventional cutting methods. Thus, in vitro propagation techniques have been developed [12]. The present study reports a mass propagation system for *Pyracantha angustifolia*, an economically valuable ornamental shrub. The optimum concentration of cytokinins for maximum shoot multiplication is different for each species. This might be due to the different contents of endogenous phytohormones. Moreover, the response to exogenous PGRs varies between cultivar and explant type [11,17,18]. Several studies have reported that the use of 0.1–2 mg·L^{-1} BAP is optimal for shoot proliferation in tree and shrub species [13,23,24]. The current study confirmed that BAP is effective for the proliferation of micro-shoots in *P. angustifolia* when used in moderately high concentrations (2.5 mg·L^{-1}), as higher contents of this cytokinin inhibited callus induction. Research on several species of the Rosaceae family has revealed that the maximum shoot multiplication rates were obtained in media supplemented with both BAP and IBA [16,20,25]. Several researchers have evaluated the effect of other cytokinins such as zeatin (ZEA), thidiazuron (TDZ), N6-(Δ^2-isopentenyl) adenine (2-iP), and meta-Topolin (mT) on the shoot multiplication rate of some woody plants [26,27], although BAP was the most cost-effective [26].

Jagiełło-Kubiec et al. [26] found a positive effect of gibberellic acid (GA_3) on shoot proliferation in ninebark (*Physocarpus opulifolius* L. Maxim.). In our study, GA_3 was used for hormonal balance in the multiplication step. Similar results were reported for other members of the Rosaceae family [14,17]. In peach (*Prunus persica* (L.) Batsch.), BAP (0.5 $mg \cdot L^{-1}$) together with IBA (0.01 $mg \cdot L^{-1}$) and GA_3 (0.5 $mg \cdot L^{-1}$) was found to be the optimal combination treatment for shoot initiation (100%) from nodal explants. Likewise, 2 $mg \cdot L^{-1}$ BAP together with 0.01 $mg \cdot L^{-1}$ IBA and 0.5 $mg \cdot L^{-1}$ GA_3 was the best treatment for shoot proliferation (7.67) in *Prunus persica* (L.) Batsch. 'Garnem' [14]. Zare Khafri et al. [16] demonstrated that the highest number of lateral shoots induction in three Iranian apricot cultivars (*Prunus armenica* L.) was obtained in a WPM medium supplemented with 4 $mg \cdot L^{-1}$ GA_3 and 1 $mg \cdot L^{-1}$ BAP.

The proliferation of axillary shoots is strongly influenced by cytokinins, and the improper concentration of these PGRs often causes hyperhydricity of the tissues [12], as observed in the present study. Hyperhydricity (previously known as vitrification) is a morphological and physiological disorder of plants that results in excessive hydration, low lignification, impaired stomatal, and reduced mechanical strength of plants vegetatively propagated in vitro [28]. It can be a source of serious financial loss to commercial laboratories and ought to be avoided. The relatively low share of hyperhydrated shoots in the present study (2.053–3.716%) confirms the suitability of the described protocol in the micropropagation of firethorn.

In the present study, among the three basal media types used, the MS medium was most effective both for shoot multiplication and rooting. Likewise, in the dwarf rootstock of apple (*Malus domestica* Borkh 'Gami Almasi'), MS was better than Chu's N6 basal medium [29] during shoot proliferation [30]. As for *Prunus africana* (Hook f.) Kalkman and *Mandevilla guanabarica* Casar. ex MFSales, an endemic plant from Brazil with pharmacological and ornamental potential, the WPM medium was more effective than the MS medium [31,32]. The addition of coconut water or other natural supplements could further increase shoot formation in *P. angustifolia*, as reported in a study on sandalwood by Solle and Semiar [33].

The current study confirmed the positive effect of IBA on root induction and growth in *P. angustifolia*. IBA is usually the more effective and more frequently used auxin for rooting than other auxins, as reported for several other woody species [34–37]. However, root formation can be significantly affected not only by the auxin type, but also by its concentration [12]. Some investigations have demonstrated that the optimum root initiation and development were obtained when actively growing axillary shoots were cultured on a medium enriched with a high concentration of IBA [37,38]. Zare Khafri et al. [16] showed that the highest rooting percentage in three Iranian apricot cultivars was obtained in half-strength QL medium [39] supplemented with 4 $mg \cdot L^{-1}$ of IBA. In peach, IBA at a concentration of 1.5 $mg \cdot L^{-1}$ induced the maximum rooting rate (42.86%), the maximum root number (6.33 per shoot), and the longest roots (7.17 cm) [14]. These findings were similar to our research regarding root number.

Our study showed that the maximum root induction and root length in *P. angustifolia* were obtained in the MS medium supplemented with moderately low IBA concentrations (1 and 1.5 $mg \cdot L^{-1}$). Likewise, in ninebark, the best rooting was at 1 $mg \cdot L^{-1}$ IBA in half-strength MS medium [26]. It is worth mentioning that in the study by Dong et al. [12], an increased rooting percentage from 0% to 77% in 1.5 months was reported for *P. coccinea* by adding 18.9 $mg \cdot L^{-1}$ IBA to the culture medium. In the present study, we were able to obtain successful rooting of this species by using much lower concentrations of PGRs, which makes the here-described micropropagation protocol more suitable for both scientific and commercial laboratories. Differences between the species and the content of endogenous PGRs in each cultivar are the most important factors responsible for this phenomenon. For example, some researchers have reported successful rooting after using a combination of auxin and cytokinin [11]. The present study revealed that BAP treatment is effective for inducing root formation when applied with IBA. On the other hand, in some historical roses, optimal rooting was found on half-strength control MS medium, without IBA [40]. Therefore, the optimal concentration of PGRs for root production and growth is different between tree and shrub species. The obtained high acclimatization efficiency (84.08–100%), which is one of the most critical steps in the micropropagation protocol [32], confirms the functionality of the root system and the good quality of the micro-shoots. It is also worth mentioning that since the micro-shoots were produced from a meristematic explant without a callus phase, their genetic fidelity should be maintained [31]. Ex vitro rooting, including the treatments with IBA and abscisic acid (ABA), could significantly shorten the production time, as reported by Jagiełło-Kubiec et al. [26], and could be considered in future studies with *P. angustifolia*.

5. Conclusions

The in vitro propagation of woody plants is a difficult task. Therefore, more research in this area is necessary. This maiden study aimed to establish and propagate *P. angustifolia* in vitro by using single-node segments (axillary buds) as explants, as well as BAP and IBA as PGRs through the direct organogenesis method. The highest micro-shoot proliferation, rooting efficiency, and root number rates were obtained in MS medium augmented with 2.5 $mg{\cdot}L^{-1}$ BAP, 0.3 $mg{\cdot}L^{-1}$ GA_3, and 0.1 $mg{\cdot}L^{-1}$ IBA (caulogenesis), and in MS medium with 1 or 1.5 $mg{\cdot}L^{-1}$ IBA together with 0.1 $mg{\cdot}L^{-1}$ BAP (rhizogenesis). The produced plantlets were successfully acclimatized to the greenhouse conditions. The here-presented micropropagation protocol allows for the large-scale (re)production of *P. angustifolia*.

Supplementary Materials: The following supporting information can be downloaded at: https://www.mdpi.com/article/10.3390/horticulturae8100964/s1, Figure S1: effect of different culture media types on the measured characteristics of *Pyracantha angustifolia* plants grown in vitro in the multiplication step; Figure S2: effect of different concentrations of BAP on the measured characteristics of *Pyracantha angustifolia* plants grown in vitro in the multiplication step; Figure S3: effect of different culture media types on the measured characteristics of *Pyracantha angustifolia* plants grown in vitro in the rooting step; Figure S4: effect of different concentrations of IBA on the measured characteristics of *Pyracantha angustifolia* plants grown in vitro in the rooting step.

Author Contributions: All authors made a significant contribution to the work presented. All authors have read and agreed to the published version of the manuscript.

Funding: This research received no external funding.

Data Availability Statement: Data are available by e-mail upon reasonable request.

Conflicts of Interest: The authors declare no conflict of interest.

References

1. Potter, D.; Eriksson, T.; Evans, R.C.; Oh, S.; Smedmark, J.; Morgan, D.R.; Kerr, M.; Robertson, K.R.; Arsenault, M.; Dickinson, T.A.; et al. Phylogeny and classification of Rosaceae. *Plant Syst. Evol.* **2007**, *266*, 5–43. [CrossRef]
2. Chari, L.D.; Martin, G.D.; Steenhuisen, S.; Adams, L.D.; Clark, V.R. Biology of invasive plants 1. *Pyracantha angustifolia* (Franch.) CK Schneid. *Invasive Plant Sci. Manag.* **2020**, *13*, 120–142. [CrossRef]
3. Yang, Y.; Lee, J.A. Antioxidant and anti-inflammatory effect of *Pyracantha angustifolia* fruit extracts. *J. Converg. Inf. Technol.* **2019**, *12*, 294–301. [CrossRef]
4. Jocou, A.I.; Gandullo, R. Synopsis of *Pyracantha* (Rosaceae, Maloideae) species naturalized in Argentina. *Bol. Soc. Argent. Bot.* **2019**, *54*, 599–616. [CrossRef]
5. Retief, E.; Meyer, N.L. *Plants of the Free State: Inventory and Identification Guide*; South African National Biodiversity Institute: Pretoria, South Africa, 2017; p. 1236, ISBN 9781928224150.

6. Weber, E. *Invasive Plant Species of the World: A Reference Guide to Environmental Weeds*; CABI Publishing: Wallingford, UK, 2017; p. 581, ISBN 9781780643861.
7. Nunes, S.; Sousa, D.; Pereira, V.T.; Correia, S.; Marum, L.; Santos, C.; Dias, M.C. Efficient protocol for in vitro mass micropropagation of slash pine. *Vitr. Cell. Dev. Biol. Plant* **2018**, *54*, 175–183. [CrossRef]
8. Ramirez-Estrada, K.; Vidal-Limon, H.; Hidalgo, D.; Moyano, E.; Golenioswki, M.; Cusidó, R.; Palazon, J. Elicitation, an effective strategy for the biotechnological production of bioactive high-added value compounds in plant cell factories. *Molecules* **2016**, *21*, 182. [CrossRef]
9. Sahari Moghaddam, A.; Kaviani, B.; Mohammadi Torkashvand, A.; Abdousi, V.; Eslami, A.R. Micropropagation of English yew, an ornamental-medicinal tree. *J. Ornamen. Plants* **2022**, *12*, 1–9.
10. Gaidamashvili, M.; Benelli, C. Threatened woody plants of Georgia and micropropagation as a tool for in vitro conservation. *Agronomy* **2021**, *11*, 1082. [CrossRef]
11. Adibi Baladeh, D.; Kaviani, B. Micropropagation of medlar (*Mespilus germanica* L.), a Mediterranean fruit tree. *Intl. J. Fruit Sci.* **2021**, *21*, 242–254. [CrossRef]
12. Dong, C.; Li, X.; Xi, Y. Micropropagation of *Pyracantha coccinea*. *HortScience* **2017**, *52*, 271–273. [CrossRef]
13. Nasri, A.; Baklouti, E.; Ben Romdhane, A.; Maalej, M.; Schumacher, H.M.; Drira, N.; Fki, L. Large-scale propagation of Myrobolan (*Prunus cerasifera*) in RITA® bioreactors and ISSR-based assessment of genetic conformity. *Sci. Hortic.* **2019**, *9*, 144–153. [CrossRef]
14. Felek, W.; Mekibib, F.; Admassu, B. Micropropagation of peach, *Prunus persica* (L.) Batsch. cv. Garnem. *Afr. J. Biotechnol.* **2017**, *16*, 490–498. [CrossRef]
15. Fan, S.; Jian, D.; Wei, X.; Chen, J.; Beeson, R.C.; Zhou, Z.; Wang, X. Micropropagation of blueberry 'Bluejay' and 'Pink Lemonade' through in vitro shoot culture. *Sci. Hortic.* **2017**, *226*, 277–284. [CrossRef]
16. Zare Khafri, A.; Solouki, M.; Zarghami, R.; Fakheri, B.; Mahdinezhad, N.; Naderpour, M. In vitro propagation of three Iranian apricot cultivars. *Vitr. Cell. Dev. Biol. Plant* **2021**, *57*, 102–117. [CrossRef]
17. Kucharska, D.; Orlikowska, T.; Maciorowski, R.; Kunka, M.; Wójcik, D.; Pluta, S. Application of *meta*-Topolin for improving micropropagation of gooseberry (*Ribes grossularia*). *Sci. Hortic.* **2020**, *272*, 109529. [CrossRef]
18. Kudělková, M.; Pavelková, R.; Ondrušiková, E.; Vachůn, M. The issues of apricot (*Prunus armeniaca* L.) micropropagation. *Acta Univ. Agric. Silvic. Mendel. Brun.* **2017**, *65*, 67–72. [CrossRef]
19. Murashige, T.; Skoog, F. A revised medium for rapid growth and bioassays with tobacco tissue cultures. *Physiol. Plant* **1962**, *15*, 473–479. [CrossRef]
20. McCown, B.H.; Lloyd, G. Woody plant medium (WPM)—A mineral nutrient formation for microculture for woody plant species. *Hort. Sci.* **1981**, *16*, 453.
21. Linsmaier, E.M.; Skoog, F. Organic growth factor requirements of tobacco tissue cultures. *Physiol. Plant.* **1965**, *18*, 100–127. [CrossRef]
22. SAS Institute. *SAS/STAT Software. Version 9.1 for Windows*; SAS Institute: Cary, NC, USA, 2003.
23. Teixeira da Silva, J.A.; Gulyás, A.; Magyar-Tábori, K.; Wang, M.-R.; Wang, Q.-C.; Dobránszki, J. In vitro tissue culture of apple and other *Malus* species: Recent advances and applications. *Planta* **2019**, *249*, 975–1006. [CrossRef]
24. Gago, D.; Sánchez, C.; Aldrey, A.; Christie, C.B.; Bernal, M.Á.; Vidal, N. Micropropagation of plum (*Prunus domestica* L.) in bioreactors using photomixotrophic and photoautotrophic conditions. *Horticulturae* **2022**, *8*, 286. [CrossRef]
25. Vujović, T.; Jevremović, D.; Marjanović, T.; Glišić, I. In vitro propagation and medium-term conservation of autochthonous plum cultivar 'Crvena Ranka'. *Acta Agric. Serbica* **2020**, *25*, 141–147. [CrossRef]
26. Jagiełło-Kubiec, K.; Nowakowska, K.; Ilczuk, A.; Łukaszewska, A. Optimizing micropropagation conditions for a recalcitrant ninebark (*Physocarpus opulifolius* L. maxim.) cultivar. *Vitr. Cell. Dev. Biol. Plant* **2021**, *57*, 281–295. [CrossRef]
27. Lotfi, M.; Bayoudh, C.; Werbrouck, S.; Mars, M. Effects of *meta*–Topolin derivatives and temporary immersion on hyperhydricity and in vitro shoot proliferation in *Pyrus communis*. *Plant Cell Tissue Organ Cult.* **2020**, *143*, 499–505. [CrossRef]
28. Abdalla, N.; El-Ramady, H.; Seliem, M.K.; El-Mahrouk, M.E.; Taha, N.; Bayoumi, Y.; Shalaby, T.A.; Dobránszki, J. An academic and technical overview on plant micropropagation challenges. *Horticulturae* **2022**, *8*, 677. [CrossRef]
29. Chu, C.C.; Wang, C.C.; Sun, C.S.; Hsu, C.; Yin, K.C.; BI, C.V. Establishment of an efficient medium for anther culture of rice through comparative experiments on the nitrogen source. *Sci. Sin.* **1975**, *18*, 659–668. [CrossRef]
30. Mohseniazar, M.; Nazeri, S.; Ghadimzadeh, M.; Malboobi, M.A. Effect of medium type and some biochemical components on in vitro proliferation of dwarf rootstock of apple (*Malus domestica* Borkh cv Gami Almasi). *Plant Prod. Technol.* **2012**, *1*, 33–41.
31. Komakech, R.; Kim, Y.-G.; Kim, W.J.; Omujal, F.; Yang, S.; Moon, B.C.; Okello, D.; Rahmat, E.; Kyeyune, G.N.; Matsabisa, M.G.; et al. A micropropagation protocol for the endangered medicinal tree *Prunus africana* (Hook f.) Kalkman: Genetic fidelity and physiological parameter assessment. *Front. Plant Sci.* **2020**, *11*, 548003. [CrossRef]
32. Cordeiro, S.Z.; Simas, N.K.; Henriques, A.B.; Sato, A. Micropropagation and callogenesis in *Mandevilla guanabarica* (Apocynaceae), an endemic plant from Brazil. *CBAB* **2014**, *14*, 108–115. [CrossRef]
33. Solle, H.R.L.; Semiarti, E. Micropropagation of sandalwood (*Santalum album* L.) endemic plant from East Nusa Tenggara, Indonesia. *AIP Conf. Proc.* **2016**, *1744*, 020026. [CrossRef]
34. Sulusoglu, M.; Cavusoglu, A. Micropropagation of cherry laurel *Prunus laurocerasus* L. *J. Food, Agric. Environ.* **2013**, *11*, 576–579.
35. Mahipal, S.; Shekhawat, N.S.; Manokari, M. In vitro propagation, micromorphological studies and *ex vitro* rooting of cannon ball tree (*Couroupita guianensis* aubl.): A multipurpose threatened species. *Physiol. Molecul. Biol. Plants* **2016**, *22*, 131–142. [CrossRef]

36. Dinesh, R.M.; Patel, A.K.; Vibha, J.B.; Shekhawat, S.; Shekhawat, S.N. Cloning of mature pomegranate (*Punica granatum*) cv. Jalore seedless via in vitro shoot production and ex vitro rooting. *Vegetos* **2019**, *32*, 181–189. [CrossRef]
37. Nand, N.; Drew, R.A.; Ashmore, S. Micropropagation of two Australian native fruit species, *Davidsonia pruriens* and *Davidsonia jerseyana* G. Harden and J.B. Williams. *Plant Cell Tiss. Org. Cult.* **2004**, *77*, 193–201. [CrossRef]
38. Prakash, E.; Sha Valli Khan, P.S.; Vivek Sreenivasa Rao, T.J.; Meru, E.S. Micropropagation of red sanders (*Pterocarpus santalinus* L.) using mature nodal explants. *J. For. Res.* **2006**, *11*, 329–335. [CrossRef]
39. Quoirin, M.; Lepoivre, P. Improved media for in vitro culture of *Prunus* sp. *Acta Hortic.* **1977**, *78*, 437–442. [CrossRef]
40. Kwaśniewska, E.; Pawłowska, B. Efficient in vitro propagation of historical roses for biodiversity conservation. *Propag. Ornam. Plants* **2017**, *17*, 3–11.

horticulturae

MDPI

Article

Shoot Induction, Multiplication, Rooting and Acclimatization of Black Turmeric (*Curcuma caesia* Roxb.): An Important and Endangered *Curcuma* Species

Zainol Haida, Uma Rani Sinniah, Jaafar Juju Nakasha and Mansor Hakiman *

Department of Crop Science, Faculty of Agriculture, Universiti Putra Malaysia, Serdang 43400, Selangor, Malaysia
* Correspondence: mhakiman@upm.edu.my; Tel.: +60-3-9786-4903

Abstract: *Curcuma caesia* Roxb., commonly known as Kali Haldi or black turmeric, is one of the important species in the genus *Curcuma*. This species has been classified as one of the endangered *Curcuma* species due to the drastic decrement of this plant in its natural habitat. *C. caesia* has been overharvested for various purposes, including bioactive compound extraction to fulfill the pharmaceutical industry demand. Hence, this study was conducted to establish a protocol for the propagation of *C. caesia* via plant tissue culture techniques. In the shoot induction stage, three basal medium formulations, including Murashige and Skoog (MS medium), the combination of Murashige and Skoog macronutrients and B5 micronutrients (MSB5 medium) and woody plant medium (WPM medium) supplemented with 15 µM of 6-benzylaminopurine (BAP), were used. The results found that the MSB5 medium was the most suitable basal medium formulation for shoot induction of *C. caesia*. In the subsequent experiment, different types of cytokinin, including BAP, kinetin and 2-iP at concentrations of 5, 10, 15 and 20 µM, were fortified in the MSB5 medium for shoot multiplication. The shoot multiplication was further enhanced by supplementing the MSB5 medium with indole-3-butyric acid (IBA) or 1-napthaleneacetic acid (NAA) at the concentrations of 2, 4, 6 and 8 µM. The results showed that a combination of 15 µM of BAP and 6 µM of IBA significantly increased the shoot multiplication with 100% shoot induction, 3.53 shoots/explant, 10.81 cm of shoot length, 9.57 leaves, 0.486 g of leaves fresh weight and 0.039 g of leaves dry weight. After the multiplication, the rooting stage was carried out by altering the basal medium strength into half and full strength and supplementing with 2.5, 5, 7.5 and 10 µM of indole-3-acetic acid (IAA). The full strength of MSB5 medium supplemented with 5 µM of IAA exhibited the highest number of roots and length of roots, with 6.13 roots and 5.37 cm, respectively. After the rooting stage, the plantlets were successfully acclimatized in the potting medium with the combination of cocopeat and peatmoss, and the ratio of 1:1 was found to produce the highest survival rate with 77.78%. In conclusion, the protocol established in this study could be useful for large-scale raw material production, either for conservation or bioactive compound extraction.

Keywords: *Curcuma caesia*; basal medium formulation; cytokinin; auxin; acclimatization

Citation: Haida, Z.; Sinniah, U.R.; Nakasha, J.J.; Hakiman, M. Shoot Induction, Multiplication, Rooting and Acclimatization of Black Turmeric (*Curcuma caesia* Roxb.): An Important and Endangered *Curcuma* Species. *Horticulturae* **2022**, *8*, 740. https://doi.org/10.3390/horticulturae8080740

Academic Editors: Alicja Tymoszuk and Dariusz Kulus

Received: 12 July 2022
Accepted: 12 August 2022
Published: 17 August 2022

1. Introduction

Curcuma caesia Roxb. belongs to the family Zingiberaceae and genus *Curcuma*. This species is well known in India as Kali Haldi and is commonly called Indian Black Turmeric in other countries. *C. caesia* has been used since ancient times for the treatment of various ailments and diseases in the Indian community. The *C. caesia* extract is used for the treatment of asthma, cancer, inflammation, epilepsy, fever and allergies [1,2]. In addition, the rhizomes and leaves of *C. caesia* were found to contain essential oils such as camphor, eucalyptol, tropolone, ledol and camphene, which are responsible for the aromatic odor [3–5]. In the numerous pharmacological studies on *C. caesia*, the extracts of *C. caesia* possess anticancer, anti-asthmatic, anti-acne, anti-inflammatory and anti-microbial properties [6–9].

As the pharmaceutical industry keeps growing, the demand for plant raw materials has increased for the extraction of various bioactive compounds. Therefore, there is an urgent need to look for an alternative method for the large-scale production of plant raw materials to fulfill the growing demand from the industry. To date, micropropagation or in vitro propagation is a plant tissue culture technique that has many advantages in producing high amounts of plantlets in a short time, and this technique can also be used for the production of bioactive compounds [10,11]. In addition, the plant tissue culture technique is very useful for the conservation of endangered plant species, including *C. caesia,* which has a drastically reduced population due to overharvesting activity [3]. Previously, several studies related to *C. caesia* micropropagation, including regenerating the plant via direct and indirect organogenesis, have been conducted [12–18]. The studies involved the alteration of basal medium formulation, cytokinin and auxin [12–18].

Hence, the aim of this study was to develop an improved protocol for micropropagation of *C. caesia* as affected by basal medium formulation, plant growth regulators, basal medium strength and potting medium for acclimatization.

2. Materials and Methods

2.1. Plant Materials

Curcuma caesia Roxb. fresh rhizomes were obtained from the nursery located in Muar, Johor, Malaysia (Coordinates: 2.06° N, 102.58° E). The sprouted shoots (approximately 5 cm) were cut and brought to the laboratory for sterilization processes. The sterilization process was conducted by placing the shoots under running tap water for 30 min to eliminate the soils, debris and contaminants. Then, the shoots were transferred to the bottle jar containing 70% commercial bleach (Clorox®, Oakland, CA, USA) and shaken for 20 min. The shoots were rinsed with autoclaved distilled water three times and excised into approximately 1 ± 0.5 cm.

2.2. Shoot Induction of *C. caesia* as Affected by Basal Medium Formulations

For the shoot induction experiment, shoots were inoculated onto the three formulations of basal medium, namely MS medium [19], MSB5 medium (combination of macronutrients and micronutrients formulated by Murashige and Skoog [15] and vitamins formulated by Gamborg et al. [20]) and Woody Plant Medium (WPM medium) formulated by Lloyd and McCown [21]. All the chemicals used were analytical grade (R&M Chemical, Malaysia; Sigma-Aldrich, USA). All the basal medium formulations were supplemented with 30 g/L of sucrose, 3 g/L of gelrite and 5 μM of BAP. The pH of the basal mediums was adjusted to 5.75 prior to autoclave (Hirayama, Japan). The basal medium was poured into the 250 mL conical flasks, with each flask containing 50 mL of basal medium. All basal mediums were autoclaved at the temperature of 121°C for 20 min at a pressure of 1.06 kg cm^{-2}. For each conical flask, one aseptic microshoot was inoculated onto the basal medium and data were collected after six weeks of inoculation. All cultures were kept in the culture room under 16 h of light and 8 h of dark using LED white light with an irradiation of 45 μmol m^{-2} s^{-1} at a temperature of 25 ± 2 °C.

2.3. Effect of Different Concentrations of Cytokinins and Auxins on Shoot Multiplication of *C. caesia*

After shoot induction, the multiplication experiment was conducted by supplementing the basal medium with cytokinin, including BAP, kinetin and 2-iP, at the concentrations of 5, 10, 15 and 20 μM. After eight weeks of incubation, the data were recorded. In the subsequent experiment, the shoot multiplication was further enhanced by combining 15 μM of BAP with auxin (IBA and NAA) at the concentrations of 2, 4, 6 and 8 μM. The data were collected after eight weeks of inoculation.

2.4. *Effect of Basal Medium Strength and Different Concentrations of Indole-3-Acetic Acid (IAA) on Root Induction of C. caesia Plantlets*

For root induction, healthy plantlets at eight weeks old, which were maintained on the MSB5 medium supplemented with 15 μM of BAP and 6 μM of IBA, were used. The rooting medium used was full and half strength of MSB5 medium supplemented with IAA at the concentrations of 2.5, 5, 7.5 and 10 μM. The plantlets were exposed to rooting medium for four weeks, and the data were collected after four weeks of inoculation.

2.5. *Acclimatization of C. caesia Plantlets as Affected by Different Potting Mediums*

The acclimatization was conducted by transferring the healthy plantlets with at least two leaves and well rooted (more than 1.5 cm long) into the potting medium. The potting medium used was a combination of cocopeat and peatmoss, perlite and peatmoss and vermiculite and peatmoss at the ratio of 1:1 for all potting medium combinations. The data were recorded after two weeks of transplantation.

2.6. *Statistical Analysis*

All the experiments were conducted in a completely randomized design with three replications and ten explants for each replication (n = 30). The Analysis of Variance (ANOVA) was used for data analysis, and Duncan's Multiple Test (DMRT) was used for means separation. The analysis was performed using Statistical Analysis System (SAS ver. 9.4, Cary, NC, USA).

3. Results and Discussion

3.1. *Shoot Induction of C. caesia as Affected by Basal Medium Formulations*

In this experiment, the shoot induction was significantly affected by different formulations of the basal medium. The MSB5 medium formulation significantly produced the highest percentage of shoot induction with 100%, followed by MS medium and WPM medium with 93.33 and 83.33%, respectively (Table 1). The MSB5 medium also significantly produced the highest number of shoots, length of shoot and number of leaves, with 2.70 shoots, 7.88 cm and 5.33 leaves (Figure 1). Meanwhile, there was no significant difference between MS medium and WPM medium formulations on number of shoots, length of shoot and number of leaves. For the biomass of leaves, the highest fresh and dry weight of leaves were accumulated from the MSB5 medium formulation with 0.179 and 0.018 g, respectively.

Table 1. Shoot induction of *C. caesia* as affected by different formulations of basal medium.

Basal Medium	Shoot Induction (%)	No. of Shoot	Shoot Length (cm)	No. of Leaves	Leaves Fresh Weight (g)	Leaves Dry Weight (g)	Root Induction (%)	No. of Root	Root Length (cm)
MSB5	100 a	2.70 a	7.88 a	5.33 a	0.179 a	0.018 a	100 a	8.40 a	3.34 a
MS	93.33 a	1.50 b	5.53 b	3.00 b	0.105 b	0.008 b	86.67 b	3.85 b	2.19 a
WPM	83.33 b	1.37 b	4.86 b	2.97 b	0.085 c	0.006 b	80 b	3.67 b	2.02 a

Data were collected after six weeks of incubation. Means (n = 30) followed by the same letter within the columns were not significantly different at $p < 0.05$ using Duncan's Multiple Range Test.

By conducting the plant tissue culture technique, finding the most suitable basal medium formulation is a crucial step for producing healthy plantlets. Since a long time ago, numerous basal medium formulations have been established for propagating various species. In micropropagation of *Curcuma* species, MS medium is the most frequently used for propagating *C. longa*, *C. zedoaria*, *C. aromatica* and also *C. caesia* [22–25], in contrast with this study, which found that MSB5 medium formulation was more efficient than MS medium formulation. The high efficiency of MSB5 medium formulation as compared to MS medium might be due to higher vitamin concentrations in the MSB5 formulation that lead to increased plant growth. Based on the MSB5 medium composition, the pyridoxine HCl and nicotinic acid concentrations were two-fold higher, and thiamine HCl concentration was 100-fold higher than the MS medium vitamin formulations. According to Saad and

Elshahed [26], vitamins such as thiamine HCl, nicotinic acid and pyridoxine HCL at low concentrations are required by plants as a catalyst for various metabolic processes. In addition, the amount needed varies by plant species as some plants are able to synthesize their own vitamins [27].

Figure 1. Establishment of *C. caesia* microshoots in MS medium, MSB5 medium and WPM medium after six weeks of inoculation. Scale bar represents 1 cm of actual size.

3.2. *Effect of Different Concentrations of Cytokinins on Shoot Multiplication of C. caesia*

Based on the previous experiment, the MSB5 medium formulation was found to be the most suitable basal medium for *C. caesia*. In this experiment, the shoots were multiplied by using different concentrations of cytokinin (Table 2). After eight weeks of inoculations, the treatments of 10 and 15 µM BAP and 10, 15 and 20 µM kinetin were recorded at 100% of shoot induction. The lowest percentage of shoot induction was significantly produced by the treatment of 20 µM of 2-iP with 30% of induction. The treatment of 15 µM of BAP significantly produced the highest number of shoots and number of leaves, with 2.77 shoots and 7.00 leaves, respectively. Meanwhile, the highest length of shoots was exhibited by the treatments of 15 µM of BAP, 15 and 20 µM of kinetin with 10.86, 11.10 and 10.14 cm, respectively (Figure 2). As the highest number of shoots and number of leaves were produced from the treatment of 15 µM of BAP, the leaves biomass accumulation was also recorded from the same treatment with 0.343 and 0.029 g of fresh and dry leaves. In this experiment, callus formation was observed in all treatments of 2-iP. The treatment of 5 µM of 2-iP significantly produced the highest callus induction percentage and callus fresh weight with 23.22% and 0.425 g, respectively (Figure 3). In contrast, no callus formation was observed in other treatments.

Figure 2. Shoot multiplication of *C. caesia* in MSB5 medium supplemented with 15 µM BAP at eight weeks after inoculation. Scale bar represents 1 cm of actual size.

Table 2. Effect of different concentrations of cytokinins in MSB5 medium on shoot multiplication of *C. caesia*.

Cytokinin (μM)		Shoot Induction (%)	No. of Shoot	Length of Shoot (cm)	No. of Leaves	Leaves Fresh Weight (g)	Leaves Dry Weight (g)	Callus Induction (%)
Control		93.33 a	1.20 d	8.96 d	4.10 e	0.109 f	0.017 f	0 c
BAP	5	96.67 a	1.97 b	9.54 c	5.73 b	0.248 cd	0.025 b	0 c
	10	100 a	2.07 b	10.10 b	5.73 b	0.277 bc	0.025 b	0 c
	15	100 a	2.77 a	10.86 a	7.00 a	0.343 a	0.029 a	0 c
	20	96.67 a	1.97 b	7.94 e	4.30 de	0.195 e	0.018 ef	0 c
Kin	5	96.67 a	1.37 cd	9.22 cd	4.63 d	0.242 d	0.020 de	0 c
	10	100 a	1.27 cd	9.63 c	4.13 d	0.240 d	0.022 cd	0 c
	15	100 a	1.47 c	11.10 a	5.03 c	0.285 b	0.022 cd	0 c
	20	100 a	1.47 c	10.14 a	5.10 c	0.246 cd	0.024 bc	0 c
2-iP	5	80 b	1.20 d	4.76 f	3.03 f	0.109 f	0.008 g	23.22 a
	10	60 c	0.70 e	3.84 g	1.53 g	0.089 f	0.009 g	16.67 b
	15	36.67 d	0.47 f	2.20 h	1.01 h	0.037 g	0.004 h	16.67 b
	20	30 e	0.43 f	1.84 h	1.17 gh	0.040 g	0.005 h	3.33 c

Data were collected after eight weeks of incubation. Means ($n = 30$) followed by the same letter within the columns were not significantly different at $p < 0.05$ using Duncan's Multiple Range Test.

Figure 3. (**a**) Callus produced from the treatment MSB5 medium supplemented with 15 μM BAP and 5 μM 2-iP at eight weeks after inoculation. (**b**) Regenerated shoot (in circle) from the treatment MSB5 medium supplemented with 15 μM BAP and 5 μM 2-iP at eight weeks after inoculation. Scale bars represent 1 cm of actual size.

Based on the results of this experiment, 15 μM of BAP was more prominent than the other treatments. Among the BAP, kinetin and 2-iP treatment, BAP was the most superior, followed by kinetin and 2-iP. The finding was in agreement with Fong and Sani [28], which found that BAP was more effective for shoot production of *C. caesia* compared to kinetin and TDZ. In addition, the efficiency of BAP was also reported on other species, including *Calotropis procera* [29], *Artemisia arborescens* [30], *Boerhaavia diffusa* [31] and *Andrographis alata* [32]. The suitability of BAP over other types of cytokinin in shoot multiplication might be due to the ability of the plant to metabolize BAP more readily as nucleotides, and ribosides stability is present naturally in BAP [33].

3.3. Enhancement of Shoot Multiplication by Supplementation of 15 μM of BAP with Different Concentrations of Auxins

The supplementation of cytokinins in the previous experiment found that 15 μM of BAP was the most prominent in the multiplication of *C. caesia*. Hence, the shoot multiplication was further enhanced by supplementing 15 μM of BAP with different concentrations of auxins. By supplementing the MSB5 medium with auxins, 100% of shoot induction was obtained in all the treatments (Table 3). In terms of the number of shoots, the treatment of 6 and 8 μM of IBA produced the highest number of shoots, with 3.53 shoots. However, the number of shoots recorded was statistically non-significant with all NAA treatments.

However, the highest length of shoots was significantly exhibited by the treatment of 6 µM of IBA with 10.81 cm. In addition, the same treatment also produced the highest number of leaves and leaves biomass accumulation with 9.57 leaves, 0.486 g of leaves fresh weight and 0.039 g of leaves dry weight, respectively. This experiment showed that the application of cytokinin and auxin produced a positive effect on *C. caesia* growth.

Table 3. Effect of 15 µM of BAP with auxins in the MSB5 medium on enhancement of shoot multiplication of *C. caesia*.

PGR	Conc (µM)	Shoot Induction (%)	No. of Shoot	Shoot Length (cm)	No. of Leaves	Leaves Fresh Weight (g)	Leaves Dry Weight (g)
Control	0	100 a	2.57 c	10.00 b	5.90 f	0.294 e	0.022 ef
IBA	2	100 a	2.97 b	9.19 c	7.13 de	0.311 d	0.023 de
	4	100 a	3.00 b	9.74 bc	7.37 cd	0.362 b	0.028 b
	6	100 a	3.53 a	10.81 a	9.57 a	0.486 a	0.039 a
	8	100 a	3.53 a	9.58 bc	8.10 b	0.320 d	0.026 bc
NAA	2	100 a	3.27 ab	8.00 de	7.00 e	0.254 g	0.020 f
	4	100 a	3.40 a	8.57 d	7.57 c	0.333 c	0.025 cd
	6	100 a	3.20 ab	7.68 e	6.83 e	0.295 e	0.021 ef
	8	100 a	3.30 ab	8.13 de	7.00 e	0.278 f	0.022 def

Data were collected after eight weeks of incubation. Means (n = 30) followed by the same letter within the columns were not significantly different at $p < 0.05$ using Duncan's Multiple Range Test.

In plant tissue culture, supplementation of auxin in the basal medium is frequently used for the rooting stage. However, the addition of auxin can also be conducted for the enhancement of shoot multiplication as the synergistic effect between the auxin and cytokinin will enhance the cell division and elongation of the shoot [34,35]. Several studies have reported that the combination of auxin and cytokinin significantly increased plant growth. A study by Hailu et al. [36] found that the combination of BAP and IBA significantly increased the shoot induction percentage, number of leaves and number of shoots of *Aframomum corrorima*. Besides that, the synergistic effect between auxin and cytokinin on shoot growth was also reported on *Petunia hybrida* [37]. The integration of auxin and cytokinin is important to support plant growth and development [38]. The increased plant growth might be due to the different nutrient uptakes, translocation rates, metabolic processes and ability of plants to regulate the level of plant hormones [39].

3.4. *Effect of Basal Medium Strength and Different Concentrations of Indole-3-Acetic Acid (IAA) on Root Induction of* C. caesia

In micropropagation of Zingiberaceae plant species, including those from the genus *Curcuma*, the roots are commonly produced simultaneously with the shoots. However, the number of roots produced per shoot is not enough to support the plantlets during acclimatization. Hence, the rooting stage was conducted to produce adequate amounts of roots for high survival rates during acclimatization. In this study, the shoots were initially cultured on the MSB5 medium supplemented with 15 µM of BAP and 6 µM of IBA for the first eight weeks. After that, the healthy plantlets with more than two leaves and growth of more than 5 cm were separated from the clumps and inoculated individually on rooting medium. The results in Table 4 showed that full-strength MSB5 medium supplemented with 5 µM of IAA significantly produced the highest number of roots and length of roots, with 6.13 roots and 5.37 cm, respectively. Meanwhile, the treatment of full-strength MSB5 medium supplemented with 10 µM of IAA significantly produced the lowest number of roots and length of roots, with 3.20 roots and 1.79 cm, respectively. In this study, full-strength MSB5 fortified with 5 µM of IAA gave the best results for adventitious root formation of *C. caesia* (Figure 4). The suitability of full-strength basal medium over half-strength medium might be due to a sufficient amount of nutrients to stimulate the root formation. Meanwhile, IAA, which is a frequently used auxin for root induction, alongside IBA, was reported to produce a good rooting response in *Nardostachys jatamansi* and *Alkanna tinctoria* [40,41].

Table 4. Effect of MSB5 medium strength and IAA on root induction of *C. caesia* plantlets.

Strength	IAA (μM)	Number of Roots	Length of Roots (cm)
Full	0	4.40 de	3.85 e
	2.5	5.87 b	4.39 c
	5	6.13 a	5.37 a
	7.5	4.20 ef	3.38 f
	10	3.20 h	1.79 h
Half	0	3.60 g	2.81 g
	2.5	4.27 def	2.59 g
	5	5.13 c	4.88 b
	7.5	4.53 d	4.15 cd
	10	4 f	4.01 de

Data were collected after twelve weeks of incubation. Means (n = 24) followed by the same letter within the columns were not significantly different at $p < 0.05$ using Duncan's Multiple Range Test.

Figure 4. Regeneration of adventitious roots of *C. caesia* in the full strength of MSB5 medium supplemented with 5 μM of IAA.

3.5. Acclimatization of *C. caesia* as Affected by Different Potting Mediums

Acclimatization is an important step in the micropropagation of commercially important plants. After the rooting stage, the healthy plantlets with at least two leaves, more than 5 cm in plant height, more than 1.5 cm in root length and without any morphology abnormalities were acclimatized on different potting mediums. Among the combination of cocopeat, perlite and vermiculite with peatmoss, cocopeat significantly produced the highest percentage of plantlet survival, with 77.78%, followed by perlite (61.11%) and vermiculite (44.43%) after two weeks of acclimatization (Table 5) (Figure 5). The higher survival percentage of plantlets in the combination of cocopeat and peatmoss could be due to the higher porosity and air space between the cocopeat and peatmoss, which provided better aeration for the roots to grow and, consequently, produced a higher survival percentage compared to the other treatments. The efficiency of cocopeat as a potting medium for acclimatization was reported on *Bacopa monnieri* and *Ficus carica* [42,43].

Table 5. Effect of potting medium on acclimatization of *C. caesia plantlets.*

Medium	Survival (%)
	Day 14
Cocopeat + peatmoss	77.78 a
Perlite + peatmoss	61.11 b
Vermiculite + peatmoss	44.43 c

Data were collected after two weeks of acclimatization. Means (n = 18) followed by the same letter within the columns were not significantly different at $p < 0.05$ using Duncan's Multiple Range Test.

Figure 5. Plantlet during acclimatization in the potting medium of cocopeat and peatmoss.

4. Conclusions

An efficient protocol for micropropagation of *C. caesia* was established in this study with the aim to mass produce the raw materials of *C. caesia* in a short time for the conservation of wild populations of this species. The basal medium formulation and plant growth regulators were tested, and results showed that the MSB5 medium supplemented with 15 μM BAP and 6 μM of IBA was the optimum formulation for shoot induction and multiplication of *C. caesia*. At the rooting stage, full-strength MSB5 medium supplemented with 5 μM IAA was the best concentration for the root formation, and acclimatization was successfully carried out using the potting medium of cocopeat with peatmoss (1:1). Hence, the protocol developed in this study could be used for the large-scale production of *C. caesia*.

Author Contributions: Conceptualization: Z.H. and M.H.; methodology, Z.H.; software, Z.H.; validation, Z.H., U.R.S., J.J.N. and M.H.; formal analysis, Z.H.; investigation, Z.H.; data curation, Z.H.; writing—original draft preparation, Z.H.; writing—review and editing, Z.H., U.R.S., J.J.N. and M.H.; supervision, U.R.S., J.J.N. and M.H. All authors have read and agreed to the published version of the manuscript.

Funding: This research received no external funding.

Institutional Review Board Statement: Not applicable.

Informed Consent Statement: Not applicable.

Data Availability Statement: Not applicable.

Acknowledgments: Authors would like to thank the Faculty of Agriculture, Universiti Putra Malaysia, for providing facilities and materials for conducting this research.

Conflicts of Interest: The authors declare no conflict of interest.

References

1. Fozia, I.; Fouzia, H.; Baqir, S.N.; Iqbal, A.; Sabahat, J.; Syed, M.F.H. Studies on antibacterial activity of some traditional medicinal plants used in folk medicine. *Pak. J. Pharm. Sci.* **2012**, *25*, 669–674.
2. Amalraj, V.A.; Velayudhan, K.C.; Muralidharan, V.K. A note on the anomalous flowering behaviour in *Curcuma caesia* (Zingiberaceae). *J. Bombay Nat. Hist. Soc.* **1989**, *86*, 278–279.
3. Borah, A.; Paw, M.; Gogoi, R.; Loying, R.; Sarma, N.; Munda, S.; Pandey, S.K.; Lal, M. Chemical composition, antioxidant, anti-inflammatory, anti-microbial and in vitro cytotoxic efficacy of essential oil of *Curcuma caesia* Roxb. leaves: An endangered medicinal plant of North East India. *Ind. Crops Prod.* **2019**, *129*, 448–454. [CrossRef]

4. Kumar, A.; Gautam, S.S. Volatile constituents of *Curcuma caesia* Roxb. rhizome from North India. *Nat. Acad. Sci. Lett.* **2020**, *43*, 607–610. [CrossRef]
5. Singh, S.; Sahoo, B.C.; Ray, A.; Jena, S.; Dash, M.; Nayak, S.; Kar, B.; Sahoo, S. Intraspecific chemical variability of essential oil of *Curcuma caesia* (Black Turmeric). *Arab. J. Sci. Eng.* **2021**, *46*, 191–198. [CrossRef]
6. Paliwal, P.; Pancholi, S.S.; Patel, R.K. Pharmacognostic parameters for evaluation of the rhizomes of *Curcuma caesia*. *J. Adv. Pharm. Technol. Res.* **2011**, *2*, 56–61. [CrossRef]
7. Karmakar, I.; Dolai, N.; Suresh, K.R.B.; Kar, B.; Roy, S.N.; Haldar, P.K. Antitumor activity and antioxidant property of *Curcuma caesia* against Ehrlich's ascites carcinoma bearing mice. *Pharm. Biol.* **2013**, *51*, 753–759. [CrossRef]
8. Kaur, R.; Kaur, B.; Suttee, A.; Kalsi, V. Comparative assessment of in vitro antimicrobial activity of *Curcuma caesia* Roxb. and *Curcuma amada* Roxb. *Asian J. Pharm. Clin. Res.* **2018**, *11*, 94–97. [CrossRef]
9. Rai, N.; Shukla, T.P.; Loksh, K.R.; Karole, S. Synthesized silver nanoparticle loaded gel of *Curcuma caesia* for effective treatment of acne. *J. Drug Deliv. Ther.* **2020**, *10*, 75–82. [CrossRef]
10. Chandran, H.; Meena, M.; Barupal, T.; Sharma, K. Plant tissue culture as a perpetual source for production of industrially important bioactive compounds. *Biotechnol. Rep.* **2020**, *26*, e00450. [CrossRef]
11. Khan, T.; Khan, M.A.; Karam, K.; Ullah, N.; Mashwani, Z.U.R.; Nadhman, A. Plant in vitro culture technologies; A promise into factories of secondary metabolites against COVID-19. *Front. Plant Sci.* **2021**, *12*, 356. [CrossRef] [PubMed]
12. Zuraida, A.R. Improved in vitro propagation of *Curcuma caesia*, a valuable medicinal plant. *J. Trop. Agric. Food Sci.* **2013**, *41*, 273–281.
13. Jose, S.; Thomas, T.D. High-frequency callus organogenesis, large-scale cultivation and assessment of clonal fidelity of regenerated plants of *Curcuma caesia* Roxb., an important source of camphor. *Agrofor. Syst.* **2015**, *89*, 779–788. [CrossRef]
14. Singh, W.R.; Singh, H.B.; Devi, S.S.; Singh, W.N.; Singh, N.M.; Devi, Y.P. Conservation of *Curcuma caesia* by in vitro techniques. *Helix* **2015**, *2*, 708–713.
15. Sarma, I.; Deka, A.C. Conservation of *Curcuma caesia* Roxb.-A critically endangered species via in vitro plant regeneration from organogenic callus. *Asian J. Conserv. Biol.* **2020**, *9*, 151–155.
16. Sarma, I.; Deka, A.C.; Sarma, T.C. A protocol for rapid clonal propagation and microrhizome production of *Curcuma caesia* Roxb (Zingiberaceae): A critically endangered medicinal plant of North East India. *Indian J. Agric. Res.* **2021**, *55*, 13–22. [CrossRef]
17. Neha, B.; Tiwari, K.L.; Jadhav, S.K. Effect of explant type in development of in vitro micropropagation protocol of an endangered medicinal plant: *Curcuma caesia* Roxb. *Biotechnology* **2014**, *13*, 22–27.
18. Shahinozzaman, M.; Ferdous, M.; Faruq, M.; Azad, M.; Amin, M. Micropropagation of black turmeric (*Curcuma caesia* Roxb.) through in vitro culture of rhizome bud explants. *J. Cent. Eur. Agric.* **2013**, *14*, 110–115. [CrossRef]
19. Murashige, T.; Skoog, F. A revised medium for rapid growth and bio assays with tobacco tissue cultures. *Physiol. Plant.* **1962**, *15*, 473–497. [CrossRef]
20. Gamborg, O.L.; Miller, R.; Ojima, K. Nutrient requirements of suspension cultures of soybean root cells. *Exp. Cell Res.* **1968**, *50*, 151–158. [CrossRef]
21. Lloyd, G.; McCown, B. Commercially-feasible micropropagation of mountain laurel, *Kalmia latifolia*, by use of shoot-tip culture. *Comb. Proc. Int. Plant Propag. Soc.* **1980**, *30*, 421–427.
22. Chowdhury, S.; Pal, K.; Chakraborty, M.; Chakraborty, S.; Mandal, S.; Pandit, M.S.; Sahana, N. Conservation and in vitro propagation of an endangered wild turmeric (*Curcuma caesia* Roxb.) species from Sub-Himalayan Terai region of West Bengal. *Int. J. Curr. Microbiol. Appl. Sci.* **2020**, *9*, 2132–2140. [CrossRef]
23. Ferrari, M.P.; dos Santos, Q.M.; de Andrade, M.M.; Trettel, J.R.; Magalhães, H.M. Growth regulators, sucrose and potassium in the growth and biochemical activity of *Curcuma longa* L. micropropagated. *Sci. Agrar. Paranaensis* **2020**, *1*, 18–26. [CrossRef]
24. Jena, S.; Ray, A.; Sahoo, A.; Sahoo, S.; Dash, B.; Kar, B.; Nayak, S. Rapid plant regeneration in industrially important *Curcuma zedoaria* revealing genetic and biochemical fidelity of the regenerants. *3 Biotech* **2020**, *10*, 17. [CrossRef] [PubMed]
25. Parida, R.; Nayak, S. Rapid in vitro leaf biomass production of genetically stable *Curcuma aromatica*-An under exploited medicinal plant. *J. Biol. Act. Prod. Nat.* **2021**, *11*, 497–504. [CrossRef]
26. Saad, A.I.; Elshahed, A.M. Plant tissue culture media. *Rec. Adv. Plant In Vitro Cult.* **2012**, 30–40.
27. Phillips, G.C.; Garda, M. Plant tissue culture media and practices: An overview. *In Vitro Cell Dev. Biol. Plant* **2019**, *3*, 242–257. [CrossRef]
28. Fong, Y.M.; Sani, H.B. Studies on micropropagation of *Curcuma caesia* Roxb. (Kunyit Hitam). *Int. J. Innov. Sci. Eng. Technol. Technol.* **2019**, *6*, 101–106.
29. Sethy, R.; Kullu, B. Micropropagation of ethnomedicinal plant *Calotropis* sp. and enhanced production of stigmasterol. *Plant Cell Tissue Organ Cult.* **2022**, *149*, 147–158. [CrossRef]
30. Riahi, L.; Chograni, H.; Ben Rejeb, F.; Ben Romdhane, M.; Masmoudi, A.S.; Cherif, A. Efficient in vitro regeneration of the endangered species *Artemisia arborescens* L. through direct organogenesis and impact on secondary metabolites production. *Hortic. Environ. Biotechnol.* **2022**, *63*, 439–450. [CrossRef]
31. Majumder, S.; Biswas, A.; Rahman, M.M. In vitro propagation and cytological analysis of *Boerhaavia diffusa* L.—An important medicinal plant species of Bangladesh. *Am. J. Plant Sci.* **2021**, *12*, 1347–1360. [CrossRef]
32. Kadapatti, S.S.; Murthy, H.N. Rapid plant regeneration, analysis of genetic fidelity, and neoandrographolide content of micro-propagated plants of *Andrographis alata* (Vahl) Nees. *J. Genet. Eng. Biotechnol.* **2021**, *19*, 20. [CrossRef] [PubMed]

33. Khajuria, A.K.; Bisht, N.S.; Bhagat, N. In vitro organogenesis and plant regeneration of *Thymus serpyllum* L.: An important aromatic medicinal plant. *In Vitro Cell Dev. Biol. Plant* **2020**, *56*, 652–661. [CrossRef]
34. Gaspar, T.; Kevers, C.; Penel, C.; Greppin, H.; Reid, D.M.; Thorpe, T.A. Plant hormones and plant growth regulators in plant tissue culture. *In Vitro Cell Dev. Biol. Plant* **1996**, *32*, 272–289. [CrossRef]
35. Sultana, K.W.; Das, S.; Chandra, I.; Roy, A. Efficient micropropagation of *Thunbergia coccinea* Wall. and genetic homogeneity assessment through RAPD and ISSR markers. *Sci. Rep.* **2022**, *12*, 1683. [CrossRef] [PubMed]
36. Hailu, T.; Haileselassie, T.; Feyissa, T. In Vitro regeneration of Korarima (*Aframomum corrorima* (Braun) PCM Jansen): A threatened spice and medicinal herb from Ethiopia. *Scientifica* **2022**, *2022*, 8237723. [CrossRef]
37. Farooq, I.; Qadri, Z.A.; Rather, Z.A.; Nazki, I.T.; Banday, N.; Rafiq, S.; Masoodi, K.Z.; Noureldeen, A.; Mansoor, S. Optimization of an improved, efficient and rapid in vitro micropropagation protocol for *Petunia hybrida* Vilm. Cv. "Bravo". *Saudi J. Biol. Sci.* **2021**, *28*, 3701–3709. [CrossRef]
38. Kudikala, H.; Jogam, P.; Sirikonda, A.; Mood, K.; Allini, V.R. In vitro micropropagation and genetic fidelity studies using SCoT and ISSR primers in *Annona reticulata* L.: An important medicinal plant. *Vegetos* **2020**, *33*, 446–457. [CrossRef]
39. Gajula, H.; Kumar, V.; Vijendra, P.D.; Rajashekar, J.; Sannabommaji, T.; Basappa, G.; Anuradha, C.M. In vitro regeneration of *Psoralea corylifolia* Linn.: Influence of polyamines during in vitro shoot development. *In Vitro Cell Dev. Biol. Plant* **2022**, *58*, 103–113.
40. Dhiman, N.; Devi, K.; Bhattacharya, A. Development of low cost micropropagation protocol for *Nardostachys jatamansi*: A critically endangered medicinal herb of Himalayas. *S. Afr. J. Bot.* **2021**, *140*, 468–477. [CrossRef]
41. Cartabia, A.; Sarropoulou, V.; Grigoriadou, K.; Maloupa, E.; Declerck, S. In vitro propagation of *Alkanna tinctoria* Tausch.: A medicinal plant of the Boraginaceae family with high pharmaceutical value. *Ind. Crops Prod.* **2022**, *182*, 114860. [CrossRef]
42. Chauhan, R.; Shirkot, P. Micropropagation of endangered medicinal plant *Bacopa monnieri* (L.) Pennell. *J. Pharmacogn. Phytochem.* **2020**, *9*, 1614–1620.
43. Sriskanda, D.; Liew, Y.X.; Khor, S.P.; Merican, F.; Subramaniam, S.; Chew, B.L. An efficient micropropagation protocol for *Ficus carica* cv. Golden Orphan suitable for mass propagation. *Biocatal. Agric. Biotechnol.* **2021**, *38*, 102225. [CrossRef]

horticulturae

Article

Impact of Artificial Polyploidization in *Ajuga reptans* on Content of Selected Biologically Active Glycosides and Phytoecdysone

Božena Navrátilová [1], Vladan Ondřej [1,*], Naděžda Vrchotová [2], Jan Tříska [2], Štěpán Horník [3] and Roman Pavela [4]

1 Department of Botany, Faculty of Science, Palacký University Olomouc, 783 71 Olomouc, Czech Republic; bozena.navratilova@upol.cz
2 Laboratory of Metabolomics and Isotopic Analyses, Global Change Research Institute, Academy of Sciences of the Czech Republic, Bělidla 986/4a, 603 00 Brno, Czech Republic; vrchotova.n@czechglobe.cz (N.V.); triska.j@czechglobe.cz (J.T.)
3 Department of Analytical Chemistry, Institute of Chemical Process Fundamentals, Academy of Sciences of the Czech Republic, Rozvojová 135, 165 05 Praha, Czech Republic; hornik@icpf.cas.cz
4 Crop Research Institute, Drnovská 507/73, 161 06 Praha, Czech Republic; pavela@vurv.cz
* Correspondence: vladan.ondrej@upol.cz; Tel.: +420-585-634-825

Abstract: Polyploidization in plants, which involves doubling or further multiplying of genome, has the potential to improve the constituents that make medicinal plants, like *Ajuga reptans*, attractive to the pharmaceutical, cosmetics, and food production industries; botanical pesticide effects could also be derived. The aim of this study was to determine how artificial polyploidization in *A. reptans* plants affected the composition and quantity of biologically active substances from the glycoside and phytoecdysone families. Diploids and artificial tetraploids of *A. reptans* were analyzed. Changes in the contents of *trans*-teupolioside, *trans*-verbascoside, and 20-hydroxyecdysone were evident in the aboveground parts of the cultivated plants (e.g., leaves and flowers). The tetraploid lines of *Ajuga* plants displayed variability in, and increased levels of, *trans*-teupolioside and *trans*-verbascoside content. The 20-hydroxecdysone content was slightly higher in tetraploids. These findings indicated that *Ajuga* tetraploids could be used in breeding programs to enhance the yield of substances with potential medicinal and industrial applications.

Keywords: *Ajuga reptans*; artificial polyploidy; *trans*-teupolioside; *trans*-verbascoside; 20-hydro xyecdysone

Citation: Navrátilová, B.; Ondřej, V.; Vrchotová, N.; Tříska, J.; Horník, Š.; Pavela, R. Impact of Artificial Polyploidization in *Ajuga reptans* on Content of Selected Biologically Active Glycosides and Phytoecdysone. *Horticulturae* **2022**, *8*, 581. https://doi.org/10.3390/horticulturae8070581

Academic Editors: Alicja Tymoszuk and Dariusz Kulus

Received: 25 May 2022
Accepted: 25 June 2022
Published: 27 June 2022

1. Introduction

Artificial polyploidization in plants is a breeding technique used to modify the genome, mainly by doubling (and occasionally further multiplying) chromosome sets. However, because this biotechnological approach affects the genome, it is not classified as genetic modification. Polyploidization techniques are experiencing a renaissance; they are increasingly being incorporated into the breeding programs of many interesting crops and aromatic and medicinal plants [1–4].

Especially in the cases of aromatic and medicinal plants, polyploidization affects the quantities of a wide range of biologically active substances [3]. For secondary metabolites, in particular, these changes could allow the use of plant extracts in the pharmaceutical, cosmetics, and food processing industries, and as botanical pesticides [5]. One such medicinal plant, which is known for its biologically active substances, is *Ajuga reptans* (blue bugle), belonging to the family Lamiaceae. Plants from the genus *Ajuga* have a wide variety of biologically active substances, such as phytoecdysteroids, anthocyanins, carotenoids, diterpenes, and phenylethanoid glycosides. Extracts from this plant species contain two glycosides: teupolioside and verbascoside. Several studies have demonstrated the potential anticancer, anti-inflammatory, and antioxidant effects of this species, which could, therefore, have cosmeceutical and food production applications [6,7]. Moreover, the presence of phytoecdysteroids (20-hydroxyecdysone) suggests the possible use of *Ajuga* extract as a botanical pesticide. Phytoecdysteroids have the same structural features as the ecdysteroids found in insects and other arthropods. It is believed that plant ecdysteroids have antifeedant effects on herbivores and suppress insect molting [8,9].

Although some researchers have attempted to obtain the abovementioned substances from in vitro or cell suspension cultures [7,10], breeding programs could provide new genotypes for field production and biomass harvesting. This study aimed to address how artificial tetraploids of *Ajuga reptans* obtained from previous experiments differ from wild progenitor plants, in terms of the contents of three important substances: *trans*-teupolioside, *trans*-verbascoside, and 20-hydroxyecdysone.

2. Materials and Methods

2.1. Plant Material

The plants of two diploid progenitor strains of *A. reptans* (denoted as 4 and 7), originating from the Crop Protection section of the Crop Research Institute (CRI), Prague-Ruzyně, Czech Republic, were used in this study. Tetraploid plants of three strains of clones (sc4, sc28, and sc12) were also analyzed. Tetraploid strains sc4 and sc28 were derived from progenitor strain 7, and sc12 (as well as sc5, for which only the flowers and leaves were analyzed) was derived from strain 4, to represent the lines of viable plants with the characteristic morphological and cytological features of tetraploids, as reported in Švécarová et al. (2018) [11]. The plants of all of these strains were cultivated in field condition, harvested at the maximum flowering stage, and dried at room temperature under enhanced air circulation. Plants growing in field condition were also morphologically characterized. The surface areas of dried, prepared and scanned leaves were measured using Image J software (https://imagej.nih.gov/ij/index.html (accessed on 23 June 2022). Only two pairs of leaves, located exactly under inflorescence, were taken for analysis. Two datasets, one for diploid leaf areas and one for tetraploid leaf areas, were statistically evaluated by *t*-test (GraphPad software, San Diego, CA, USA).

2.2. Extraction of the Plant Materials

The dried above-ground plant materials were powdered and extracted with methanol for 2 h at room temperature. Three parallel samples were prepared from each plant material. The extracts were stored at −18 °C. Methanol LiChrosolv, gradient grade for LC, was purchased from Merck (Prague, Czech Republic).

2.3. HPLC and LC/MS Analyses

The extracts were analyzed using HPLC (Hewlett Packard 1050), column Phenomenex Luna C18(2) (150 × 2 mm and 3 μm), diode array detector (DAD Agilent G1315B) and fluorescence detector (FLD Agilent G1321A).

Mobile phase A: 5% acetonitrile + 0.1% of o-phosphoric acid; mobile phase B: 80% acetonitrile + 0.1% of o-phosphoric acid. The gradient: 0. min 0% B, 10. min 10% B, 20. min 30% B, 25. min 40% B, 30. min 60% B. The flow rate was 0.25 mL/min, the column temperature was 35° C. The injection volume was 5 μL.

The substances were also analyzed by LC-MS (APCI-LC-MS, LCQ Accela Fleet), with the same column, but, instead of o-phosphoric acid, formic acid was used. Complete identification was performed by LC-NMR.

Quantification was performed by HPLC using calibration curves for 20-hydroxyecdysone and *trans*-verbascoside (syn. acteoside); *trans*-teupoloside was quantified according to the calibration curve for verbascoside.

Acetonitrile for LC/MS was purchased from Merck (Prague, Czech Republic), *o*-phosporic acid p.a. from Fluka (Prague, Czech Republic), formic acid from Sigma-Aldrich

(Prague, Czech Republic) and standard of *trans*-verbascoside and 20-hydroxy-ecdysone from Sigma-Aldrich (Prague, Czech Republic).

Datasets of each measured substance for each pair (progenitor diploid strain and its derived tetraploid strain) were statistically evaluated by unpaired t-test ($p < 0.01$) (GraphPad software).

2.4. NMR Analysis-Identification of trans-Verbascoside and trans-Teupolioside

For the preparation of *trans*-verbascoside and *trans*-teupolioside, a commercial HPLC system (Dionex UltiMate 3000, Thermo Fisher Scientific) with a 4.6 × 250 mm HPLC column (Luna C18 (2), Phenomenex, 5 µm, 100 Å pore size) was employed. The separation was performed by isocratic elution using acetonitrile-deuterium oxide system (22% ACN-78% D2O) and monitored at 254 nm. The flow rate was 0.5 mL/min. The concentrated methanol solution (50 µL) was injected multiple times into the HPLC system and fractions of individual chromatographic peaks were collected. The individual fractions were evaporated to dryness and subsequently dissolved in methanol-d4 for NMR spectroscopy analysis.

The NMR data were acquired using a Varian INOVA 500 MHz spectrometer (Varian Inc., Palo Alto, CA, USA) operating at 499.87 MHz for ^{1}H and 125.70 MHz for ^{13}C. The ^{1}H, COSY and HSQC spectra were used for structure elucidation of *trans*-teupolioside, while only ^{1}H NMR spectrum was recorded for *trans*-verbascoside. NMR spectra were referenced to the line of the solvent (methanol-d4, δ = 3.31 ppm for 1H and δ = 49.00 ppm for ^{13}C), see Supplementary Materials. The identification was based on comparison of NMR data of isolated compounds with available literature [12].

3. Results

Diploid and tetraploid *A. reptans* plants cultivated in field conditions and originated from in vitro cultures showed typical features for their ploidy levels. Plants from tetraploid strains were characterized by more robust growth of the shoots in comparison with diploid progenitor plants (Figure 1A,B). Moreover, the size of tetraploid flowers and leaves was also bigger than in diploid ones (Figure 1C,D). Measured areas of tetraploid leaves were significantly larger (1.7-fold) than for diploid leaves.

Figure 1. Differences between diploid and tetraploid plants of *A. reptans*. (**A**) Tetraploid plants displayed more robust growth than diploid plants in field conditions, even though they were planted at the same time. (**B**) Sizes of diploid and tetraploid plants at the maximum flowering stage. (**C**) Leaf size and shape of tetraploid and diploid plants. Leaves were collected from the basal part of shoots. (**D**) Tetraploid flowers were bigger than diploid ones. Flowers and shoots flowering were collected at the same stage.

The biochemical analyses focused on whole green and flowering shoots, and especially on leaves and flowers. In all studied strains of *A. reptans* plants, both diploids and tetraploids were identified in the same spectrum, respectively, in the presence of the analyzed substances of *trans*-teupolioside, *trans*-verbascoside and 20-hydroxyecdysone. However, differences in their content were revealed. In terms of the 20-hydroxyecdysone content in whole shoots, the differences between diploid progenitor strains were negligible. The comparison of 20-hydroxyecdysone content in tetraploids and diploids (Figure 2) also showed mostly insignificant differences. Similar results were obtained for analyzed leaves and flowers. Only leaves and flowers of sc4 tetraploids contained a significantly higher amount of 20-hydroxyecdysone in relation to the measured content in the diploid progenitor (Figure 3).

On the other hand, changes were found in the contents of phenylpropanoids. In the case of *trans*-teupolioside content, there was no significant difference among diploid progenitor strains. In addition, the content of *trans*-teupolioside in whole shoots showed no significant difference between progenitor strain 4 and its derived tetraploid strain sc12 (Figure 2). A non-significantly higher level of *trans*-teupolioside was detected in tetraploid flowers and leaves compared with diploids (Figure 3). Moreover, the sc28 and sc4 tetraploid strains had higher and lower *trans*-teupolioside contents, respectively, than progenitor strain 7 (Figure 2).

The *trans*-verbascoside content in whole shoots showed differences among tetraploid strains derived from progenitor strain 7; sc4 also differed from diploid progenitor strain 7 (Figure 2). Regarding sc4, it showed a 1.87-fold higher *trans*-verbascoside content than progenitor strain 7. Furthermore, sc12 tetraploids showed markedly and significantly higher *trans*-verbascoside content (1.7-fold) than diploid progenitor strain 4 (Figure 2). Tetraploid strain sc4 showed higher content of *trans*-verbascoside not only in whole shoots, but also separately in flowers and leaves (Figure 3).

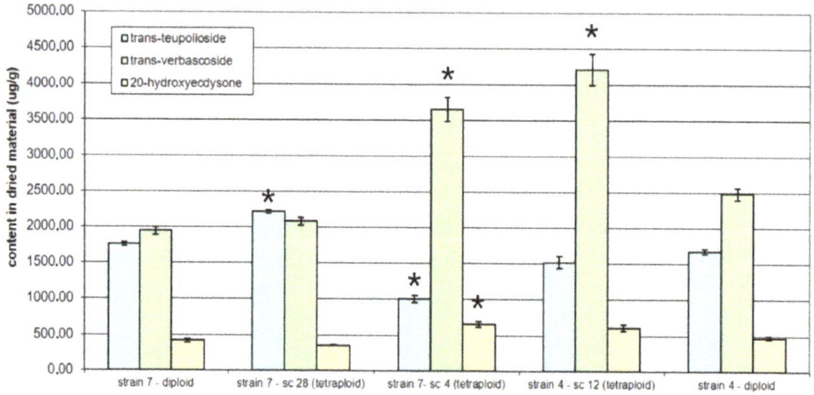

Figure 2. Contents of *trans*-teupolioside, *trans*-verbascoside and 20-hydroxyecdysone measured in dried shoots of two progenitor strains (7 and 4) and in tetraploid strains derived from those progenitors. Asterisks (*) mark significant differences of substance content in tetraploids in relation to diploid progenitors (p value < 0.01).

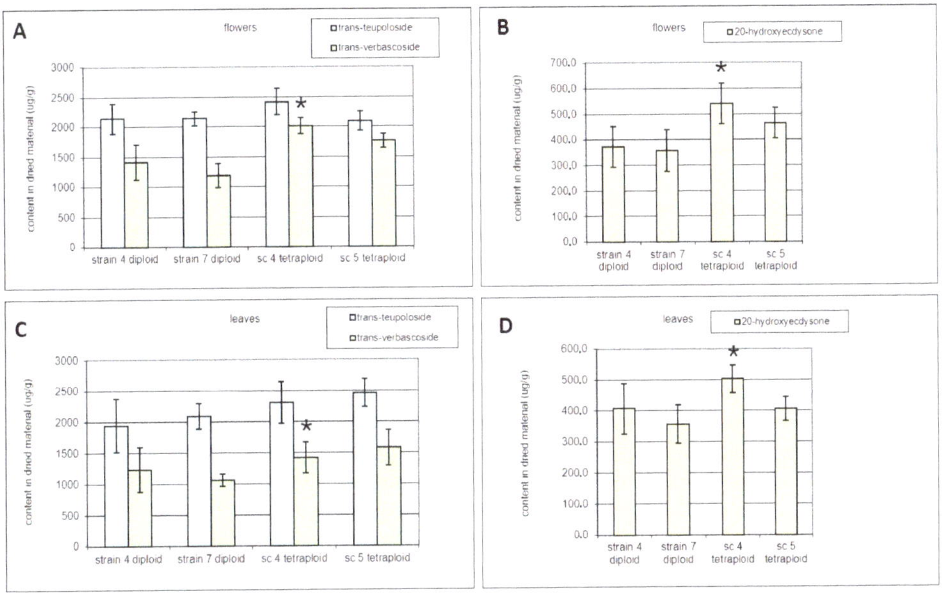

Figure 3. Contents of glycosides *trans*-teupolioside, *trans*-verbascoside (**A**,**C**) and 20-hydroxyecdy sone (**B**,**D**) measured in dried flowers (**A**,**B**) and leaves (**C**,**D**) of two progenitor strains (7 and 4) and in tetraploid strains derived from those progenitors. Asterisks (*) mark significant differences of substances content in tetraploids in relations to its diploid progenitors (p value < 0.01).

4. Discussion

The enlarged shoots and flowers of tetraploid plants grown in greenhouses and field conditions accord with previous findings for *A. reptans* artificial tetraploids cultivated in vitro [12]. The experimental plants were derived from the tetraploids of in vitro cultures. Enlarged leaves, stems, and flowers in artificially polyploidized plants have been documented for many plant species [3,13,14], as also discussed in the review of Trojak-Goluch et al. (2021) [15].

Artificial polyploidization not only enhances biomass production, but also alters the content of various phytochemicals, including secondary metabolites [16–18]. We analyzed *trans*-teupolioside, *trans*-verbascoside, and 20-hydroxyecdysone, which were also identified in plant extracts from the genus *Ajuga* [7,19].

Polyploidization can influence plant metabolism, both qualitatively and quantitatively. Tetraploids of several plant species produce terpenoids [13,17], morphine (in the case of *Papaver somniferum*) [20], and/or secondary metabolites, such as tetrahydrocannabinol and cannabidiol [21]. In phenylpropanoids, secondary metabolites include teupolioside and verbascoside. Studies of the tetraploids of *Solanum commersonii* [22] and *Solanum bulbocastanum* [23] revealed higher phenylpropanoid content in *S. commersonii* compared with the diploid progenitors, However, the tetraploids of *S. bulbocastaneum* displayed similar (or lower) phenylpropanoid content to those of their diploid parents.

This study demonstrated differences in phenylpropanoid contents in the tetraploid strain. The phytochemical most affected by polyploidization was *trans*-verbascoside, the content of which was higher in tetraploids compared with diploid progenitors. Verbascoside exerts numerous biological effects, such as anti-inflammatory effects on the skin, and in the intestines and lungs. It also exerts cytoprotective effects through free radical scavenging, which could be useful for treating neurodegenerative diseases, as reviewed by Alipieva et al. (2014) [24]. The nearly two-fold higher *trans*-verbascoside content in tetraploids, compared to the diploid progenitors, observed in this study indicated that *A. reptans* could be used in the pharmacological industry. Changes in 20-hydroxyecdysone contents were also examined, and our results suggested that genome doubling affects the expression of genes involved in complex biosynthetic pathways; this applies to both phenylpropanoids and all of the secondary metabolites. New tetraploid genomes are typically unique, and strains with altered secondary metabolite contents can be produced; such genotypes have potential for breeding programs.

5. Conclusions

In this work, it was demonstrated that the method of artificial polyploidization in *A. reptans* that utilizes in vitro techniques is able not only to double the genome, but also to change the content of bioactive substances in the plant bodies. Here, new tetraploid materials were tested for increases in their phytochemical contents. The phytochemical most increased by polyploidization was *trans*-verbascoside. Tetraploid plants which are displaying robust features could also represent new genotypes for usage in gardening and landscaping, as a quick ground covering plant for empty areas.

Supplementary Materials: The following supporting information can be downloaded at: https://www.mdpi.com/article/10.3390/horticulturae8070581/s1. NMR data of *trans*-teupolioside, *trans*-verbascoside.

Author Contributions: All co-authors worked on presented research and manuscript in following parts: B.N.—preparing polyploid plants and transfer them from in vitro cultures, plant morphology analysis, writing manuscript; V.O.—design of experiments with polyploids, writing manuscript; N.V. and J.T.—HPLC and LC/MS analysis, writing manuscript; Š.H.—NMR analysis; R.P.—donor plants, writing of manuscript. All authors have read and agreed to the published version of the manuscript.

Funding: This research was supported by the grants QK1910103 (NAZV, Ministry of Agriculture, Czech Republic) and the Ministry of Education, Youth and Sports of the Czech Republic, project SustES–Adaptation strategies for sustainable ecosystem services and food security under adverse environmental conditions (CZ.02.1.01/0.0/0.0/16_019/0000797).

Institutional Review Board Statement: Not applicable.

Informed Consent Statement: Not applicable.

Data Availability Statement: The data presented in this study are available on request from the corresponding author.

Acknowledgments: We thank to technical staff of Department of Botany, Palacký University Olomouc, M. Klíčová for taking care about plants.

Conflicts of Interest: The authors declare no conflict of interest.

References

1. van Duren, M.; Morpurgo, R.; Dolezel, J.; Afza, R. Induction and verification of autotetraploids in diploid banana (*Musa acuminata*) by in vitro techniques. *Euphytica* **1996**, *88*, 25–34. [CrossRef]
2. Xing, S.H.; Guo, X.B.; Wang, Q.; Pan, Q.F.; Tian, Y.S.; Liu, P.; Zhao, J.Y.; Wang, G.F.; Sun, X.F.; Tang, K.X. Induction and flow cytometry identification of tetraploids from seed-derived explants through colchicine treatments in *Catharanthus roseus* (L.) G. don. *BioMed. Res. Internat.* **2011**, *2011*, 10. [CrossRef]
3. Tavan, M.; Mirjalili, M.H.; Karimzadeh, G. In vitro polyploidy induction: Changes in morphological, anatomical and phytochemical characteristics of *Thymus persicus* (Lamiaceae). *Plant Cell Tissue Org. Cult.* **2015**, *122*, 573–583. [CrossRef]
4. Švécarová, M.; Navrátilová, B.; Hašler, P.; Ondřej, V. Artificial induction of tetraploidy in *Humulus lupulus* L. using oryzalin. *Acta Agrobot.* **2019**, *72*, 1764. [CrossRef]
5. Pavela, R. Acute toxicity and synergistic and antagonistic effects of the aromatic compounds of some essential oils against *Culex quinquefasciatus* Say larvae. *Parasitol. Res.* **2015**, *114*, 3835–3853. [CrossRef] [PubMed]
6. Di Paola, R.; Esposito, E.; Mazzon, E.; Riccardi, L.; Caminiti, R.; Dal Toso, R.; Pressi, G.; Cuzzocrea, S. Teupolioside, a phenylpropanoid glycosides of *Ajuga reptans*, biotechnologically produced by IRBN22 plant cell line, exerts beneficial effects on a rodent model of colitis. *Biochem. Pharmacol.* **2009**, *77*, 845–857. [CrossRef]
7. Marchev, A.S.; Georgiev, M.I. Plant In vitro Systems as a sustainable source of sctive ingredients for cosmeceutical application. *Molecules* **2020**, *25*, 2006. [CrossRef]

8. Dinan, L. Phytoecdysteroids: Biological aspects. *Phytochemistry* **2001**, *57*, 325–339. [CrossRef]
9. Aly, R.; Ravid, U.; Abu-Nassar, J.; Botnick, I.; Lebedev, G.; Gal, S.; Ziadna, H.; Achdari, G.; Smirov, E.; Meir, A.; et al. Biological activity of natural phytoecdysteroids from Ajuga iva against the sweetpotato whitefly Bemisia tabaci and the persea mite *Oligonychus perseae*. *Pest Manag. Sci.* **2011**, *67*, 1493–1498. [CrossRef]
10. Gubser, G.; Vollenweider, S.; Eibl, D.; Eibl, R. Food ingredients and food made with plant cell and tissue cultures: State-of-the art and future trends. *Eng. Life Sci.* **2021**, *21*, 87–98. [CrossRef]
11. Švécarová, M.; Navrátilová, B.; Ondřej, V. In vitro polyploidization of *Ajuga reptans* L. using oryzalin. *Acta Biol. Crac. Bot.* **2018**, *60*, 69–73.
12. Rungsimakan, S.; Rowan, M.G. Terpenoids, flavonoids and caffeic acid derivatives from *Salvia viridis* L. cvar. Blue Jeans. *Phytochemistry* **2014**, *108*, 177–188. [CrossRef] [PubMed]
13. Gao, S.L.; Chen, B.J.; Zhu, D.N. In vitro production and identification of autotetraploids of *Scutellaria baicalensis*. *Plant Cell Tissue Org. Cult.* **2002**, *70*, 289–293. [CrossRef]
14. Navrátilová, B.; Švécarová, M.; Bednář, J.; Ondřej, V. In vitro polyploidization of *Thymus vulgaris* L. and its effect on composition of essential oils. *Agronomy* **2021**, *11*, 596. [CrossRef]
15. Trojak-Goluch, A.; Kawka-Lipińska, M.; Wielgusz, K.; Praczyk, M. Polyploidy in industrial crops: Applications and perspectives in plant breeding. *Agronomy* **2021**, *11*, 2574. [CrossRef]
16. Dhawan, O.P.; Lavania, U.C. Enhancing the productivity of secondary metabolites via induced polyploidy: A review. *Euphytica* **1996**, *87*, 81–89. [CrossRef]
17. Banyai, W.; Sangthong, R.; Karaket, N.; Inthima, P.; Mii, M.; Supaibulwatana, K. Overproduction of artemisinin in tetraploid *Artemisia annua* L. *Plant Biotechnol. J.* **2010**, *27*, 427–433. [CrossRef]
18. Xu, C.G.; Tang, T.X.; Chen, R.; Liang, C.H.; Liu, X.Y.; Wu, C.L.; Yang, Y.S.; Yang, D.P.; Wu, H. A comparative study of bioactive secondary metabolite production in diploid and tetraploid *Echinacea purpurea* (L.) Moench. *Plant Cell Tissue Org. Cult.* **2014**, *116*, 323–332. [CrossRef]
19. Taha-Salaime, L.; Lebedev, G.; Abo-Nassar, J.; Marzouk, S.; Inbar, M.; Ghanim, M.; Aly, R. Activity of *Ajuga iva* extracts against the African cotton leafworm *Spodoptera littoralis*. *Insects* **2020**, *11*, 726. [CrossRef]
20. Mishra, B.K.; Pathak, S.; Sharma, A.; Trivedi, P.K.; Shukla, S. Modulated gene expression in newly synthesized autotetraploid of *Papaver somniferum* L. *South Afr. J. Bot.* **2010**, *76*, 447–452. [CrossRef]
21. Parson, J.L.; Martin, S.L.; Golenia, G.; James, T. Polyploidization for the genetic improvement of *Cannabis sativa*. *Front. Plant Sci.* **2019**, *10*, 476. [CrossRef] [PubMed]
22. Caruso, I.; Lepore, L.; De Tommasi, N.; Dal Piaz, F.; Frusciante, L.; Aversano, R.; Garramone, R.; Carputo, D. Secondary metabolite profile in induced tetraploids of wild *Solanum commersonii* Dun. *Chem. Biodivers.* **2011**, *8*, 2226–2237. [CrossRef] [PubMed]
23. Caruso, I.; Dal Piaz, F.; Malafronte, N.; De Tommasi, N.; Aversano, R.; Wulff Zottele, C.; Scarano, M.-T.; Carputo, D. Impact of ploidy change on secondary metabolites and photochemical efficiency in *Solanum bulbocastanum* Dun. *Nat. Prod. Commun.* **2013**, *8*, 1387–1392. [PubMed]
24. Alipieva, K.; Korkina, L.; Orhan, I.; Georgiev, M. Verbascoside—A review of its occurrence, (bio)synthesis and pharmacological significance. *Biotech. Adv.* **2014**, *32*, 1065–1076. [CrossRef] [PubMed]

horticulturae

Review

In Vitro Propagation of Caper (*Capparis spinosa* L.): A Review

Meriyem Koufan [1,*], Ilham Belkoura [2] and Mouaad Amine Mazri [3,*]

[1] Natural Resources and Local Products Research Unit, Regional Center of Agricultural Research of Agadir, National Institute of Agricultural Research, Avenue Ennasr, BP 415 Rabat Principale, Rabat 10090, Morocco
[2] In Vitro Culture Laboratory, Department of Basic Sciences, National School of Agriculture, BP S/40, Meknes 50001, Morocco
[3] Agro-Biotechnology Research Unit, Regional Center of Agricultural Research of Marrakech, National Institute of Agricultural Research, Avenue Ennasr, BP 415 Rabat Principale, Rabat 10090, Morocco
* Correspondence: meriyem.koufan@inra.ma (M.K.); mouaadamine.mazri@inra.ma (M.A.M.)

Abstract: Caper (*Capparis spinosa* L.) is a shrubby plant species recalcitrant to vegetative propagation and generally difficult to propagate by seeds. This is due to the difficulties associated with seed germination, root induction from stem cuttings, and plant hardening. Propagation by tissue culture would be a good alternative and promising approach to overcome the limitations of conventional propagation. Tissue culture methods can be used for the clonal propagation of caper plants. Indeed, in many plant species, micropropagation has played a decisive role in the rapid and large-scale production of uniform and genetically stable plants. Tissue culture methods can also be used in genetic improvement and conservation programs. In this review, we first provided an overview on caper and its conventional means of propagation, then we described the different methods of caper micropropagation, i.e., in vitro seed germination and seedling development, propagation by nodal segmentation of elongated shoots (i.e., microcuttings), and adventitious organogenesis. These micropropagation methods can make it possible to overcome all the obstacles preventing large-scale propagation and genetic improvement of caper. Thus, the most updated information on the progress made in the field of caper micropropagation is reported and future perspectives are outlined.

Keywords: caper; in vitro germination; microcuttings; micropropagation; organogenesis

Citation: Koufan, M.; Belkoura, I.; Mazri, M.A. In Vitro Propagation of Caper (*Capparis spinosa* L.): A Review. *Horticulturae* **2022**, *8*, 737. https://doi.org/10.3390/horticulturae8080737

Academic Editors: Barbara Ruffoni, Alicja Tymoszuk and Dariusz Kulus

Received: 17 May 2022
Accepted: 27 June 2022
Published: 17 August 2022

1. Introduction

Caper (*Capparis spinosa* L.) is a shrubby plant native to the Mediterranean region. It belongs to the family Capparidaceae and genus Capparis, which includes more than 250 species generally used for ornamental, culinary, cosmetic, pharmaceutical, and medicinal purposes [1]. *Capparis spinosa* is characterized by high morphological and ecological diversity, which led some authors to differentiate several intraspecific variants and taxa [2–5]. It plays important socio-economic roles in the arid regions of many countries, and is well adapted to high temperatures, intense sunlight, and fluctuating climates [6–8].

Caper is cultivated for both unopened flowers and young fruits that are used in many traditional dishes [9]. The fruits of caper are harvested from both wild and cultivated plants. The main producers are Morocco, Spain, and Turkey [10,11].

Caper is a shrub with high medicinal values. This species is rich in bioactive compounds such as flavonoids, glucosinolates, phenolic acids, and alkaloids that can be used for medicinal, culinary, and ornamental purposes. Along this line, many health-promoting properties of caper extracts were scientifically demonstrated, in particular anti-cancer and antioxidant activities [12]. Caper is also used for ecological purposes since it helps in preventing soil erosion and preserving biodiversity and soil water [13,14].

Caper is commonly propagated by seeds. This method allows the maintenance of high genetic diversity within its populations. However, propagation by seeds cannot be used for the production of true-to-type plants [12,15]. Due to the high socio-economic and ecological importance of caper, it becomes preparatory to genetically characterize the existing material in order to identify the best and most promising genotypes, with superior agronomic characteristics. Such genotypes would be subjected to rapid and large-scale propagation, and then used for different ecological and industrial purposes. Propagation by stem cuttings may be a good alternative to sexual propagation. Nevertheless, this vegetative propagation method is associated with serious rooting problems [12]. Therefore, developing efficient propagation methods through tissue culture is needed today for caper.

In vitro propagation is a very promising approach for rapid production of caper plants. Tissue culture of caper was first reported in 1984 [16]. Since, several studies have been published describing micropropagation systems for caper, but with extremely variable rooting and acclimatization rates [9,12,17–23]. The high variability of findings reported in the literature may be related to several factors such as the genotype, age, and physiological state of the donor plant, explant, plant growth regulators (PGRs), and culture conditions [16]. Today, it is extremely important to develop a reproducible and efficient micropropagation system for the best-performing genotypes.

The present review provides the most updated data on the progress made in the field of caper micropropagation. This includes propagation by in vitro seed germination and seedling development, propagation by microcuttings, and through adventitious organogenesis. This review also describes the micropropagation steps of caper and discusses the difficulties encountered in each in vitro culture method.

1.1. Geographic Distribution, Botanical Classification, and Ecological Characteristics of Caper

Caper is a xerophytic and heliophilic species endowed with a great adaptation capacity to difficult environmental conditions [24]. Caper is mainly located in the arid and semi-arid areas of the Mediterranean region. It is found in north and east Africa, Madagascar, southwest and central Asia (China and Iran), the Middle East, southern Europe (Greece, Italy), Australia, and Oceania [3,25]. It grows naturally from the Atlantic coasts of the Canary Islands and Morocco to the Black Sea, and to the shore of the Caspian Sea [3,26]. The wild and cultivated caper plants are used throughout its entire area of growth, from west Africa to the Norfolk Islands in the Pacific [27].

Caper belongs to the genus Capparis and family Capparidaceae [3,28–30]. It is a perennial species, generally thorny and hairy, with a height of about 50–80 cm. Caper is characterized by a very high level of heterogeneity within its populations [3,29,31]. The taxonomic position of *C. spinosa* is still difficult to assess due to the several taxa of different ranks that have been described [32]. Therefore, the classification of this species remains ambiguous and controversial. A morphological study of some caper ecotypes from north-central Morocco revealed the existence of three genotypic clusters that correspond to three species: *Capparis spinosa*, *C. ovata*, and *C. cartilaginea* [31]. *C. spinosa* is by far the most economically important one and is mainly found in the Mediterranean region [1,3,29]. It is also one of the most common medicinal and aromatic plants that grow along roads, on slopes, and in rocky and stony areas [33].

Flower buds (i.e., capers) are the most sought-after parts of caper. The firmness of capers is a key indicator of their commercial quality (i.e., an indicator of good storability). The firmness index of capers varies depending on the genotype and area of growth [34]. Caper fruit is a dehiscent berry of 2–4 cm long (Figure 1). The number of seeds per fruit is on average 130, with a minimum of 15 seeds in the smallest fruits and a maximum of 400 seeds in large fruits [13]. The roots of caper are very deep but not highly branched [13]. Indeed, the caper has a powerful root system that can mobilize large volumes of soil. These characteristics make caper highly tolerant to drought and able to grow even on the poorest soils and on steep slopes. Hence, caper has a great ecological interest as it can prevent erosion in arid and semi-arid regions [35].

Figure 1. Caper (*Capparis spinosa* L.) fruits.

As a xerophytic plant, caper has morphological and physiological characteristics that make it tolerant to the most severe climatic conditions of arid and semi-arid regions [11]. Caper is found in sites with temperatures as low as −4 °C, but generally the annual average temperature of its natural regions of growth exceeds 14 °C. It withstands temperatures of up to 40 °C and is also drought tolerant. It can survive with no rain for several months. In its area of growth, the rainfall varies from 200 to 550 mm per year [13].

Caper is a rupicolous species that can be cultivated on different types of soils. It grows on loamy clay, sandy, rocky, and gravelly soils. It also grows on clayey and poorly draining soils and on light sandy loam soils with alkaline pH (Figure 2). However, it seems that caper prefers light, well-draining soils with a neutral to alkaline pH [13,14].

Figure 2. Wild caper (*Capparis spinosa* L.) plants grown naturally in Morocco.

1.2. Economic Importance and Medicinal Uses

The economic importance of caper lies in the value of its young shoots, tender young fruits, and flower buds, and they have a high demand in the international market [24]. Flower buds are generally harvested before flowering and used for aromatic purposes in the cosmetic industry. The wild buds are harvested by seasonal pickers then stored in salt before packaging [36].

Since ancient times, different parts of the caper have been used as traditional herbal remedies due to their beneficial effects on human health [37]. In recent years, the biological activities of caper bud extracts have been evaluated and numerous health promoting properties were reported. Caper extracts possess a high inhibitory activity against acetylcholinesterase (AChE) and butyrylcholinesterase (BChE), as well as antioxidant and metal chelating properties [38]. Caper leaves, stems, flowers, roots, and fruits were also reported to have antidiabetic, anti-obesity, antimicrobial, anti-inflammatory, and antihypertensive activities [39–43]. This is due to their content of flavonoids, phenols, alkaloids, tannins, and glycosides [44].

In Greek folk medicine, herbal teas made from the young shoots of caper are used against rheumatism. The bark and leaves of caper have anticarcinogenic activities. Indeed, they contain rutin, selenium, and quercetin, which contribute to the prevention of certain forms of cancer [14].

2. Conventional Propagation of Caper

2.1. Propagation by Seeds

Propagation by seeds is the most widely used method for caper propagation. However, many factors may hamper caper seed germination, among which the most important are the hardness of seed coat, seed dormancy, the hormonal balance within seeds, and embryo immaturity. The seed coat and mucilage surrounding the seeds are ecological adaptations to prevent water loss and maintain seed viability during dry seasons [14,45,46]. Generally, viable embryos germinate within 3 to 4 days after partial removal of the lignified seed coat [47].

Several studies have been conducted to overcome plant seed dormancy. Treatments such as plant hormones, sulfuric acid (H_2SO_4), methanol, potassium nitrates, boiling water, and stratification were suggested to break seed dormancy [48]. In the case of caper, H_2SO_4, cold stratification, scarification, and treatment with gibberellic acid (GA_3) or nitric acid (KNO_3) have been suggested [16,49–53]. However, the germination rate of caper seeds was not high enough for rapid and large-scale propagation even after the application of these treatments.

2.2. Propagation by Stem Cuttings

Vegetative propagation by stem cuttings is a practicable and efficient method for the clonal multiplication of plants. However, in many species, this method is hampered by rooting difficulties [48]. Caper is a plant species difficult to root. Root induction from stem cuttings depends on many factors such as the type and age of cutting, time of cuttings' collection, the substrate used for planting, and other seasonal and environmental parameters. Moghaddasi et al. [14] observed a root induction rate of up to 55% in one-year-old twigs of caper. Some authors used PGRs to improve the rooting ability of caper cuttings [16,46,54,55]. Bahrani et al. [46] evaluated the effects of indole-3-butyric acid (IBA) on root induction from growing season (softwood) and one year old (semi-hardwood) stems of caper. IBA did not promote root induction from softwood cuttings, while it showed a very low rooting percentage (16.1%) in semi-hardwood cuttings [46]. According to Saifi et al. [54], treating caper cuttings with 3000 ppm IBA for one minute resulted in a rhizogenesis rate of 50%. On the other hand, the use of leafy cuttings showed a slightly higher rhizogenesis rate (67.1%) than leafless cuttings (61.4%) [15]. The ability of caper cuttings to induce roots is also influenced by other endogenous and exogenous factors such as the genotype, cutting thickness, and planting season [45,55,56].

Caper propagation by stem cuttings allows for the production of true-to-type plants. However, as mentioned above, this method is hindered by rooting problems. Additionally, caper plants produced by stem cuttings are highly susceptible to drought during the first years of planting.

Based on the above, it seems that the use of in vitro culture methods would considerably help in the rapid and large-scale propagation of this medicinal plant. Up to date, few studies have been undertaken on caper micropropagation. Three different methods were described: (i) in vitro seed germination and seedling development; (ii) nodal segmentation of elongated shoots (i.e., microcuttings); and (iii) organogenesis. These micropropagation methods are described and discussed in the following sections.

3. Caper Propagation by Tissue Culture

3.1. In Vitro Seed Germination and Seedling Development

Propagation by in vitro seed germination allows the conservation of caper diversity and ensures the sustainable use of this species. Under natural conditions, the propagation of caper by seeds is hampered by many factors as previously mentioned. In vitro seed germination may considerably help to overcome these constraints. Indeed, seeds are cultured on appropriate culture media and under controlled conditions that provide embryos with the optimal nutrients and conditions to promote germination and subsequent growth [48]. Accordingly, in vitro seed germination (Figure 3) can be considered the best approach to produce a large number of caper plants while preserving the genetic diversity within this species. In vitro seed germination is also a powerful tool for genetic improvement.

Figure 3. In vitro seed germination and seedling development of caper (*Capparis spinosa* L.).

3.1.1. Surface Disinfection and Culture Medium

Various surface disinfection protocols were described for caper seeds. For example, Rhimi et al. [8] started the disinfection procedure by rinsing the seeds five times with sterile distilled water. The seeds were then treated with H_2SO_4 for 30 min, followed by immersion in 70% alcohol for 5 min and then in 6% sodium hypochlorite for 20 min, with continuous stirring. Finally, the seeds were rinsed five times (5 min each) with sterile distilled water. Germanà and Chiancone [57] soaked caper seeds in 70% ethyl alcohol for 3 min, followed by 20 min in a solution of 25% commercial bleach (approximately 1.5% active chlorine), and finally by three rinses with sterile distilled water.

3.1.2. Seed Germination

Several factors can affect seed germination in vitro, for example, plant genotype, culture medium components, PGRs, seed coat, pretreatments, and culture conditions [8,48,58–61]. In caper, a large variation was observed in germination rates among the different populations (Table 1). It appears that Murashige and Skoog medium (MS; [62]) is the most suitable for in vitro seed germination of caper. After scarification, Chalak et al. [17] observed a seed germination rate of 71% on PGR-free MS medium, while the use of sterile distilled water gelled with agar showed a germination rate of 64%. According to Rhimi et al. [8], a 75% germination rate was observed in caper seeds from the Nahli site (Tunisia) following

30-min H_2SO_4 and 48-h GA_3 (2000 mg L^{-1}) pretreatments. In a different work, the caper seeds harvested from a wild tree located in the Madonie Mountains and that belongs to the subspecies *rupestris* showed a germination rate of 80.4%. The seeds were subjected to a heat pretreatment at 40 °C for one hour, then cultured on modified MS medium [57]. Treating seeds with gamma rays may improve their germination ability in vitro and promote seedling growth and development [63,64]. Seed irradiation with gamma rays resulted in an in vitro germination rate of 50% [65].

Table 1. In vitro seed germination of caper (*Capparis spinosa* L.).

Pretreatment	Germination Medium	PGRs	Culture Conditions	Germination Percentage (%)	Reference
H_2SO_4 for 20 min with scratching	MS	PGR-free	Darkness, 25 °C	46%	Al-Safadi and Elias [65]
Gamma irradiation (a 100 Gray dose)	MS	PGR-free	Darkness, 25 °C	50%	
H_2SO_4 for 30 min followed by soaking in 2000 mg L^{-1} GA_3 for 48 h	MS	PGR-free	16 h photoperiod, 22 °C	75%	Rhimi et al. [8]
	Sterile distilled water	PGR-free	16 h photoperiod, 22 °C	62.5%	
Scarification	MS	PGR-free	16 h photoperiod, 26 °C	71%	Chalak et al. [17]
	Sterile distilled water	PGR-free	16 h photoperiod, 26 °C	64%	
Imbibition in 20 ppm GA_3	MS	0.4 mg L^{-1} NAA + 0.45 mg L^{-1} BAP + 0.7 mg L^{-1} GA_3	16 h photoperiod, 27 °C	32.1%	Germanà and Chiancone [57]
Hot temperature (40 °C) for 1 h	MS	0.4 mg L^{-1} NAA + 0.45 mg L^{-1} BAP + 0.7 mg L^{-1} GA_3	16 h photoperiod, 27 °C	80.4%	

Abbreviations: BAP, 6-benzylaminopurine; GA_3, gibberellic acid; H_2SO_4, sulfuric acid; MS, Murashige and Skoog medium; NAA, 1-naphthaleneacetic acid; PGR, plant growth regulators.

The effect of PGRs on seed germination has also been the subject of investigation [57,66–68]. During in vitro seed germination, shoot and root growth depends strongly on exogenous PGRs, their concentrations, as well as the metabolism of endogenous plant hormones, their concentration, and interaction with PGRs [69,70]. Based on our experiments (data not published), the use of MS medium supplemented with GA_3 and 6-benzylaminopurine (BAP) or thidiazuron (TDZ) promotes caper seed germination. This is in good agreement with results reported for other plant species [71,72]. GA_3 is a growth regulator widely used for in vitro seed germination. The exogenous GA_3 added to culture medium acts by increasing the concentration of endogenous GA_3 and lowering that of abscisic acid (ABA), which promotes seed germination [73,74]. BAP is a potent cytokinin commonly used in tissue culture due to its efficacy and affordability [75]. Regarding TDZ, it is a phenylurea derivative that has a potent cytokinin-like activity and is widely used for the micropropagation of Mediterranean species [76].

3.1.3. Seedling Growth and Plantlet Acclimatization

Research on in vitro development of caper seedlings and their acclimatization is still very limited. According to Heydariyan et al. [77], salicylic acid and GA_3 significantly improved the growth and development of caper seedlings. Germanà and Chiancone [57] noted that the combination of 1-naphthaleneacetic acid (NAA), BAP, and GA_3 inhibited apical bud growth. However, new buds emerged around the cotyledons, in the region between the epicotyl and hypocotyl. According to these authors, 40% of seedlings produced new shoots [57].

Successful acclimatization of plantlets grown in vitro is a crucial step for the mass propagation of any species. Rhimi et al. [8] reported a survival rate of 95% after the acclimatization of 15-day-old seedlings in pots containing a mixture of peat and sand (2:1, *w*/*w*).

3.2. Caper Propagation by Microcuttings

Despite the large number of caper genotypes and their reputed medicinal virtues, an efficient and reproducible propagation system through microcuttings has yet to be developed. Such system should aim not only to produce a large number of true-to-type plants, but also to harness the economic potential of capers.

Propagation by microcuttings, also known as propagation by nodal segmentation of elongated shoots, is a vegetative propagation method based on the multiplication of axillary shoots derived from microcuttings (i.e., tender stem segments bearing 1–3 buds) in vitro. Elongated shoots are cut into small segments (i.e., microshoots) containing one or two nodes then placed on a rooting medium to promote the induction of adventitious roots. The plantlets are then transferred to the glasshouse for acclimatization. This method allows for the production of true-to-type plants [48].

The findings of several studies carried out on capper micropropagation by microcuttings underlined the enormous potential of this method [9,12,17]. However, successful regeneration through microcuttings depends on many factors associated with the genotype and culture conditions.

3.2.1. Plant Material and Culture Medium

In order to establish aseptic cultures and obtain leafy shoots, nodal segments were used [9,12,17,78]. In addition, the effects of different basal media on the multiplication and rooting of caper microshoots were evaluated, for example, MS medium, Linsmaier and Skoog medium [79], Nitsch and Nitsch medium [80], Woody Plant Medium (WPM; [81]), and Rugini medium [82]. However, MS medium was found to be the most suitable for culture establishment and shoot proliferation [9,20].

3.2.2. Surface Disinfection

The development of an efficient disinfection protocol is a necessary step for the setup of a micropropagation system. However, the establishment of axenic cultures from microcuttings may be very challenging [9,18]. Various protocols have been described in the literature, most of them used sodium hypochlorite as a disinfectant agent [17,20,22,23]. The use of sodium hypochlorite coupled with ethanol was also suggested and showed convincing results. Indeed, some authors suggested to first soak microcuttings in a solution of 70–80% ethanol for 2 to 5 min before surface disinfection with sodium hypochlorite [9,12,23,78].

3.2.3. Shoot Multiplication and Elongation

Shoot multiplication and elongation are strongly affected by PGRs. High levels of endogenous cytokinins may be needed to induce cell division. Addition of exogeneous PGRs promote cell elongation and multiplication. Many authors suggested the use of a BAP-auxin combination for caper shoot multiplication and elongation (Table 2). For example, El-Mekawy et al. [20] reported that the combination of 0.5 mg L^{-1} BAP and 0.05 mg L^{-1} NAA gave an average number of 3.89 shoots per explant, with an average length of 3.24 cm.

The undeniable effect of BAP on caper shoot multiplication and elongation was also reported by Rodriguez et al. [78], who indicated that 4 μM ($\simeq$0.9 mg L^{-1}) BAP (in combination with 0.3 μM ($\simeq$0.05 mg L^{-1}) indole-3-acetic acid (IAA) and 0.3 μM ($\simeq$0.1 mg L^{-1}) GA_3) promoted shoot proliferation and growth. This was confirmed by Sottile et al. [83], who reported that the combination of 6 μM ($\simeq$1.35 mg L^{-1}) BAP and 0.12 μM ($\simeq$0.02 mg L^{-1}) IBA allowed for maximum proliferation of caper leafy shoots (91%). This combination also improved the number of shoots per explant (8.7) and shoot length (2.7 cm). Multiple shoot formation from nodal buds was observed by Chalak et al. [17] on media containing 1.5 mg L^{-1} BAP, 0.05 mg L^{-1} IBA, and 0.1 mg L^{-1} GA_3, with an average number of shoots ranging from 4.25 to 5.43.

Some other factors were reported to affect shoot multiplication and elongation of caper, such as the type and concentration of carbohydrates added to culture medium [12]. Sucrose is by far the most commonly used carbon source for the multiplication of caper shoots. Glucose promoted caper shoot growth as well and, in some cases resulted in better growth than sucrose [12]. Successive subcultures have been reported to affect the proliferation capacity of caper shoots. Shoot multiplication was enhanced by successive subcultures on fresh culture media, with an average rate exceeding 20 new shoots per explant at the end of the sixth subculture. However, the proliferation capacity decreased during the following subcultures [17].

3.2.4. In Vitro Rhizogenesis of Caper Microshoots

Adventitious root induction from microshoots is the most decisive step of the microcuttings' technique. In many plant species, it was very difficult to induce roots from in vitro cultured shoots [84]. Root formation from microshoots depends on many factors such as the genotype, culture medium and conditions, and PGRs, particularly auxins and their concentrations. Early cell division is a prerequisite for root formation. Generally, the root induction requires high levels of auxins. Addition of cytokinins to culture medium may be necessary since they are involved in the regulation of cell division and organ differentiation [12,18,69].

In many plant species, the use of the auxins IAA, IBA and NAA resulted in high in vitro rhizogenesis rates [85–87]. These auxins, used either alone or in combination, have also shown convincing results in *Capparis* species [9,18,20]. Generally, the roots appear after 15 to 20 days of culture [9]. Attia et al. [22] observed an in vitro rooting percentage of 56.7% on MS medium containing 1.5 mg L^{-1} NAA. According to Carra et al. [9] and Sottile et al. [83], the use of IBA at 5 μM ($\simeq$1 mg L^{-1}) resulted in high rhizogenesis rates (93.4–93.7%). However, increasing IBA concentration to 10 μM ($\simeq$2 mg L^{-1}) decreased the rhizogenesis rate to 68.7% [9]. On the other hand, 1 μM ($\simeq$0.18 mg L^{-1}) NAA showed a rhizogenesis rate of 68.7% [9]. The average number of roots per shoot ranged from 2.4 on the medium supplemented with 1 μM ($\simeq$0.18 mg L^{-1}) NAA to 3.5 on that containing 5 μM ($\simeq$0.87 mg L^{-1}) IAA under light conditions, whereas the highest average length of roots was 28 mm when 5 μM ($\simeq$1 mg L^{-1}) IBA was added to culture medium [9]. The combination of NAA and IBA at 0.75 mg L^{-1} and 0.25 mg L^{-1}, respectively, resulted in a rooting percentage of 67% [16]. Rodriguez et al. [78] observed 70% rooting after a 20-day incubation period in darkness on MS medium containing IAA. Chalak et al. [17] and Chalak and Elbitar [18] suggested a pulse treatment for 4 h in a solution of 100 mg L^{-1} IAA before transferring the explants to MS or half-strength Murashige and Skoog medium ($\frac{1}{2}$MS). This treatment resulted in a rhizogenesis range of 87–92%. Soaking caper microshoots in a solution of 24.6 μM ($\simeq$5 mg L^{-1}) IBA for 10 min before culture on MS medium under a 16 h photoperiod enhanced the rooting percentage to 80.5% [47].

Table 2. Caper (*Capparis spinosa* L.) propagation by microcuttings.

Bud Break/ Culture Initiation Medium	Photoperiod	Multiplication-Elongation Medium	Photoperiod	Average Number of Shoots Per Explant	Rooting Medium	Photoperiod	Rooting Percentage	Plantlet Acclimatization	Reference
MS	16 h	MS + 6 µM (≃1.35 mg L^{-1}) BAP + 0.12 µM IBA (≃0.02 mg L^{-1})	16 h	8.9	MS + 5 µM (≃1.01 mg L^{-1}) IBA	Light	93.7%	75–82%	Carra et al. [9]
MS	16 h	MS + 0.5 mg L^{-1} BAP + 0.5 mg L^{-1} IBA	16 h	5.2	Either PGR-free $\frac{1}{2}$MS or MS + 1.5 mg L^{-1} NAA	16 h	56.7%	65%	Attia et al. [22]
PGR-free MS	16 h	MS + 4 µM (≃0.9 mg L^{-1}) BAP + 0.3 µM (≃0.05 mg L^{-1}) IAA + 0.3 µM (≃0.1 mg L^{-1}) GA_3	18 h	N/A	$\frac{1}{2}$MS + 30 µM IAA	Darkness	70%	N/A	Rodriguez et al. [78]
PGR-free MS	In darkness then 16 h photoperiod	MS + 6.6 µM (≃1.59 mg L^{-1}) meta-topolin + 0.25 µM (≃0.05 mg L^{-1}) IBA	In darkness then 16 h photoperiod	5.24-7.32	MS + 0.75 mg L^{-1} NAA + 0.25 mg L^{-1} IBA	In darkness then 16 h photoperiod	67%	N/A	Gianguzzi et al. [16,23]
WPM	16 h	WPM + 0.8 mg L^{-1} kinetin + 0.05 mg L^{-1} IBA + 0.1 mg L^{-1} GA_3	16 h	4.6	$\frac{1}{2}$MS + 5 mg L^{-1} IAA	N/A	80%	63%	Musallam et al. [88]
MS + 1 mg L^{-1} zeatin	16 h	MS + 1 mg L^{-1} Zeatin	16 h	>20	4 h pulse treatment in darkness with 100 mg L^{-1} IAA solution followed by culture on $\frac{1}{2}$MS	Darkness	92%	92%	Chalak and Elbitar [18]
MS	16 h	MS + 0.50 mg L^{-1} BAP + 0.05 mg L^{-1} NAA	16 h	3.89	$\frac{1}{2}$MS + 1.5 mg L^{-1} IBA	16 h	85%	90%	El-Mekawy et al. [20]
Modified MS + 1.5 mg L^{-1} BAP + 0.05 mg L^{-1} IBA + 0.1 mg L^{-1} GA_3	N/A	Modified MS + 1.5 mg L^{-1} BAP + 0.05 mg L^{-1} IBA + 0.1 mg L^{-1} GA_3	N/A	5.43	4 h pulse treatment in darkness with 100 mg L^{-1} IAA solution followed by culture on $\frac{1}{2}$MS	Darkness	87%	40%	Chalak et al. [17]

N/A: Data not available. Abbreviations: $\frac{1}{2}$MS, half-strength Murashige and Skoog medium; BAP, 6-benzylaminopurine; GA_3, gibberellic acid; IAA, indole-3-acetic acid; IBA, indole-3-butyric acid; MS, Murashige and Skoog medium; NAA, 1-naphthaleneacetic acid; PGR, plant growth regulator; WPM, Woody Plant Medium.

3.2.5. Plantlet Acclimatization

To date, only a few studies have been conducted on the acclimatization of caper plants produced by the microcuttings technique, and divergent results were published. Some authors reported low survival rates. For example, Chalak et al. [17] observed a survival rate of 40%. On the other hand, some other studies reported high survival rates. For example, Chalak and Elbitar [18] observed a survival rate of 92%. Musallam et al. [88] reported a successful transfer of caper plants regenerated in vitro to field conditions as well as normal growth and development.

The composition and texture of culture medium used before acclimatization (e.g., elongation and/or rooting medium) may significantly influence the survival rate of the plants during acclimatization [89,90]. In *C. orientalis*, the survival rate varied from 56% to 66% depending on the hormonal composition of rooting medium [12]. The impact of rooting medium on subsequent plantlet acclimatization was observed in other plant species [91].

Based on the above, the effects of many factors on caper plantlet survival during acclimatization are still to be evaluated, such as the genotype, substrate mixtures and ratios, nutrient solutions and irrigation, pre-acclimatization, and greenhouse/glasshouse conditions. Therefore, further research should focus on these factors in order to improve the survival rate of caper plants produced by the microcuttings technique.

3.3. Caper Propagation by Organogenesis

Organogenesis is a powerful in vitro regeneration system for the mass production of genetically uniform plants [92,93]. It is based on the totipotency of plant cells and involves the development of shoot buds and roots either directly on the explant or after callus formation under aseptic and appropriate culture conditions to subsequently form complete plants [61,92,94]. Successful regeneration through organogenesis depends on many factors including the genotype and explant type as well as culture medium and conditions. Despite the enormous potential of this method, there are only few papers dealing with it in caper. Organogenesis can be considered as the best approach available today to propagate caper and can be exploited for genetic improvement and conservation purposes.

Different PGRs were evaluated to develop an in vitro regeneration system through organogenesis for caper. Movafeghi et al. [95] suggested to culture hypocotyl explants on MS medium supplemented with 0.1 mg L^{-1} NAA and 0.5 mg L^{-1} BAP. This PGR combination promoted adventitious shoot induction and proliferation. Increasing NAA concentration decreased the formation rate of adventitious shoots. Root induction was performed on MS medium containing 0.5 mg L^{-1} NAA and the regenerated plantlets were successfully acclimatized to ex vitro conditions. Elmaghrabi et al. [96] cultured leaf explants on MS medium containing 1.2 mg L^{-1} 2,4-dichlorophenoxyacetic acid (2,4-D) to induce callogenesis. For shoot and root formation, they suggested to transfer calli on MS medium supplemented with 2 mg L^{-1} BAP. The plantlets were successfully acclimatized to greenhouse conditions. Fahmideh et al. [97] suggested to use 2 mg L^{-1} kinetin and 2 mg L^{-1} NAA for shoot induction from callus culture, and 1 mg L^{-1} NAA for shoot rooting.

Al-Safadi and Elias [65] also highlighted the significant impact of PGRs on caper organogenesis. The use of 2 mg L^{-1} GA_3 gave an average of 2.2 shoots per explant while the combination of 0.1 mg L^{-1} GA_3, 1 mg L^{-1} NAA, and 2 mg L^{-1} zeatin riboside yielded an average of 5.5 shoots per explant. A survival rate of 86% was observed during acclimatization. Caglar et al. [47] pointed out the effectiveness of TDZ for the production and multiplication of caper shoots in vitro. Indeed, 4.54 μM ($\simeq$1 mg L^{-1}) TDZ gave an average number of 45.7 shoots per explant.

The effects of other factors on caper organogenesis were scarcely investigated. Al-Safadi and Elias [65] found that gamma irradiation greatly improved the growth of caper shoots induced in vitro. In fact, a 10 Gy dose of gamma irradiation promoted shoot growth by up to 200% and improved rhizogenesis to reach 100%.

4. Conclusions and Future Perspectives

Caper is a shrub that has long been considered difficult to propagate conventionally and, despite the high socioeconomic and medicinal importance of this species, research on its micropropagation is still very limited. The present review reported and discussed the main findings of the literature in the field of caper micropropagation. Based on the studies reported here, it can be seen that caper micropropagation was achieved through seed germination, microcuttings, and organogenesis. However, each method presents many difficulties and divergent results. Moreover, the organogenesis technique is not well exploited and more research is still needed in this field. Based on the information gathered in this report, it is highly important to improve the existing micropropagation methods, and to develop new regeneration systems that could be used for rapid and large-scale propagation and genetic improvement of caper, such as somatic embryogenesis. In fact, the development of the various caper industries is closely linked to the development of efficient propagation methods through tissue culture.

Somatic embryogenesis is a powerful micropropagation technique that is used not only for the production of a large number of plants in a short period of time, but also for genetic transformation, artificial seed production, and germplasm conservation. Somatic embryogenesis also offers the possibility to identify and select, under in vitro conditions, the genotypes that are resistant to biotic and abiotic stresses. It can be used for secondary metabolite production, which may be very interesting in caper. Thus, developing an in vitro regeneration system through somatic embryogenesis for caper is currently needed for the sustainable utilization and improvement of this species.

Author Contributions: Conceptualization, M.K., I.B. and M.A.M.; investigation, M.K. and M.A.M.; writing—original draft preparation, M.K.; writing—review and editing, M.A.M., M.K. and I.B.; supervision, I.B. All authors have read and agreed to the published version of the manuscript.

Funding: This research received no external funding.

Institutional Review Board Statement: Not applicable.

Informed Consent Statement: Not applicable.

Data Availability Statement: Not applicable.

Conflicts of Interest: The authors declare no conflict of interest.

References

1. Fici, S. Intraspecific variation and evolutionary trends in *Capparis spinosa* L. (Capparaceae). *Plant Syst. Evol.* **2001**, *228*, 123–141. [CrossRef]
2. Zohary, M. The species of *Capparis* in the Mediterranean and the Near Eastern countries. *Bull. Res. Counc. Isr.* **1960**, *8D*, 49–64.
3. Inocencio, C.; Rivera, D.; Concepción Obón, M.; Alcaraz, F.; Barreña, J.-A. A systematic revision of *Capparis* section *Capparis* (Capparaceae) 1, 2. *Ann. Mo. Bot. Gard.* **2006**, *93*, 122–149. [CrossRef]
4. Danin, A. *Capparis* in the East Mediterranean countries. *Fl. Medit.* **2010**, *20*, 179–185.
5. Gristina, A.S.; Fici, S.; Siragusa, M.; Fontana, I.; Garfi, G.; Carimi, F. Hybridization in *Capparis spinosa* L.: Molecular and morphological evidence from a Mediterranean island complex. *Flora Morphol. Distrib. Funct. Ecol. Plants* **2014**, *209*, 733–741. [CrossRef]
6. Jiang, H.E.; Li, X.; Ferguson, K.D.; Wang, Y.F.; Liu, C.J.; Li, C.S. The discovery of *Capparis spinosa* L. (Capparidaceae) in the Yanghai Tombs (2800 years b.p.), NW China, and its medicinal implications. *J. Ethnopharmacol.* **2007**, *113*, 409–420. [CrossRef]
7. Faran, M. Capparis spinosa—The plant on the wall. In *Medicinal and Aromatic Plants of the Middle-East (Medicinal and Aromatic Plants of the World)*; Yaniv, Z., Dudai, N., Eds.; Springer: Dordrecht, The Netherlands, 2014; pp. 59–65.
8. Rhimi, A.; Hannachi, H.; Hjaoujia, S.; Yousfi, H.; Boussaid, M. In vitro germination and seedling development of Tunisian caper (*Capparis spinosa* L.). *Int. J. Agron. Agri. Res.* **2017**, *10*, 1–8.
9. Carra, A.; Del Signore, M.B.; Sottile, F.; Ricci, A.; Carimi, F. Potential use of new diphenylurea derivatives in micropropagation of *Capparis spinosa* L. *Plant Growth Regul.* **2012**, *66*, 229–237. [CrossRef]
10. Inocencio, C.; Cowan, R.S.; Alcaraz, F.; Rivera, D.; Fay, F.M. AFLP fingerprinting in *Capparis* subgenus *Capparis* related to the commercial sources of capers. *Genet. Resour. Crop. Evol.* **2005**, *52*, 137–144. [CrossRef]
11. Kdimy, A.; El Yadini, M.; Guaâdaoui, A.; Bourais, I.; El Hajjaji, S.; Le, H.V. Phytochemistry, biological activities, therapeutic potential, and socio-economic value of caper (*Capparis spinosa* L.). *ChemRxiv* **2022**. [CrossRef]

12. Kereša, S.; Stanković, D.; Lodeta, K.B.; Jerčić, I.H.; Bolarić, S.; Barić, M.; Mihovilović, A.B. Efficient protocol for the in vitro plantlet production of caper (*Capparis orientalis* Veill.) from the East Adriatic coast. *Agronomy* **2019**, *9*, 303. [CrossRef]
13. Lakrimi, M. Le caprier. *Transfert Technol. Agri.* **1997**, *37*, 1–4.
14. Moghaddasi, S.M.; Haddad Kashani, H.; Azarbad, Z. *Capparis spinosa* L. propagation and medicinal uses. *Life Sci. J.* **2012**, *9*, 684–686.
15. Ramezani-Gask, M.; Bahrani, M.J.; Shekafandeh, A.; Salehi, H.; Taghvaei, M.; Al-Ahmadi, M.J. A comparison of different propagation methods of common Caper-bush (*Capparis spinosa* L.) as a new horticultural crop. *Int. J. Plant Dev. Biol.* **2008**, *2*, 106–110.
16. Gianguzzi, V.; Barone, E.; Sottile, A.F. In vitro rooting of *Capparis spinosa* L. as affected by genotype and by the proliferation method adopted during the multiplication phase. *Plants* **2020**, *9*, 398. [CrossRef] [PubMed]
17. Chalak, L.; Elbitar, A.; Cordahi, N.; Hage, C.; Chehade, A. In vitro propagation of *Capparis spinosa* L. *Acta Hortic.* **2003**, *616*, 335–338. [CrossRef]
18. Chalak, L.; Elbitar, A. Micropropagation of *Capparis spinosa* L. subsp. Rupestris Sibth. & Sm. by nodal cuttings. *Indian J. Biotechnol.* **2006**, *5*, 555–558.
19. Hegazi, G.; Eid, S.; Sharaf, A.E.M. Micropropagation for conservation of two rare *Capparis* species from Egypt 1. *Catrina* **2011**, *6*, 29–39.
20. El-Mekawy, M.A.; Ali, M.A.M.; Belal, A.H.; Abdallah, S.A.S. Micropropagation of caper (*Capparis spinosa*, L.) from wild plants growing in North Sinai. *Hortsci. J. Suez Canal Univ.* **2013**, *1*, 15–20. [CrossRef]
21. Mehrabani, V. In vitro micropropagation of two native *Capparis spinosa* L. cultivars from Iran. *Bio. Forum Int. J.* **2016**, *8*, 144–149.
22. Attia, A.; Dessoky, E.D.S.; Al-Sodany, Y.M.; Ismail, I.A. Ex situ preservation for some endemic and rare medicinal plants in Taif, KSA. *Biotechnol. Biotechnol. Equip.* **2017**, *31*, 912–920. [CrossRef]
23. Gianguzzi, V.; Inglese, P.; Barone, E.; Sottile, F. In vitro regeneration of *Capparis spinosa* L. by using a temporary immersion system. *Plants* **2019**, *8*, 177. [CrossRef] [PubMed]
24. Chedraoui, S.; Abi-Rizk, A.; El-Beyrouthy, M.; Chalak, L.; Ouaini, N.; Rajjou, L. *Capparis spinosa* L. in a systematic review: A xerophilous species of multi values and promising potentialities for agrosystems under the threat of global warming. *Front. Plant. Sci.* **2017**, *8*, 1845. [CrossRef] [PubMed]
25. Zarei, M.; Seyedi, N.; Maghsoudi, S.; Nejad, M.S.; Sheibani, H. Green synthesis of Ag nanoparticles on the modified grapheme oxide using *Capparis spinosa* fruit extract for catalytic reduction of organic dyes. *Inorg. Chem. Commun.* **2021**, *123*, 108327. [CrossRef]
26. Alkire, B. *Capers*; Center for New Crops and Plants Products: West Lafayette, IN, USA, 1998.
27. Rivera, D.; Inocencio, C.; Obon, C.; Alcaraz, F. Review of food and medicinal uses of *Capparis* L. subgenus Capparis (Capparidaceae). *Econ. Bot.* **2003**, *57*, 515–534. [CrossRef]
28. Hall, J.C.; Sytsma, K.J.; Iltis, H.H. Phylogeny of Capparaceae and Brassicaceae based on chloroplast sequence data. *Am. J. Bot.* **2002**, *89*, 1826–1842. [CrossRef] [PubMed]
29. Tlili, N.; Khaldi, A.; Triki, S.; Munné-Bosch, S. Phenolic compounds and vitamin antioxidants of caper (Capparis spinosa). *Plant Foods Hum. Nutr.* **2010**, *65*, 260–265. [CrossRef]
30. Bennoune, L.; Bouteldja, N.; Chebah, S.; Akkouche, Z. Etude Bibliographique de Quelques Activités Biologiques de Différentes Parties du Câprier. Master's Thesis, Jihel University, Jijel, Algeria, 2012.
31. Saifi, N.; Ibijbijen, J.; Echchgadda, G. Caractérisation morphologique et identification taxonomique du câprier (*Capparis* spp.) dans les régions de Fès, Taounate et My Idriss Zerhoune. In Proceedings of the International Congress of 'Institut National des Plantes Médicinales et Aromatiques', Fes, Morocco, 22–24 March 2007.
32. Fici, S. A taxonomic revision of the *Capparis spinosa* group (Capparaceae) from the Mediterranean to Central Asia. *Phytotaxa* **2014**, *174*, 1–24. [CrossRef]
33. Rakhimova, T.; Vaisova, G.B.; Rakhimova, N.K.; Matkarimova, A. Phytocoenotic distribution of *Capparis spinosa* L. (Capparaceae) under different ecological conditions in Uzbekistan. *Ann. Romanian Soc. Cell Biol.* **2021**, *25*, 7882–7895.
34. Ibnou Ali El Alaoui, M. Etude de la diversité génétique chez le câprier (*Capparis* spp.) au Maroc. *Périodique D'information Du Cent. Régional De La Rech. Agron. De Meknès* **2015**. Available online: https://mag.inrameknes.info/?p=886 (accessed on 15 March 2022).
35. Benseghir-Boukhari, L.A.; Seridi, R. Le câprier, une espèce arbustive pour le développement rural durable en Algérie. *Méditerranée* **2007**, *109*, 101–105. [CrossRef]
36. Aliyazicioglu, R.; Eyupoglu, O.E.; Sahin, H.; Yildiz, O.; Baltas, N. Phenolic components, antioxidant activity, and mineral analysis of *Capparis spinosa* L. *Afr. J. Biotechnol.* **2013**, *12*, 6643–6649. [CrossRef]
37. Zhang, H.; Ma, Z.F. Phytochemical and pharmacological properties of *Capparis spinosa* as a medicinal plant. *Nutrients* **2018**, *10*, 116. [CrossRef] [PubMed]
38. Mollica, A.; Stefanucci, A.; Macedonio, G.; Locatelli, M.; Luisi, G.; Novellino, E.; Zengin, G. Chemical composition and biological activity of *Capparis spinosa* L. from Lipari Island. *S. Afr. J. Bot.* **2019**, *120*, 135–140. [CrossRef]
39. Eddouks, M.; Lemhadri, A.; Michel, J.B. Hypolipidemic activity of aqueous extract of *Capparis spinosa* L. in normal and diabetic rats. *J. Ethnopharmacol.* **2005**, *98*, 345–350. [CrossRef]
40. Ali, Z.N.; Eddouks, M.; Michel, J.B.; Sulpice, T.; Hajji, L. Cardiovascular effect of Capparis spinosa aqueous extract. Part III: Antihypertensive effect in spontaneously hypertensive rats. *Am. J. Pharmacol. Toxicol.* **2007**, *2*, 111–115. [CrossRef]

41. Boga, C.; Forlani, L.; Calienni, R.; Hindley, T.; Hochkoeppler, A.; Tozzi, S.; Zanna, N. On the antibacterial activity of roots of *Capparis spinosa* L. *Nat. Prod. Res.* **2011**, *25*, 417–421. [CrossRef]
42. Gull, T.; Sultana, B.; Bhatti, I.A.; Jamil, A. Antibacterial potential of Capparis spinosa and Capparis decidua extracts. *Int. J. Agric. Biol.* **2015**, *17*, 727–733. [CrossRef]
43. Hamuti, A.; Li, J.; Zhou, F.; Aipire, A.; Ma, J.; Yang, J.; Li, J. Capparis spinosa fruit ethanol extracts exert different effects on the maturation of dendritic cells. *Molecules* **2017**, *22*, 97. [CrossRef]
44. Al-Azawi, A.H.; Ghaima, K.K.; Hawazen, H.S. Phytochemical, antibacterial and antioxidant activities of *Capparis spinosa* L. cultivated in Iraq. *Biosci. Res.* **2018**, *15*, 2611–2618.
45. Barbera, G. *Il Cappero*; Edagricole: Bologna, Italy, 1993.
46. Bahrani, M.J.; Gask, M.R.; Shekafandeh, A.; Taghvaei, M. Seed germination of wild caper (*Capparis spinosa* L., var. parviflora) as affected by dormancy breaking treatments and salinity levels. *Seed Sci. Technol.* **2008**, *36*, 776–780. [CrossRef]
47. Caglar, C.; Caglar, S.; Ergin, O.; Yarim, M. The influence of growth regulators on shoot proliferation and rooting of in vitro propagated caper. *J. Environ. Biol.* **2005**, *26*, 479–485.
48. Mazri, M.A.; Koufan, M.; Moussafir, S.; Essatte, A.; Belkoura, I. Recent advances in argan propagation: A review. *Trees* **2022**, *in press*. [CrossRef]
49. Orphanos, P.I. Germination of caper (*Capparis spinosa* L.) seeds. *J. Hortic. Sci.* **1983**, *58*, 267–270. [CrossRef]
50. Sozzi, G.O.; Chiesa, A. Improvement of caper (*Capparis spinosa* L.) seed germination by breaking seed coat-induced dormancy. *Sci. Hortic.* **1995**, *62*, 255–261. [CrossRef]
51. Suleiman, M.K.; Bhat, N.R.; Abdal, M.S.; Jacob, S.; Thomas, R.R.; Al-Dossery, S.; Bellen, R. Germination studies of *Capparis spinosa* L. *Propag. Ornam. Plants* **2009**, *9*, 35–38.
52. Arefi, I.H.; Nejad, S.K.; Kafi, M. Roles of duration and concentration of priming agents on dormancy breaking and germination of caper (*Capparis spinosa* L.) for the protection of arid degraded areas. *Pak. J. Bot.* **2012**, *44*, 225–230.
53. Labbafi, M.R.; Mehrafarin, A.; Badi, H.N.; Ghorbani, M.; Tavakoli, M. Improve germination of caper (*Capparis spinosa* L.) seeds by different induction treatments of seed dormancy breaking. *Trakia J. Sci.* **2018**, *16*, 70–74. [CrossRef]
54. Saifi, N.; Echchgadda, G.; Nassiri, L.; Ibijbijen, J. Ability to root formation and germination of some Moroccan ecotypes of caper (*Capparis* spp.). *Sci. Lib.* **2014**, *6*, 141201.
55. Ibáñez, A.J. Estudio Para la Mejora de la Propagación de la Alcaparra Mediante Estaquillas. Ph.D. Thesis, Universitat Politècnica de València, València, Spain, 2015.
56. Ghorbel, A.; Ben Salem, F.A.; Khouildi, S.; Skouri, H.; Chibani, F. Le câprier: Caractérisation et multiplication. In *Des Modèles Biologiques à l'Amélioration des Plantes*; Hamon, S., Ed.; AUF: Paris, France, 2001; pp. 157–172.
57. Germanà, M.; Chiancone, B. In vitro germination and seedling development of caper (*Capparis spinosa* L.) mature seeds. *Acta Hortic.* **2009**, *839*, 181–186. [CrossRef]
58. Koufan, M.; Belkoura, I.; Mazri, M.A.; Amarraque, A.; Essatte, A.; Elhorri, H.; Zaddoug, F.; Alaoui, T. Determination of antioxidant activity, total phenolics and fatty acids in essential oils and other extracts from callus culture, seeds and leaves of *Argania spinosa* (L.) Skeels. *Plant Cell Tissue Organ Cult.* **2020**, *141*, 217–227. [CrossRef]
59. Koufan, M.; Mazri, M.A.; Essatte, A.; Moussafir, S.; Belkoura, I.; El Rhaffari, L.; Toufik, I. A novel regeneration system through micrografting for *Argania spinosa* (L.) Skeels, and confirmation of successful rootstock-scion union by histological analysis. *Plant Cell Tissue Organ Cult.* **2020**, *142*, 369–378. [CrossRef]
60. Abbasi Khalaki, M.; Moameri, M.; Asgari Lajayer, B.; Astatkie, T. Influence of Nano-priming on seed germination and plant growth of forage and medicinal plants. *Plant Growth Regul.* **2021**, *93*, 13–28. [CrossRef]
61. Amghar, I.; Diria, G.; Boumlik, I.; Gaboun, F.; Iraqi, D.; Labhilili, M.; Mentag, R.; Meziani, R.; Mazri, M.A.; Ibriz, M.; et al. An efficient regeneration pathway through adventitious organogenesis for the endangered *Argania spinosa* (L.) Skeels. *Vegetos* **2021**, *34*, 355–367. [CrossRef]
62. Murashige, T.; Skoog, F. A revised medium for rapid growth and bioassays with tobacco tissue cultures. *Physiol. Plant* **1962**, *15*, 473–479. [CrossRef]
63. Hernández-Muñoz, S.; Pedraza-Santos, M.E.; López, P.A.; De La Cruz-Torres, E.; Martínez-Palacios, A.; Fernández-Pavía, S.P.; Chávez-Bárcenas, A.T. Estimulación de la germinación y desarrollo in vitro de *Laelia autumnalis* con rayos gamma. *Rev. Fitotec. Mex.* **2017**, *40*, 271–283. [CrossRef]
64. Ngoenngam, L.; Pongtongkam, P.; Arananant, J.; Poeaim, S.; Poeam, A. In vitro effect of gamma irradiation and plant growth regulators (PGRs) for induction and development of *Stylosanthes hamata* cv. Verano. *Int. J. Agric. Technol.* **2019**, *15*, 63–74.
65. Al-Safadi, B.; Elias, R. Improvement of caper (*Capparis spinosa* L.) propagation using in vitro culture and gamma irradiation. *Sci. Hortic.* **2011**, *127*, 290–297. [CrossRef]
66. Graeber, K.; Nakabayashi, K.; Miatton, E.; Leubner-Metzger, G.; Soppe, W.J.J. Molecular mechanisms of seed dormancy. *Plant Cell. Environ.* **2012**, *35*, 1769–1786. [CrossRef]
67. Miransari, M.; Smith, D.L. Plant hormones and seed germination. *Environ. Exp. Bot.* **2014**, *99*, 110–121. [CrossRef]
68. Kosakivska, I.V.; Voytenko, L.V.; Vasyuk, V.A.; Vedenichova, N.P.; Babenko, L.M.; Shcherbatyuk, M.M. Phytohormonal regulation of seed germination. *Fiziol. Rast. Genet.* **2019**, *51*, 187–206. [CrossRef]
69. Gaspar, T.; Kevers, C.; Penel, C.; Greppin, H.; Reid, D.M.; Thorpe, T.A. Plant hormones and plant growth regulators in plant tissue culture. In Vitro Cell. Develop. *Biol. Plant* **1996**, *32*, 272–289. [CrossRef]

70. Gentile, A.; Jàquez Gutiérrez, M.; Martinez, J.; Frattarelli, A.; Nota, P.; Caboni, E. Effect of meta-Topolin on micropropagation and adventitious shoot regeneration in Prunus rootstocks. *Plant Cell Tissue Organ Cult.* **2014**, *118*, 373–381. [CrossRef]
71. Deb, G.; Sultana, S.; Bhuiyan, M.S.U.; Sarker, K.K.; Papry, A.S. In vitro plant regeneration of wild eggplant (*Solanum sisymbriifolium*) to produce large number of rootstocks for tomato grafting. *J. Adv. Biotechnol. Exp. Ther.* **2019**, *2*, 65–72. [CrossRef]
72. Ahmad, A.; Ahmad, N.; Anis, M.; Alatar, A.A.; Abdel-Salam, E.M.; Qahtan, A.A.; Faisal, M. Gibberellic acid and thidiazuron promote micropropagation of an endangered woody tree (Pterocarpus marsupium Roxb.) using in vitro seedlings. *Plant Cell Tiss. Organ Cult.* **2021**, *144*, 449–462. [CrossRef]
73. Bicalho, E.M.; Pintó-Marijuan, M.; Morales, M.; Müller, M.; Munné-Bosch, S.; Garcia, Q.S. Control of macaw palm seed germination by the gibberellin/abscisic acid balance. *Plant Biol.* **2015**, *17*, 990–996. [CrossRef]
74. Song, Q.; Cheng, S.; Chen, Z.; Nie, G.; Xu, F.; Zhang, J.; Zhou, M.; Zhang, W.; Liao, Y.; Ye, J. Comparative transcriptome analysis revealing the potential mechanism of seed germination stimulated by exogenous gibberellin in Fraxinus hupehensis. *BMC Plant Biol.* **2019**, *19*, 199. [CrossRef]
75. Van Staden, J.; Zazimalova, E.; George, E.F. Plant growth regulators II: Cytokinins, their analogues and antagonists. In *Plant Propagation by Tissue Culture*; George, E.F., Hall, M.A., De Klerk, G.-J., Eds.; Springer: Basel, Switzerland, 2008; Volume 1, pp. 205–226.
76. Mazri, M.A.; Belkoura, I.; Meziani, R. Use of TDZ for micropropagation of some Mediterranean crop species. In *Thidiazuron: From Urea Derivative to Plant Growth Regulator*; Ahmad, N., Faisal, M., Eds.; Springer: Singapore, 2018; pp. 115–137.
77. Heydariyan, M.; Basirani, N.; Sharifi-Rad, M.; Khmmari, I.; Rafat Poor, S. Effect of seed priming on germination and seedling growth of the caper (*Capparis Spinosa*) under drought stress. *Int. J. Adv. Biol. Biomed. Res.* **2014**, *2*, 2381–2389.
78. Rodriguez, R.; Rey, M.; Cuozzo, L.; Ancora, G. In vitro propagation of caper (*Capparis spinosa* L.). *In Vitro Cell. Develop. Biol. Plant* **1990**, *26*, 531–536. [CrossRef]
79. Linsmaier, E.M.; Skoog, F. Organic growth factor requirements of tobacco tissue cultures. *Physiol. Plant.* **1965**, *18*, 100–127. [CrossRef]
80. Nitsch, J.P.; Nitsch, C. Haploid plants from pollen grains. *Science* **1969**, *163*, 85–87. [CrossRef] [PubMed]
81. Lloyd, G.; McCown, B. Commercially-feasible micropropagation of mountain laurel, Kalmia latifolia, by use of shoot-tip culture. *Comb. Proc. Int. Plant Propagators Soc.* **1980**, *30*, 421–427.
82. Rugini, E.; Verma, D.C. Micropropagation of difficult-to-propagate almond (*Prunus amygdalus*, Batsch.) cultivar. *Plant Sci. Lett.* **1983**, *28*, 273–281. [CrossRef]
83. Sottile, F.; Giuggioli, N.R.; Marinoni, D.T.; Peano, C.; Del Signore, M.B. Selection and micropropagation of valuable caper genotypes. *Hortic. Sci.* **2020**, *47*, 110–116. [CrossRef]
84. Koufan, M.; Belkoura, I.; Alaoui, T. The multiplication of the argane tree by microcutting (*Argania spinosa* L. Skeels). *Eur. J. Biotechnol. Biosci.* **2018**, *6*, 47–52.
85. Parra, R.; Amo-Marco, J.B. Effect of plant growth regulators and basal media on in vitro shoot proliferation and rooting of *Myrtus communis* L. *Biol. Plant* **1996**, *38*, 161–168. [CrossRef]
86. Nowakowska, K.; Pacholczak, A.; Tepper, W. The effect of selected growth regulators and culture media on regeneration of *Daphne mezereum* L. 'Alba'. *Rend. Lincei. Sci. Fis. Nat.* **2019**, *30*, 197–205. [CrossRef]
87. Kulus, D. Influence of growth regulators on the development, quality, and physiological state of in vitro-propagated *Lamprocapnos spectabilis* (L.) Fukuhara. *In Vitro Cell. Develop. Biol.-Plant* **2020**, *56*, 447–457. [CrossRef]
88. Musallam, I.; Duwayri, M.; Shibli, R.A. Micropropagation of caper (*Capparis spinosa* L.) from wild plants. *Funct. Plant Sci. Biotechnol.* **2011**, *5*, 17–21.
89. Mazri, M.A. Effect of liquid media and in vitro pre-acclimatization stage on shoot elongation and acclimatization of date palm (*Phoenix dactylifera* L.) cv. Najda. *J. Ornament. Hortic. Plants* **2012**, *2*, 225–231.
90. Mazri, M.A.; Meziani, R. An improved method for micropropagation and regeneration of date palm (*Phoenix dactylifera* L.). *J. Plant Biochem. Biotechnol.* **2013**, *22*, 176–184. [CrossRef]
91. Meziani, R.; Mazri, M.A.; Arhazzal, M.; Belkoura, I.; Alem, C.; Jaiti, F. Evaluation of in vitro shoot elongation and rooting of date palm, and determination of physiological characteristics of regenerated plantlets. *Not. Sci. Biol.* **2019**, *11*, 77–85. [CrossRef]
92. Mazri, M.A.; Meziani, R.; Elmaataoui, S.; Alfeddy, M.N.; Jait, F. Assessment of genetic fidelity, biochemical and physiological characteristics of in vitro grown date palm cv. *Al-Fayda*. *Vegetos* **2019**, *32*, 333–344. [CrossRef]
93. Mazri, M.A.; Meziani, R.; El Bakouri, Z. Cost analysis of date palm (cv. Mejhoul) plantlets produced by organogenesis in Morocco. *Plant Cell Tissue Organ. Cult.* **2021**, *146*, 409–415. [CrossRef]
94. Amghar, I.; Ibriz, M.; Ibrahimi, M.; Boudra, A.; Gaboun, F.; Meziani, R.; Iraqi, D.; Mazri, M.A.; Diria, G.; Abdelwahd, R. In vitro root induction from argan (*Argania spinosa* (L.) Skeels) adventitious shoots: Influence of ammonium nitrate, auxins, silver nitrate and putrescine, and evaluation of plantlet acclimatization. *Plants* **2021**, *10*, 1062. [CrossRef] [PubMed]
95. Movafeghi, A.; Ghader, H.; Aliasgharpour, M. Regeneration of *Capparis spinosa* L. using hypocotyl explants. *Iran. J. Biol.* **2008**, *21*, 289–297.
96. Elmaghrabi, A.M.; Abunghnia, E.; Hamoud, S. In vitro propagation of the wild medicinal plant, caper (*Capparis spinosa* L.). *Afr. J. Biotechnol.* **2017**. Available online: https://www.researchgate.net/profile/Adel-Elmaghrabi/publication/319687304_In_vitro_propagation_via_wild_medicinal_plants_of_Caper_Capparis_spinosa_L/links/59b97ed0a6fdcc6872314444/In-vitro-propagation-via-wild-medicinal-plants-of-Caper-Capparis-spinosa-L.pdf (accessed on 15 March 2022).

97. Fahmideh, L.; Sheikhi, M.; Benakashani, F.; Solouki, M. Callus induction and organogenesis from various explants of *Capparis spinosa* L. plant under in vitro conditions. *J. Plant Prod.* **2019**, *26*, 75–88. [CrossRef]

MDPI
St. Alban-Anlage 66
4052 Basel
Switzerland
www.mdpi.com

Horticulturae Editorial Office
E-mail: horticulturae@mdpi.com
www.mdpi.com/journal/horticulturae

www.ingramcontent.com/pod-product-compliance
Lightning Source LLC
Chambersburg PA
CBHW061509210926
51850CB00008B/1859
9783725802791